Atlas of Human Cross-Sectional Anatomy

With CT and MR Images

Second Edition

Atlas of Human Cross-Sectional Anatomy

With CT and MR Images

Second Edition

Donald R. Cahill, Ph.D.
Department of Anatomy
Mayo Medical School
Rochester, Minnesota

Matthew J. Orland, M.D.
Department of Medicine
Washington University
St. Louis, Missouri

Carl C. Reading, M.D.
Department of Diagnostic Radiology
Mayo Medical School
Rochester, Minnesota

WILEY-LISS

A JOHN WILEY & SONS, INC., PUBLICATION
New York • Chichester • Brisbane • Toronto • Singapore

Address all Inquiries to the Publisher
Wiley-Liss, Inc., 41 East 11th Street, New York, NY 10003

Printed in the United States of America

Library of Congress Cataloging-in-Publication Data

Cahill, Donald R.
    Atlas of human cross-sectional anatomy / Donald R. Cahill, Matthew
J. Orland, Carl C. Reading. – 2nd ed.
      p.   cm.
    Bibliography: p.
    Includes index.
    ISBN 0-471-50988-4
    1. Anatomy, Human–Atlases.   1. Orland, Matthew J.   II. Reading,
Carl C.   III. Title.
QM25.C24 1990
611'.022'2–dc19

89-30897
CIP

Contents

Preface

The second edition of the *Atlas of Human Cross-Sectional Anatomy* is considerably revised and expanded from its parent edition. Presented in this edition are the serial anatomic sections of the trunk and of the head and neck illustrated with the use of meticulously prepared pen and ink drawings as well as photographs from the original format. In addition, new chapters feature 1) drawings of supplemental transverse sections of the abdomen demonstrating variations of normal anatomical relationships, 2) drawings of sections of the upper and lower limbs including the hand and foot, and 3) drawings of coronal and sagittal sections of the head and neck.

By displaying all parts of the body, the second edition of the *Atlas* is a more complete reference work. Furthermore, the scope of the original *Atlas* is broadened in the second edition by including computed tomograms (CT scans) and magnetic resonance images (MR images) of normal living subjects depicting anatomy analogous to the cadaver sections of the head, limbs, and trunk. These provide a clinical, radiographic correlation with classic cross-sectional anatomy.

The carefully detailed pen and ink drawings by Cahill were made to illustrate the 115 sections in this work. The anatomic sections were prepared by freezing and sectioning. Photographs were made of the frozen sections. The sections were then thawed and fixed, and dissection of the sections was performed to authenticate and complement the features shown on the surface photographs. The pen and ink drawings were made during dissection using the surface photographs as a guide. The drawings depict the basic histologic characteristics of the tissues and organs. Care has been taken not only to illustrate the surface image of a section, but also to demonstrate the important relationships occurring within a particular section not immediately apparent without the benefit of dissection. Such attention to detail complements radiographic correlations by demonstrating the thickness of a section rather than simply the surfaces of the sections. Orientation drawings were made from reassembled cadaver sections and simplified to basic features for easy reference.

The current level of technologic development in computed tomography and magnetic resonance imaging allows for the demonstration of the anatomic features of some regions of the body better with computed tomography and other regions better with magnetic resonance. Regions of the body that are in involuntary motion, such as the trunk, are depicted primarily with CT scans, because the fast (2 second) CT scanning time produces minimal distortion of the image. On the other hand, magnetic resonance produces images of exquisite resolution of the head and limbs, which are regions that can remain motionless for the 5–10 minutes of necessary MR scanning time. Because magnetic resonance imaging has the unique capacity to depict anatomy in a variety of planes, it has been used to demonstrate the coronal and sagittal views of the head in this *Atlas*. We have also elected to demonstrate the zero degree views of the head with magnetic resonance imaging because this plane is more commonly imaged in the clinical setting using MR rather than CT. The technical note provides additional information on CT scanning and MR imaging.

The *Atlas* is prepared in chapters, including (1) The Male Thorax (9 sections with drawings of both surfaces of the sections, photographs of the sections and 8 matching CT scans), (2) The Male Abdomen (9 sections with drawings of both surfaces, photographs and 9 matching CT scans), and an Upper Abdominal Supplement (4 drawings from a second cadaver, chosen to illustrate a common variation in upper abdominal anatomy, with 8 correlative CT scans including 4 scans that illustrate variations in liver morphology), (3) The Male Pelvis (8 sections with drawings of both surfaces, photographs and 8 matching CT scans), (4) The Female Pelvis (14 sections, with drawings of both surfaces of the anatomic sections, photographs and 14 matching CT scans), (5) The Lower Limb (22 illustrated sections with 22 correlative MR images), (6) The Upper Limb (17 illustrated sections with 17 matching MR images), (7) The Head—20 degrees from Orbitomeatal Plane (7 sections with drawings of both surfaces and 14 matching CT scans), (8) The Head—0 degrees from the Orbitomeatal plane (14 sections with drawings of both surfaces and 28 matching MR images), (9) The Head in Sagittal Planes (6 sections with drawings and 15 correlative MR images), (10) The Head in Coronal Planes (6 sections with drawings and 12 matching MR images). We have routinely arranged the anatomic drawings with approximately 40 labels per drawing adjacent to matching CT scans or MR images which are supplied with fewer labels, and have attempted to use a standard anatomical nomenclature in English that closely follows the *Nomina Anatomica* to give uniformity to the labeled drawings and scans.

It is a pleasure to thank production editor, Sonny Fritz; head of illustrations, Michael O'Connor; the senior staff members, Eric Swanson, Ann Epner, and Doug McLaurine; and, indeed, the entire staff of Wiley-Liss, Inc. for their part in bringing this book to fruition.

This *Atlas* is intended for the study of sectional anatomy and the interpretation of CT scans and MR images obtained in clinical medicine. As such, this work is intended for a wide audience within the medical and paramedical fields including radiologists, anatomists, internists, surgeons, medical students, graduate students in the biomedical sciences and radiologic technologists.

Donald R. Cahill, Ph.D.
Matthew J. Orland, M.D.
Carl C. Reading, M.D.

Technical Note: CT and MR Images in This Atlas

All MR images were obtained using a 1.5 Tesla Superconducting Magnetic Resonance Imager (GE Signa). The majority of the scans were obtained utilizing a 256 × 256 matrix size, two signal acquisitions, and thin sections of 5 mm thickness. Volume coils were used to obtain images of the limbs. Because MR signal contrast is multiparametric and depends on proton density, T1 and T2 relaxation times, and blood flow, different pulse sequences were chosen to highlight specific anatomic aspects of the different regions of the body. Short repetition times (TR) and echo times (TE) were used to emphasize T1 relaxation time characteristics, whereas long repetition and echo times were used to highlight T2 relaxation time characteristics. Structures with a large number of mobile hydrogen protons, such as adipose tissue, generate a strong signal and appear bright on T1 and somewhat less bright on T2-weighted images. Other structures such as cortical bone and tendons, which have few mobile hydrogen protons, generate a weak signal and appear dark on both T1 and T2-weighted images. Fluids such as cerebrospinal fluid, bile, and urine have T1 and T2 characteristics which cause the fluid to appear dark on T1-weighted images but bright on T2-weighted images. Blood vessels are unique structures in that they contain a rapidly flowing liquid which has the potential to generate a very strong signal. In this atlas imaging techniques were used to suppress this signal and therefore, generally, blood vessels appear as dark structures. There are many additional factors which contribute to the MR image which are beyond the scope of this brief introduction to the subject. The interested reader is referred to the bibliography at the end of this section.

The CT images were obtained using fourth generation CT scanners (GE 9800 and Picker 1200). Scans of ten mm slice thickness were obtained of the head and neck after the administration of iodinated intravenous contrast material. Similarly, scans of ten mm slice thickness were obtained of the abdomen and pelvis following the administration of both iodinated intravenous and oral contrast material. Window settings were adjusted to demonstrate the soft tissue components of the sections, except in the base of the skull where the window settings were chosen to highlight the bony structures. The appearance of the anatomic structures in the CT images depends upon the differential degree of attenuation of the x-ray beam by the different tissue types. Adipose, for example, causes very little attenuation of x-rays and therefore appears dark on CT scans. Bone, which is very dense, on the other hand, attenuates a considerable portion of the x-ray beam and appears white on the CT scans. Iodinated contrast material in the intestine and blood vessels also attenuates the beam to cause these structures to appear whiter than would be seen without contrast material.

Considerable effort went into acquiring the best possible CT and MR images for this atlas at the time it was sent to press. Already, however, new advances in both CT and MR imaging are on the horizon which promise to provide even better anatomic detail than is presently available. We look forward to these new advances in technology as a means to improve our understanding and to provide a greater insight into the cross-sectional anatomy of the human body.

References

Berquist TH, Ehman RL, Richardson ML: Magnetic resonance of the musculoskeletal system. New York, New York, Raven Press, 1987.

Brant-Zawadzki M, Norman D: Magnetic resonance imaging of the central nervous system. New York, New York, Raven Press, 1987.

Stark DD, Bradley W: Magnetic resonance imaging. St. Louis, Missouri, CV Mosby Company, 1988.

Young, SW: Magnetic resonance imaging—basic principles. New York, New York, Raven Press, 1988.

Carl C. Reading

The Male Thorax

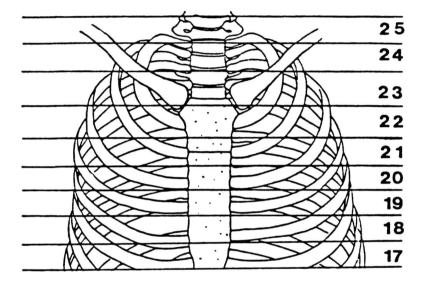

25
24
23
22
21
20
19
18
17

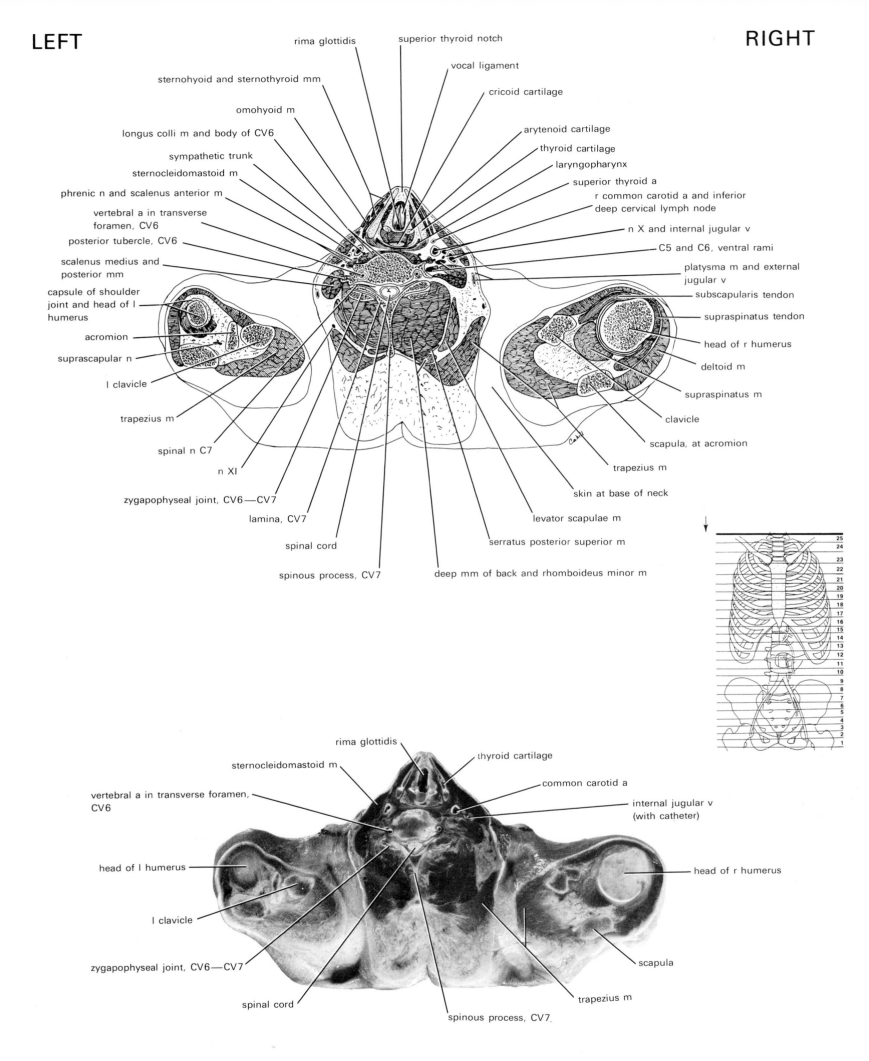

LEFT   RIGHT

rima glottidis
superior thyroid notch
vocal ligament
sternohyoid and sternothyroid mm
cricoid cartilage
omohyoid m
arytenoid cartilage
longus colli m and body of CV6
thyroid cartilage
sympathetic trunk
laryngopharynx
sternocleidomastoid m
superior thyroid a
phrenic n and scalenus anterior m
r common carotid a and inferior
deep cervical lymph node
vertebral a in transverse
foramen, CV6
n X and internal jugular v
posterior tubercle, CV6
C5 and C6, ventral rami
scalenus medius and
posterior mm
platysma m and external
jugular v
capsule of shoulder
joint and head of l
humerus
subscapularis tendon
supraspinatus tendon
acromion
head of r humerus
suprascapular n
deltoid m
l clavicle
supraspinatus m
trapezius m
clavicle
scapula, at acromion
spinal n C7
trapezius m
n XI
skin at base of neck
zygapophyseal joint, CV6—CV7
levator scapulae m
lamina, CV7
serratus posterior superior m
spinal cord
spinous process, CV7
deep mm of back and rhomboideus minor m

rima glottidis
sternocleidomastoid m
thyroid cartilage
common carotid a
vertebral a in transverse foramen,
CV6
internal jugular v
(with catheter)
head of l humerus
head of r humerus
l clavicle
zygapophyseal joint, CV6—CV7
scapula
spinal cord
trapezius m
spinous process, CV7

Section 25 from above.

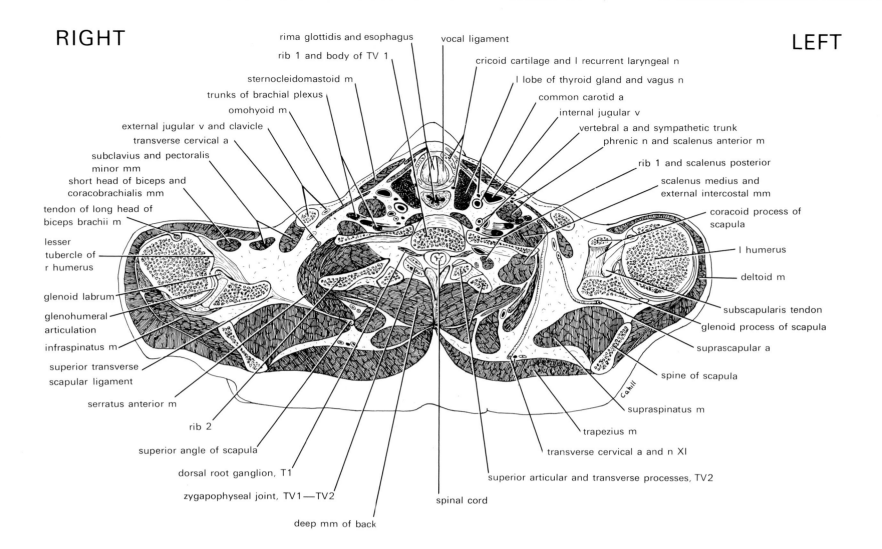

rima glottidis and esophagus
rib 1 and body of TV 1
sternocleidomastoid m
trunks of brachial plexus
omohyoid m
external jugular v and clavicle
transverse cervical a
subclavius and pectoralis minor mm
short head of biceps and coracobrachialis mm
tendon of long head of biceps brachii m
lesser tubercle of r humerus
glenoid labrum
glenohumeral articulation
infraspinatus m
superior transverse scapular ligament
serratus anterior m
rib 2
superior angle of scapula
dorsal root ganglion, T1
zygapophyseal joint, TV1—TV2
deep mm of back

vocal ligament
cricoid cartilage and l recurrent laryngeal n
l lobe of thyroid gland and vagus n
common carotid a
internal jugular v
vertebral a and sympathetic trunk
phrenic n and scalenus anterior m
rib 1 and scalenus posterior
scalenus medius and external intercostal mm
coracoid process of scapula
l humerus
deltoid m
subscapularis tendon
glenoid process of scapula
suprascapular a
spine of scapula
supraspinatus m
trapezius m
transverse cervical a and n XI
superior articular and transverse processes, TV2
spinal cord

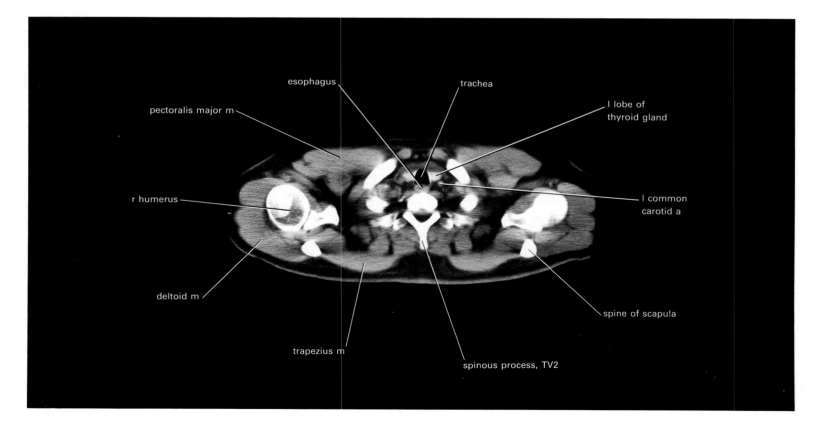

esophagus
pectoralis major m
r humerus
deltoid m
trachea
l lobe of thyroid gland
l common carotid a
spine of scapula
trapezius m
spinous process, TV2

Section 25 from below.

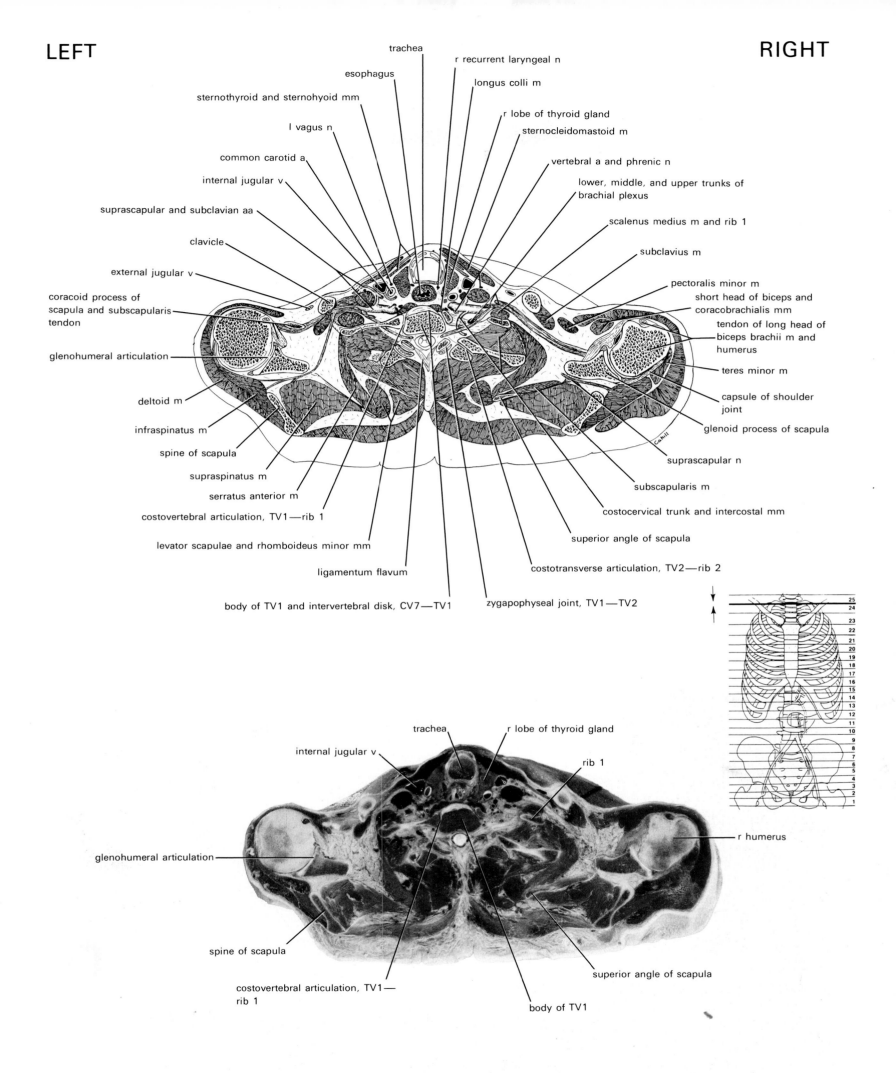

LEFT   RIGHT

trachea
esophagus
r recurrent laryngeal n
sternothyroid and sternohyoid mm
longus colli m
l vagus n
r lobe of thyroid gland
common carotid a
sternocleidomastoid m
internal jugular v
vertebral a and phrenic n
suprascapular and subclavian aa
lower, middle, and upper trunks of brachial plexus
clavicle
scalenus medius m and rib 1
external jugular v
subclavius m
coracoid process of scapula and subscapularis tendon
pectoralis minor m
short head of biceps and coracobrachialis mm
glenohumeral articulation
tendon of long head of biceps brachii m and humerus
teres minor m
deltoid m
capsule of shoulder joint
infraspinatus m
glenoid process of scapula
spine of scapula
suprascapular n
supraspinatus m
serratus anterior m
subscapularis m
costovertebral articulation, TV1—rib 1
costocervical trunk and intercostal mm
levator scapulae and rhomboideus minor mm
superior angle of scapula
ligamentum flavum
costotransverse articulation, TV2—rib 2
body of TV1 and intervertebral disk, CV7—TV1
zygapophyseal joint, TV1—TV2

trachea
r lobe of thyroid gland
internal jugular v
rib 1
r humerus
glenohumeral articulation
spine of scapula
superior angle of scapula
costovertebral articulation, TV1—rib 1
body of TV1

Section 24 from above.

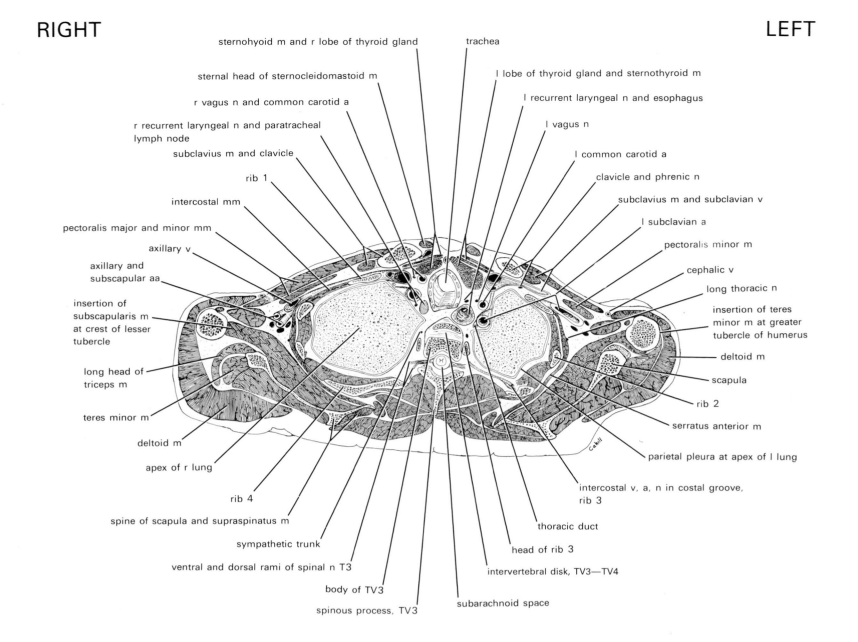

sternohyoid m and r lobe of thyroid gland

trachea

sternal head of sternocleidomastoid m

l lobe of thyroid gland and sternothyroid m

r vagus n and common carotid a

l recurrent laryngeal n and esophagus

r recurrent laryngeal n and paratracheal lymph node

l vagus n

subclavius m and clavicle

l common carotid a

rib 1

clavicle and phrenic n

intercostal mm

subclavius m and subclavian v

pectoralis major and minor mm

l subclavian a

axillary v

pectoralis minor m

axillary and subscapular aa

cephalic v

insertion of subscapularis m at crest of lesser tubercle

long thoracic n

insertion of teres minor m at greater tubercle of humerus

deltoid m

long head of triceps m

scapula

teres minor m

rib 2

deltoid m

serratus anterior m

apex of r lung

parietal pleura at apex of l lung

rib 4

intercostal v, a, n in costal groove, rib 3

spine of scapula and supraspinatus m

thoracic duct

sympathetic trunk

head of rib 3

ventral and dorsal rami of spinal n T3

intervertebral disk, TV3—TV4

body of TV3

subarachnoid space

spinous process, TV3

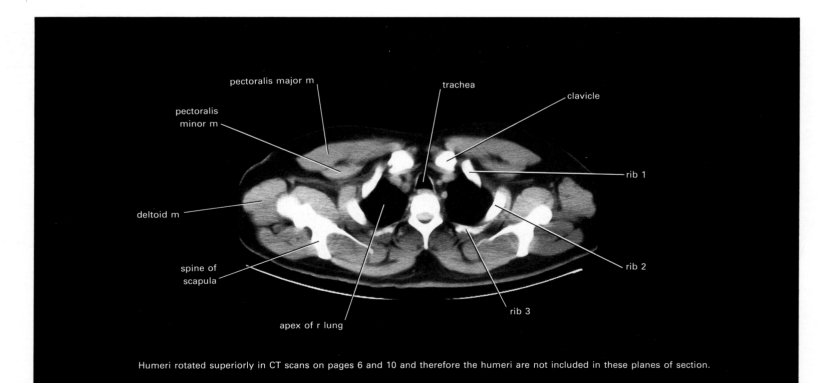

pectoralis major m

trachea

clavicle

pectoralis minor m

rib 1

deltoid m

spine of scapula

rib 2

rib 3

apex of r lung

Humeri rotated superiorly in CT scans on pages 6 and 10 and therefore the humeri are not included in these planes of section.

Section 24 from below.

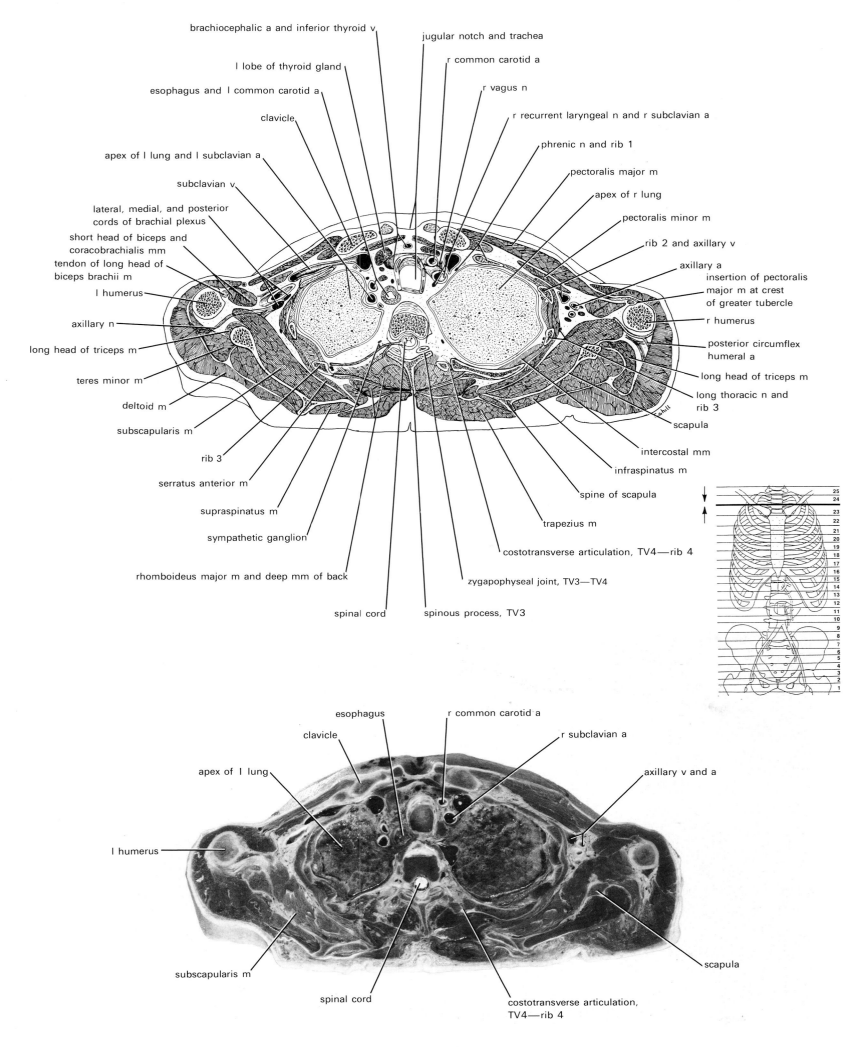

brachiocephalic a and inferior thyroid v

jugular notch and trachea

r common carotid a

l lobe of thyroid gland

r vagus n

esophagus and l common carotid a

r recurrent laryngeal n and r subclavian a

clavicle

phrenic n and rib 1

apex of l lung and l subclavian a

pectoralis major m

subclavian v

apex of r lung

lateral, medial, and posterior
cords of brachial plexus

pectoralis minor m

short head of biceps and
coracobrachialis mm

rib 2 and axillary v

tendon of long head of
biceps brachii m

axillary a
insertion of pectoralis
major m at crest
of greater tubercle

l humerus

r humerus

axillary n

posterior circumflex
humeral a

long head of triceps m

long head of triceps m

teres minor m

long thoracic n and
rib 3

deltoid m

scapula

subscapularis m

intercostal mm

rib 3

infraspinatus m

serratus anterior m

spine of scapula

supraspinatus m

trapezius m

sympathetic ganglion

costotransverse articulation, TV4—rib 4

rhomboideus major m and deep mm of back

zygapophyseal joint, TV3—TV4

spinal cord

spinous process, TV3

25
24
23
22
21
20
19
18
17
16
15
14
13
12
11
10
9
8
7
6
5
4
3
2
1

esophagus

r common carotid a

clavicle

r subclavian a

apex of l lung

axillary v and a

l humerus

subscapularis m

spinal cord

costotransverse articulation,
TV4—rib 4

scapula

Section 23 from above.

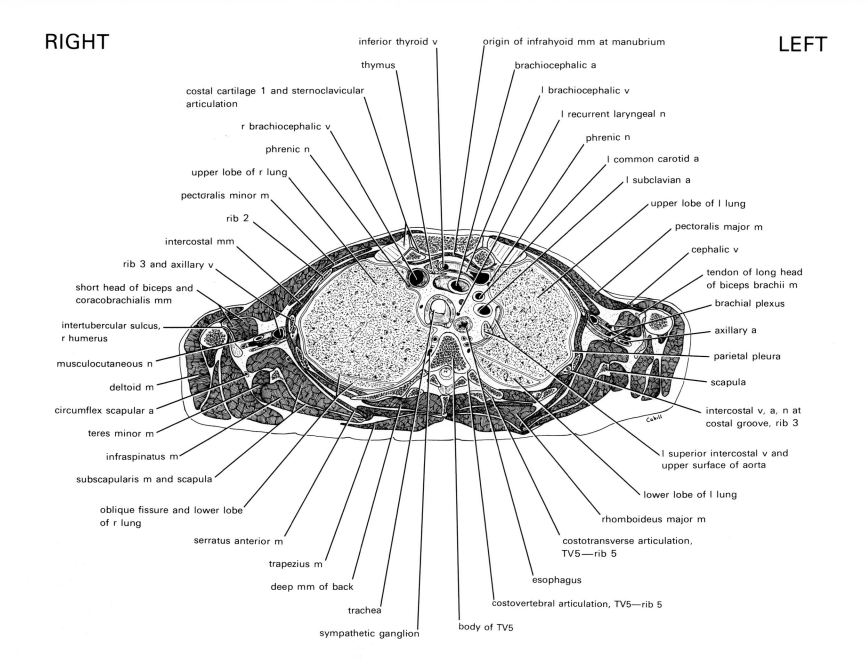

inferior thyroid v

thymus

costal cartilage 1 and sternoclavicular
articulation

r brachiocephalic v

phrenic n

upper lobe of r lung

pectoralis minor m

rib 2

intercostal mm

rib 3 and axillary v

short head of biceps and
coracobrachialis mm

intertubercular sulcus,
r humerus

musculocutaneous n

deltoid m

circumflex scapular a

teres minor m

infraspinatus m

subscapularis m and scapula

oblique fissure and lower lobe
of r lung

serratus anterior m

trapezius m

deep mm of back

trachea

sympathetic ganglion

origin of infrahyoid mm at manubrium

brachiocephalic a

l brachiocephalic v

l recurrent laryngeal n

phrenic n

l common carotid a

l subclavian a

upper lobe of l lung

pectoralis major m

cephalic v

tendon of long head
of biceps brachii m

brachial plexus

axillary a

parietal pleura

scapula

intercostal v, a, n at
costal groove, rib 3

l superior intercostal v and
upper surface of aorta

lower lobe of l lung

rhomboideus major m

costotransverse articulation,
TV5—rib 5

esophagus

costovertebral articulation, TV5—rib 5

body of TV5

Cahill

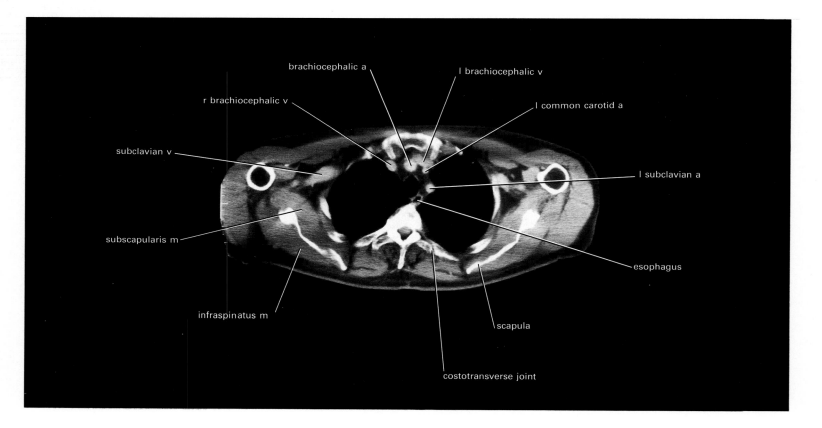

brachiocephalic a

l brachiocephalic v

r brachiocephalic v

l common carotid a

subclavian v

l subclavian a

subscapularis m

esophagus

infraspinatus m

scapula

costotransverse joint

Section 23 from below.

LEFT   RIGHT

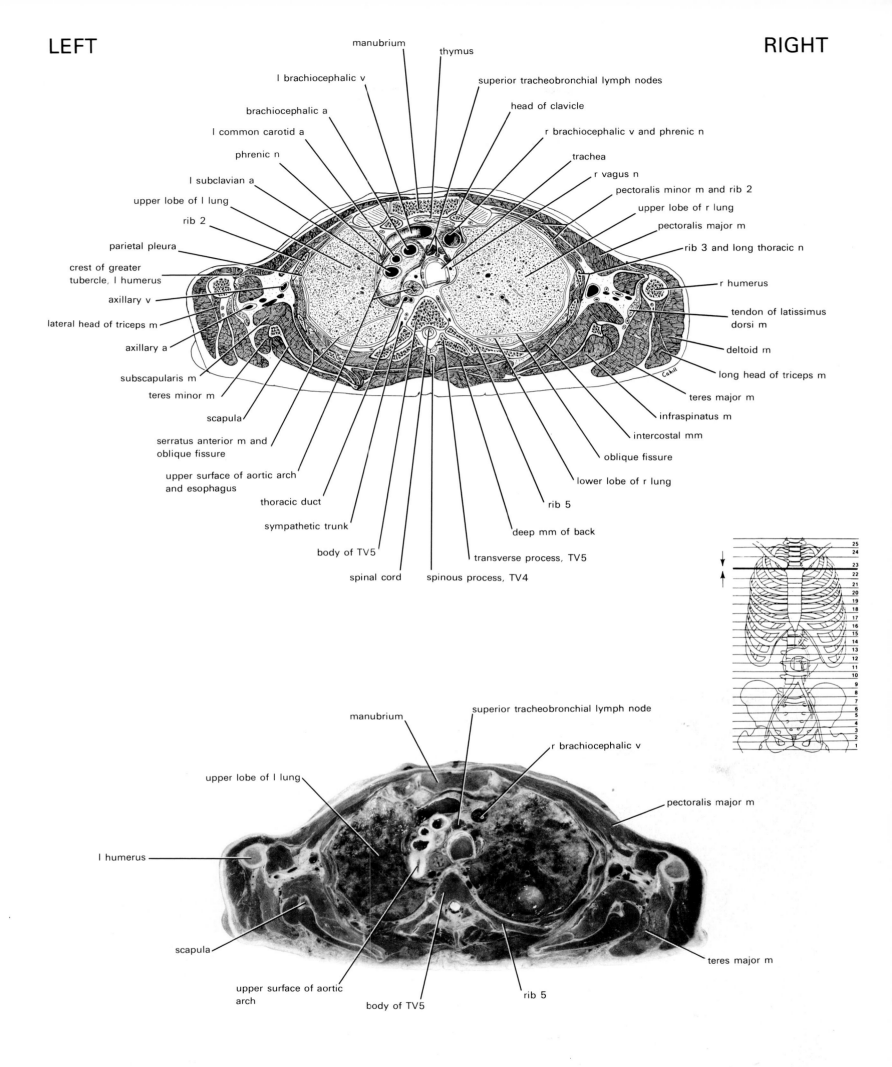

manubrium
thymus
l brachiocephalic v
superior tracheobronchial lymph nodes
brachiocephalic a
head of clavicle
l common carotid a
r brachiocephalic v and phrenic n
phrenic n
trachea
l subclavian a
r vagus n
upper lobe of l lung
pectoralis minor m and rib 2
rib 2
upper lobe of r lung
parietal pleura
pectoralis major m
crest of greater
tubercle, l humerus
rib 3 and long thoracic n
axillary v
r humerus
lateral head of triceps m
tendon of latissimus
dorsi m
axillary a
deltoid m
subscapularis m
long head of triceps m
teres minor m
teres major m
scapula
infraspinatus m
serratus anterior m and
oblique fissure
intercostal mm
upper surface of aortic arch
and esophagus
oblique fissure
thoracic duct
lower lobe of r lung
sympathetic trunk
rib 5
body of TV5
deep mm of back
transverse process, TV5
spinal cord   spinous process, TV4

manubrium
superior tracheobronchial lymph node
upper lobe of l lung
r brachiocephalic v
l humerus
pectoralis major m
scapula
teres major m
upper surface of aortic
arch
rib 5
body of TV5

Section 22 from above.

25 24 23 22 21 20 19 18 17 16 15 14 13 12 11 10 9 8 7 6 5 4 3 2 1

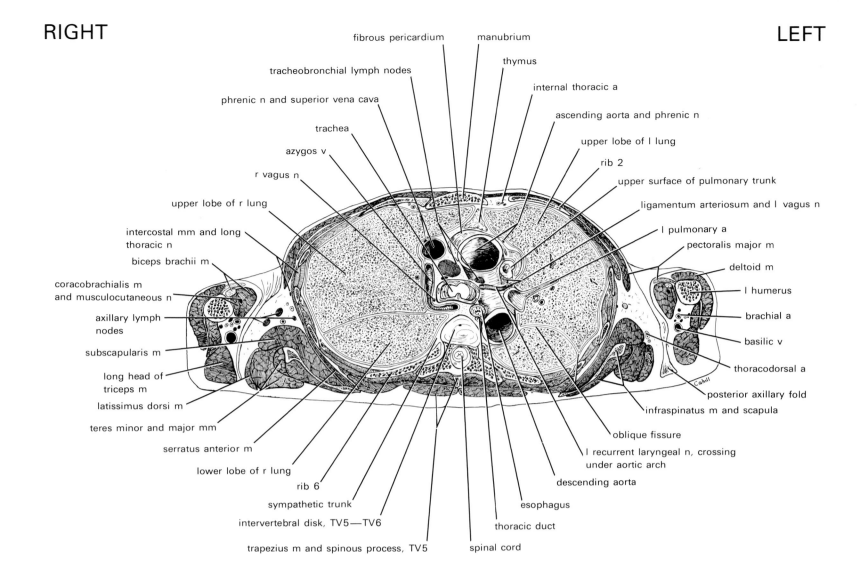

fibrous pericardium
manubrium
thymus
tracheobronchial lymph nodes
internal thoracic a
phrenic n and superior vena cava
ascending aorta and phrenic n
trachea
upper lobe of l lung
azygos v
rib 2
r vagus n
upper surface of pulmonary trunk
upper lobe of r lung
ligamentum arteriosum and l vagus n
l pulmonary a
intercostal mm and long
thoracic n
pectoralis major m
biceps brachii m
deltoid m
coracobrachialis m
and musculocutaneous n
l humerus
brachial a
axillary lymph
nodes
basilic v
subscapularis m
thoracodorsal a
long head of
triceps m
posterior axillary fold
latissimus dorsi m
infraspinatus m and scapula
teres minor and major mm
oblique fissure
serratus anterior m
l recurrent laryngeal n, crossing
under aortic arch
lower lobe of r lung
descending aorta
rib 6
sympathetic trunk
esophagus
intervertebral disk, TV5—TV6
thoracic duct
trapezius m and spinous process, TV5
spinal cord

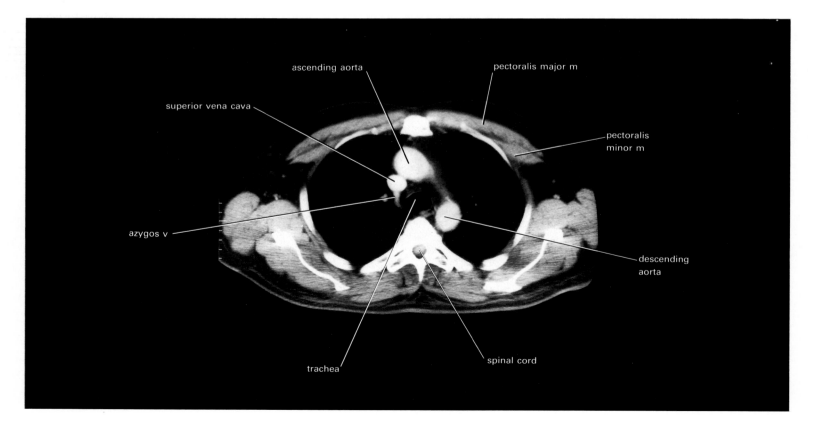

ascending aorta
pectoralis major m
superior vena cava
pectoralis
minor m
azygos v
descending
aorta
trachea
spinal cord

Section 22 from below.

LEFT   RIGHT

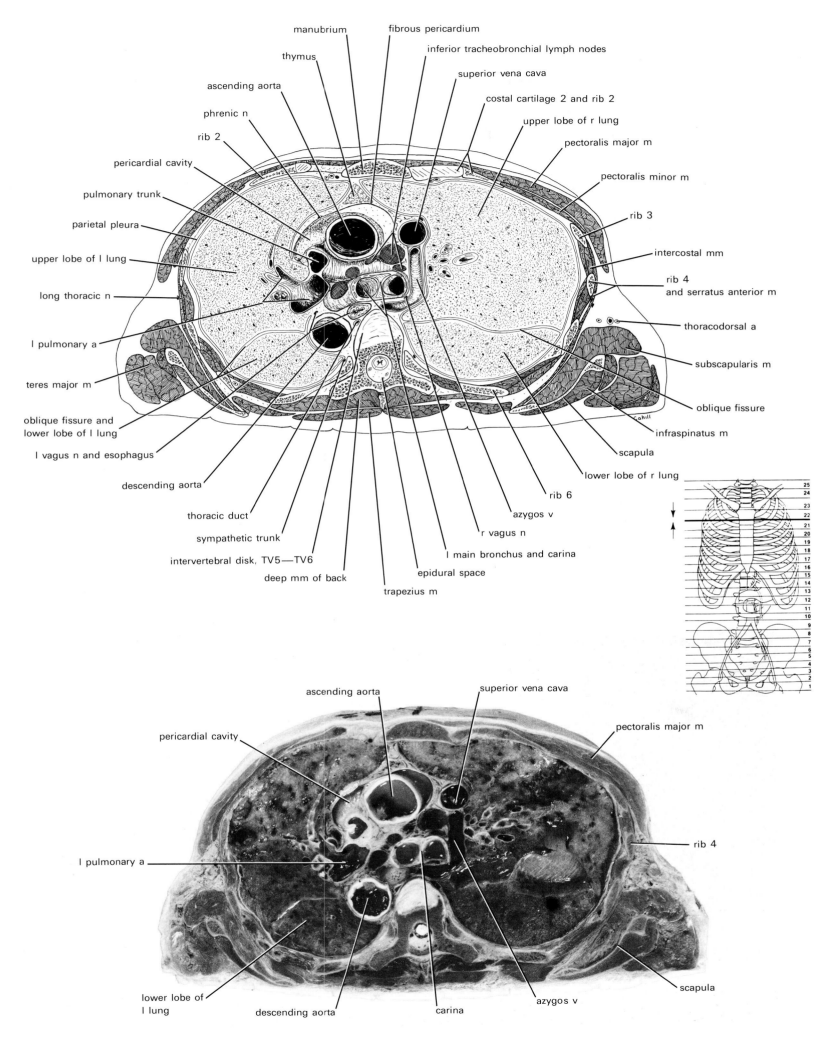

manubrium   fibrous pericardium

thymus   inferior tracheobronchial lymph nodes

ascending aorta   superior vena cava

phrenic n   costal cartilage 2 and rib 2

rib 2   upper lobe of r lung

pericardial cavity   pectoralis major m

pulmonary trunk   pectoralis minor m

parietal pleura   rib 3

upper lobe of l lung   intercostal mm

long thoracic n   rib 4
                                                              and serratus anterior m

                                                              thoracodorsal a

l pulmonary a   subscapularis m

teres major m   oblique fissure

oblique fissure and   infraspinatus m
lower lobe of l lung

l vagus n and esophagus   scapula

descending aorta   lower lobe of r lung

thoracic duct   rib 6

sympathetic trunk   azygos v

intervertebral disk, TV5—TV6   r vagus n

deep mm of back   l main bronchus and carina

trapezius m   epidural space

ascending aorta   superior vena cava

pericardial cavity   pectoralis major m

l pulmonary a   rib 4

lower lobe of   descending aorta   carina   azygos v   scapula
l lung

Section 21 from above.

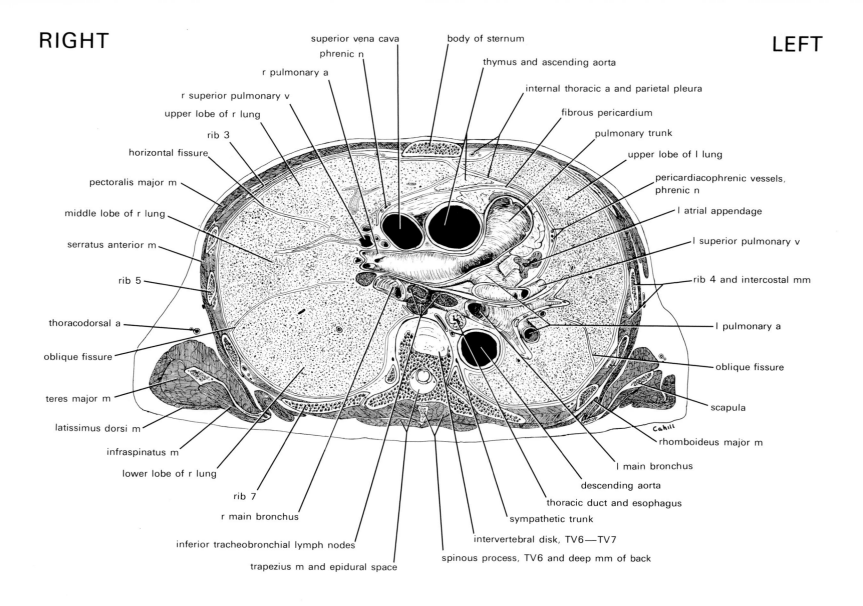

superior vena cava
phrenic n
r pulmonary a
r superior pulmonary v
upper lobe of r lung
rib 3
horizontal fissure
pectoralis major m
middle lobe of r lung
serratus anterior m
rib 5
thoracodorsal a
oblique fissure
teres major m
latissimus dorsi m
infraspinatus m
lower lobe of r lung
rib 7
r main bronchus
inferior tracheobronchial lymph nodes
trapezius m and epidural space

body of sternum
thymus and ascending aorta
internal thoracic a and parietal pleura
fibrous pericardium
pulmonary trunk
upper lobe of l lung
pericardiacophrenic vessels, phrenic n
l atrial appendage
l superior pulmonary v
rib 4 and intercostal mm
l pulmonary a
oblique fissure
scapula
rhomboideus major m
l main bronchus
descending aorta
thoracic duct and esophagus
sympathetic trunk
intervertebral disk, TV6—TV7
spinous process, TV6 and deep mm of back

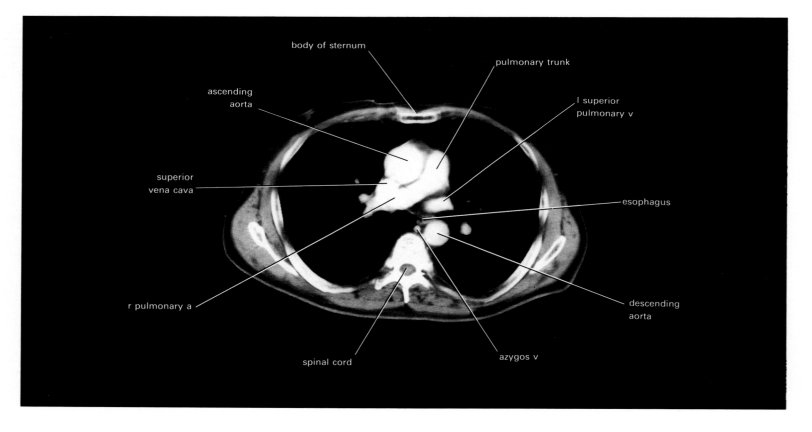

body of sternum
ascending aorta
pulmonary trunk
l superior pulmonary v
superior vena cava
esophagus
r pulmonary a
descending aorta
spinal cord
azygos v

Section 21 from below.

LEFT   RIGHT

thymus and body of sternum   fibrous pericardium
                                    epicardial fat
ascending aorta   internal thoracic a and v
upper lobe of l lung   phrenic n
r cusp of pulmonary valve   superior vena cava
pectoralis major m   r superior pulmonary v
anterior cusp of pulmonary valve   upper lobe of r lung and rib 3
rib 3 and pectoralis minor m
l cusp of pulmonary valve   horizontal fissure
phrenic n   r pulmonary a
l atrial appendage   middle lobe of r lung
l upper lobe bronchus   oblique fissure
serratus anterior m   thoracodorsal a
oblique fissure   latissimus dorsi m
                                            scapula and teres
                                            major m
l superior pulmonary v   infraspinatus m
l lower lobe bronchus   lower lobe of r lung
lower lobe of l lung   rhomboideus major m
descending aorta   rib 7
esophagus and thoracic duct   r main bronchus
costovertebral articulation, TV7—rib 7   r lung in retroesophageal recess
accessory hemiazygos v   spinal cord

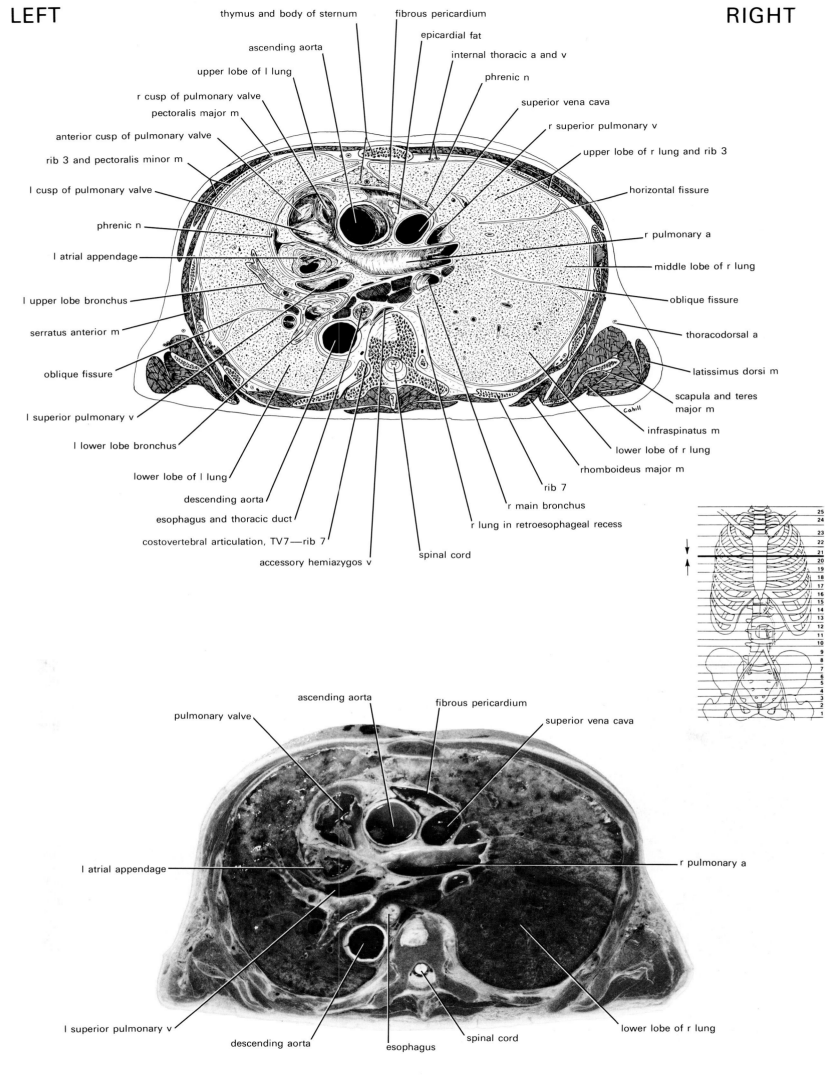

ascending aorta   fibrous pericardium
pulmonary valve   superior vena cava
l atrial appendage   r pulmonary a
l superior pulmonary v   lower lobe of r lung
descending aorta   spinal cord
                        esophagus

Section 20 from above.

RIGHT   LEFT

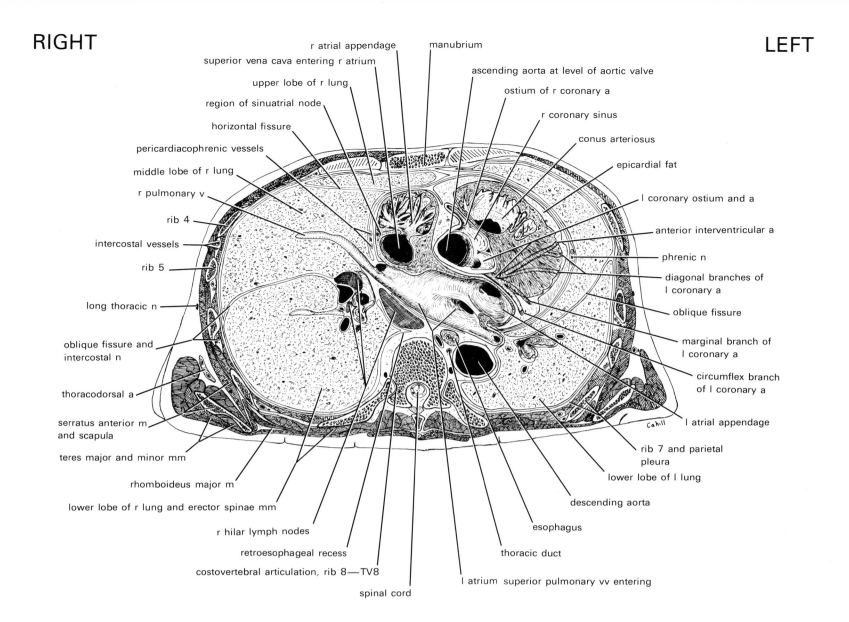

r atrial appendage — manubrium
superior vena cava entering r atrium
ascending aorta at level of aortic valve
upper lobe of r lung
ostium of r coronary a
region of sinuatrial node
r coronary sinus
horizontal fissure
conus arteriosus
pericardiacophrenic vessels
epicardial fat
middle lobe of r lung
l coronary ostium and a
r pulmonary v
anterior interventricular a
rib 4
phrenic n
intercostal vessels
diagonal branches of
rib 5
l coronary a
long thoracic n
oblique fissure
marginal branch of
oblique fissure and
l coronary a
intercostal n
circumflex branch
of l coronary a
thoracodorsal a
l atrial appendage
serratus anterior m
and scapula
rib 7 and parietal
pleura
teres major and minor mm
lower lobe of l lung
rhomboideus major m
descending aorta
lower lobe of r lung and erector spinae mm
esophagus
r hilar lymph nodes
thoracic duct
retroesophageal recess
costovertebral articulation, rib 8—TV8
l atrium superior pulmonary vv entering
spinal cord

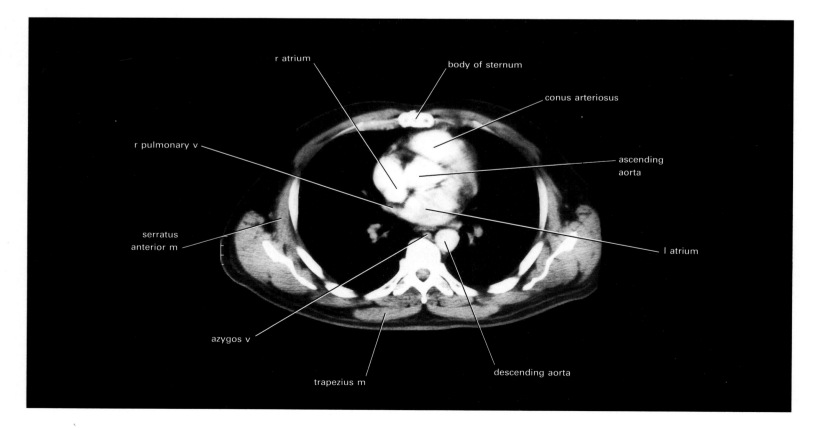

r atrium — body of sternum
conus arteriosus
r pulmonary v
ascending
aorta
serratus
anterior m
l atrium
azygos v
descending aorta
trapezius m

Section 20 from below.

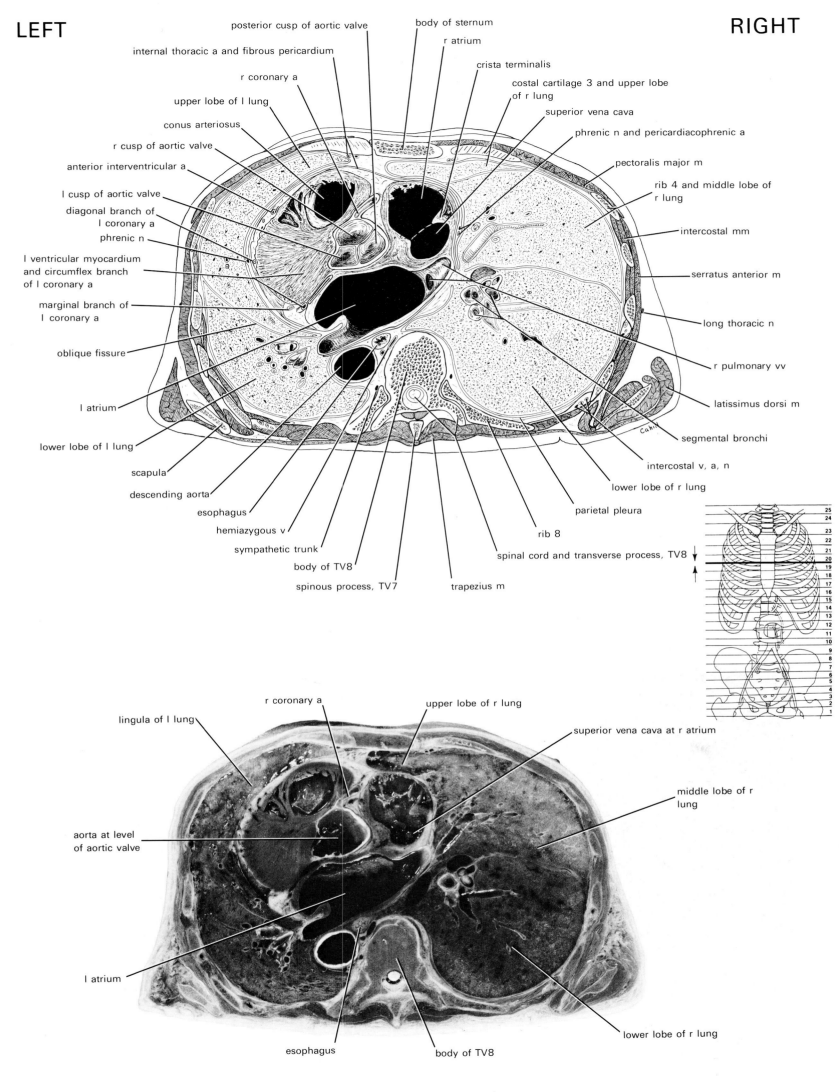

LEFT   RIGHT

posterior cusp of aortic valve
body of sternum
r atrium
internal thoracic a and fibrous pericardium
crista terminalis
r coronary a
costal cartilage 3 and upper lobe
of r lung
upper lobe of l lung
superior vena cava
conus arteriosus
phrenic n and pericardiacophrenic a
r cusp of aortic valve
pectoralis major m
anterior interventricular a
rib 4 and middle lobe of
r lung
l cusp of aortic valve
diagonal branch of
l coronary a
intercostal mm
phrenic n
l ventricular myocardium
and circumflex branch
of l coronary a
serratus anterior m
marginal branch of
l coronary a
long thoracic n
oblique fissure
r pulmonary vv
l atrium
latissimus dorsi m
lower lobe of l lung
segmental bronchi
scapula
intercostal v, a, n
descending aorta
lower lobe of r lung
esophagus
parietal pleura
hemiazygous v
rib 8
sympathetic trunk
spinal cord and transverse process, TV8
body of TV8
spinous process, TV7
trapezius m

Section 19 from above.

r coronary a
upper lobe of r lung
lingula of l lung
superior vena cava at r atrium
middle lobe of r
lung
aorta at level
of aortic valve
l atrium
lower lobe of r lung
esophagus
body of TV8

25
24
23
22
21
20
19
18
17
16
15
14
13
12
11
10
9
8
7
6
5
4
3
2
1

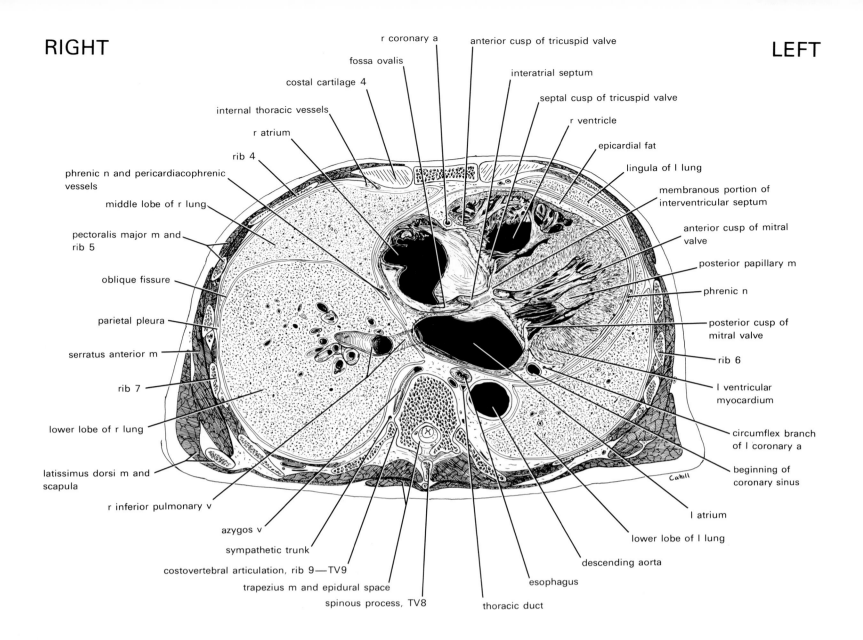

RIGHT   LEFT

r coronary a
fossa ovalis
anterior cusp of tricuspid valve
costal cartilage 4
interatrial septum
internal thoracic vessels
septal cusp of tricuspid valve
r atrium
r ventricle
rib 4
epicardial fat
phrenic n and pericardiacophrenic
vessels
lingula of l lung
middle lobe of r lung
membranous portion of
interventricular septum
pectoralis major m and
rib 5
anterior cusp of mitral
valve
oblique fissure
posterior papillary m
parietal pleura
phrenic n
serratus anterior m
posterior cusp of
mitral valve
rib 7
rib 6
lower lobe of r lung
l ventricular
myocardium
latissimus dorsi m and
scapula
circumflex branch
of l coronary a
r inferior pulmonary v
beginning of
coronary sinus
azygos v
l atrium
sympathetic trunk
lower lobe of l lung
costovertebral articulation, rib 9—TV9
descending aorta
trapezius m and epidural space
esophagus
spinous process, TV8
thoracic duct

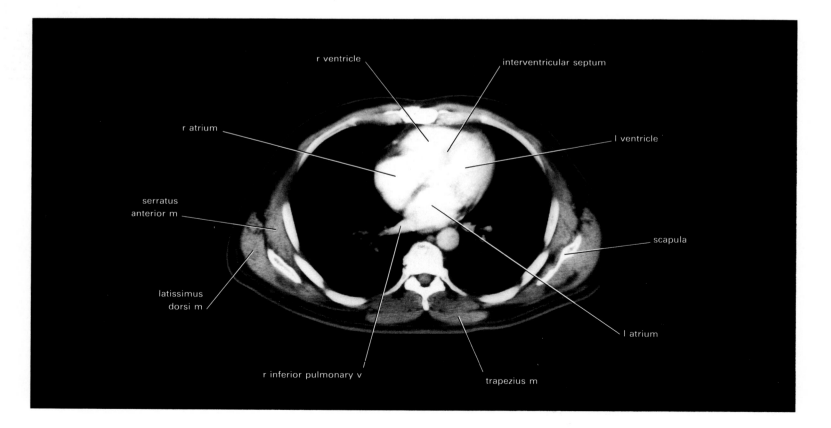

r ventricle
interventricular septum
r atrium
l ventricle
serratus
anterior m
scapula
latissimus
dorsi m
l atrium
r inferior pulmonary v
trapezius m

Section 19 from below.

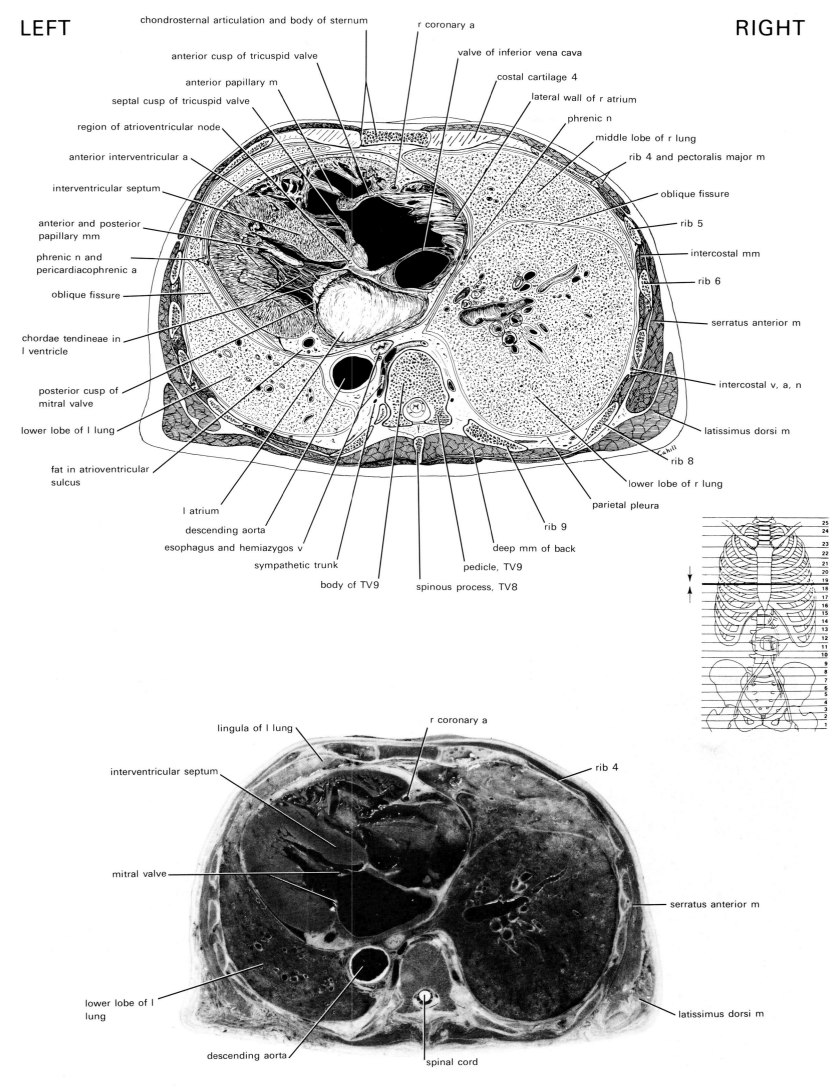

chondrosternal articulation and body of sternum
r coronary a
anterior cusp of tricuspid valve
valve of inferior vena cava
anterior papillary m
costal cartilage 4
septal cusp of tricuspid valve
lateral wall of r atrium
region of atrioventricular node
phrenic n
middle lobe of r lung
anterior interventricular a
rib 4 and pectoralis major m
interventricular septum
oblique fissure
anterior and posterior
papillary mm
rib 5
intercostal mm
phrenic n and
pericardiacophrenic a
rib 6
oblique fissure
serratus anterior m
chordae tendineae in
l ventricle
intercostal v, a, n
posterior cusp of
mitral valve
latissimus dorsi m
lower lobe of l lung
rib 8
fat in atrioventricular
sulcus
lower lobe of r lung
parietal pleura
l atrium
descending aorta
rib 9
esophagus and hemiazygos v
deep mm of back
sympathetic trunk
pedicle, TV9
body of TV9
spinous process, TV8

lingula of l lung
r coronary a
interventricular septum
rib 4
mitral valve
serratus anterior m
lower lobe of l
lung
latissimus dorsi m
descending aorta
spinal cord

Section 18 from above.

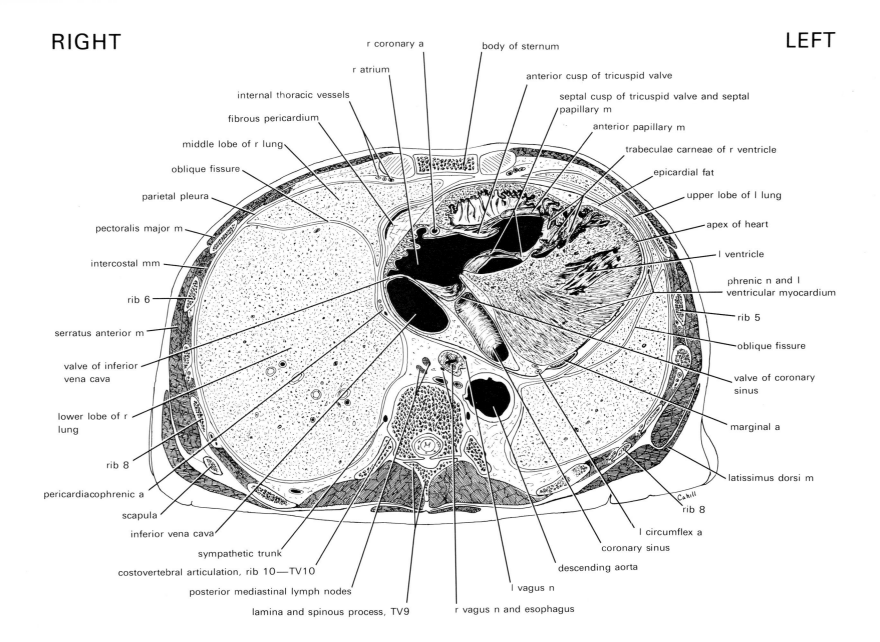

r coronary a
r atrium
internal thoracic vessels
fibrous pericardium
middle lobe of r lung
oblique fissure
parietal pleura
pectoralis major m
intercostal mm
rib 6
serratus anterior m
valve of inferior vena cava
lower lobe of r lung
rib 8
pericardiacophrenic a
scapula
inferior vena cava
sympathetic trunk
costovertebral articulation, rib 10—TV10
posterior mediastinal lymph nodes
lamina and spinous process, TV9

body of sternum
anterior cusp of tricuspid valve
septal cusp of tricuspid valve and septal papillary m
anterior papillary m
trabeculae carneae of r ventricle
epicardial fat
upper lobe of l lung
apex of heart
l ventricle
phrenic n and l ventricular myocardium
rib 5
oblique fissure
valve of coronary sinus
marginal a
latissimus dorsi m
rib 8
l circumflex a
coronary sinus
descending aorta
l vagus n
r vagus n and esophagus

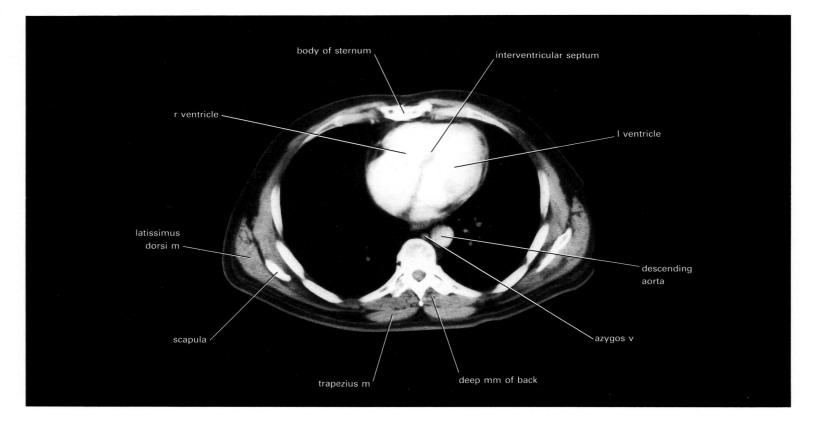

body of sternum
interventricular septum
r ventricle
l ventricle
latissimus dorsi m
descending aorta
scapula
azygos v
trapezius m
deep mm of back

Section 18 from below.

LEFT   RIGHT

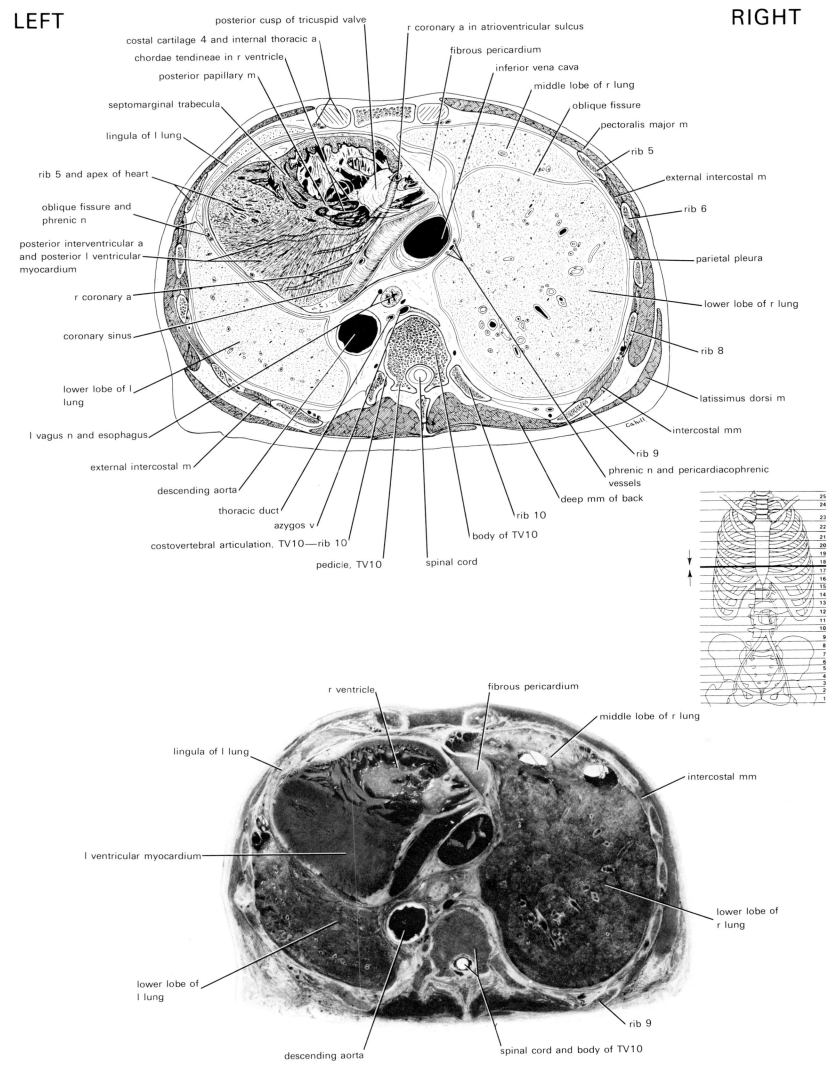

posterior cusp of tricuspid valve
costal cartilage 4 and internal thoracic a
chordae tendineae in r ventricle
posterior papillary m
septomarginal trabecula
lingula of l lung
rib 5 and apex of heart
oblique fissure and
phrenic n
posterior interventricular a
and posterior l ventricular
myocardium
r coronary a
coronary sinus
lower lobe of l
lung
l vagus n and esophagus
external intercostal m
descending aorta
thoracic duct
azygos v
costovertebral articulation, TV10—rib 10
pedicle, TV10

r coronary a in atrioventricular sulcus
fibrous pericardium
inferior vena cava
middle lobe of r lung
oblique fissure
pectoralis major m
rib 5
external intercostal m
rib 6
parietal pleura
lower lobe of r lung
rib 8
latissimus dorsi m
intercostal mm
rib 9
phrenic n and pericardiacophrenic
vessels
deep mm of back
rib 10
body of TV10
spinal cord

r ventricle
lingula of l lung
l ventricular myocardium
lower lobe of
l lung
descending aorta

fibrous pericardium
middle lobe of r lung
intercostal mm
lower lobe of
r lung
rib 9
spinal cord and body of TV10

Section 17 from above.

The Male Abdomen

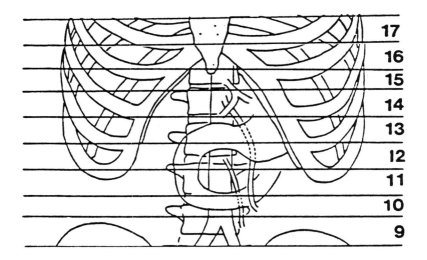

17
16
15
14
13
12
11
10
9

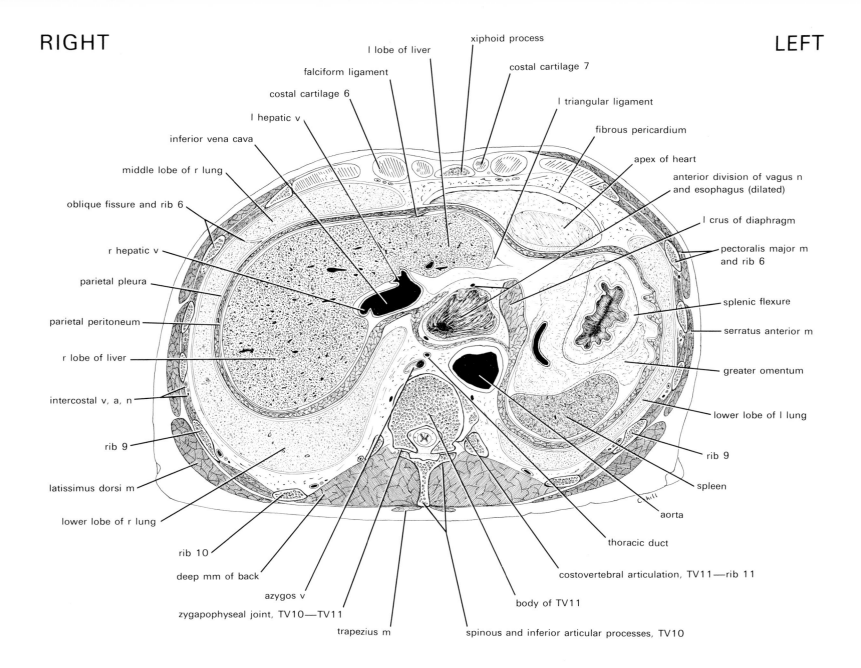

xiphoid process
l lobe of liver
costal cartilage 7
falciform ligament
costal cartilage 6
l triangular ligament
l hepatic v
fibrous pericardium
inferior vena cava
apex of heart
middle lobe of r lung
anterior division of vagus n
and esophagus (dilated)
oblique fissure and rib 6
l crus of diaphragm
r hepatic v
pectoralis major m
and rib 6
parietal pleura
splenic flexure
parietal peritoneum
serratus anterior m
r lobe of liver
greater omentum
intercostal v, a, n
lower lobe of l lung
rib 9
rib 9
latissimus dorsi m
spleen
lower lobe of r lung
aorta
rib 10
thoracic duct
deep mm of back
costovertebral articulation, TV11—rib 11
azygos v
body of TV11
zygapophyseal joint, TV10—TV11
spinous and inferior articular processes, TV10
trapezius m

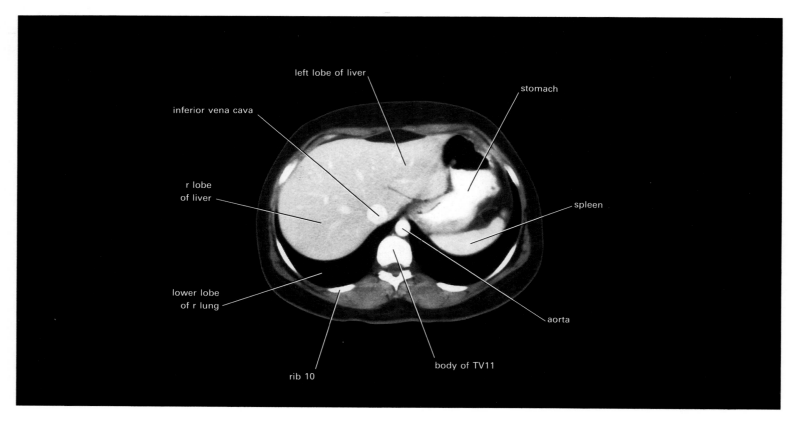

left lobe of liver
stomach
inferior vena cava
r lobe
of liver
spleen
lower lobe
of r lung
aorta
rib 10
body of TV11

Section 17 from below.

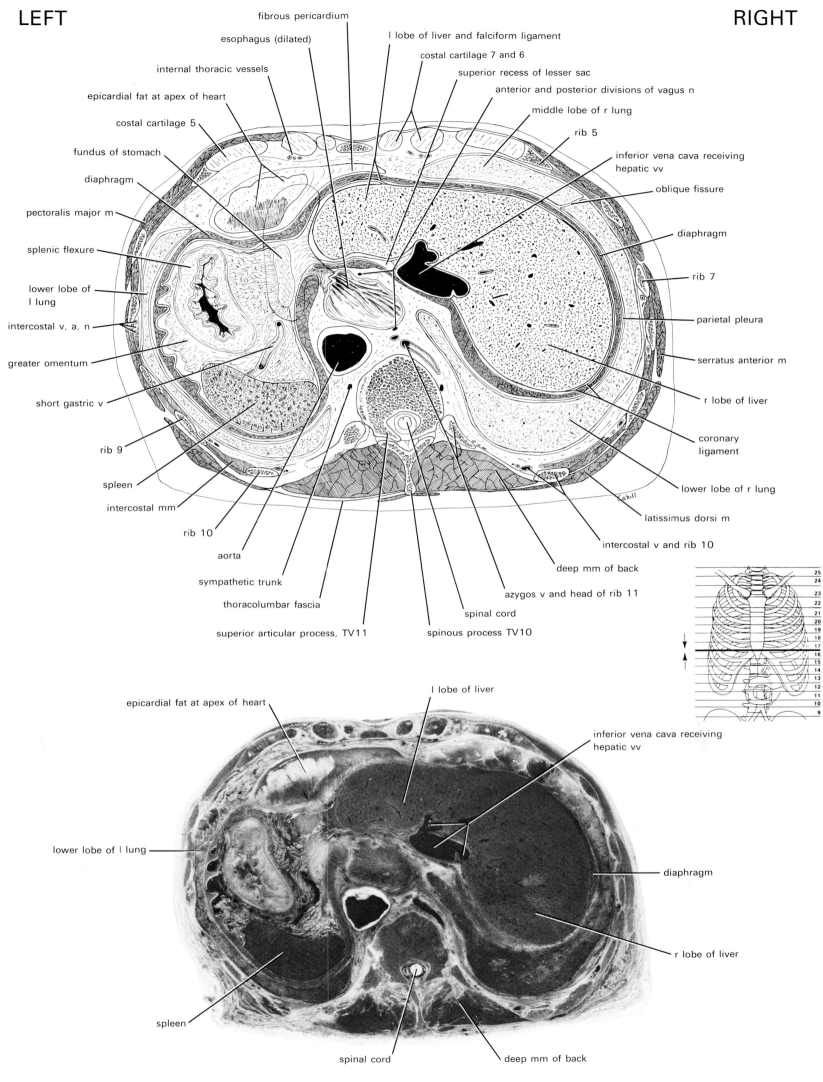

LEFT   RIGHT

fibrous pericardium
esophagus (dilated)
l lobe of liver and falciform ligament
internal thoracic vessels
costal cartilage 7 and 6
epicardial fat at apex of heart
superior recess of lesser sac
costal cartilage 5
anterior and posterior divisions of vagus n
fundus of stomach
middle lobe of r lung
diaphragm
rib 5
pectoralis major m
inferior vena cava receiving hepatic vv
splenic flexure
oblique fissure
lower lobe of l lung
diaphragm
intercostal v, a, n
rib 7
greater omentum
parietal pleura
short gastric v
serratus anterior m
rib 9
r lobe of liver
spleen
coronary ligament
intercostal mm
lower lobe of r lung
rib 10
latissimus dorsi m
aorta
intercostal v and rib 10
sympathetic trunk
deep mm of back
thoracolumbar fascia
azygos v and head of rib 11
superior articular process, TV11
spinal cord
spinous process TV10

epicardial fat at apex of heart
l lobe of liver
inferior vena cava receiving hepatic vv
lower lobe of l lung
diaphragm
r lobe of liver
spleen
spinal cord
deep mm of back

Section 16 from above.

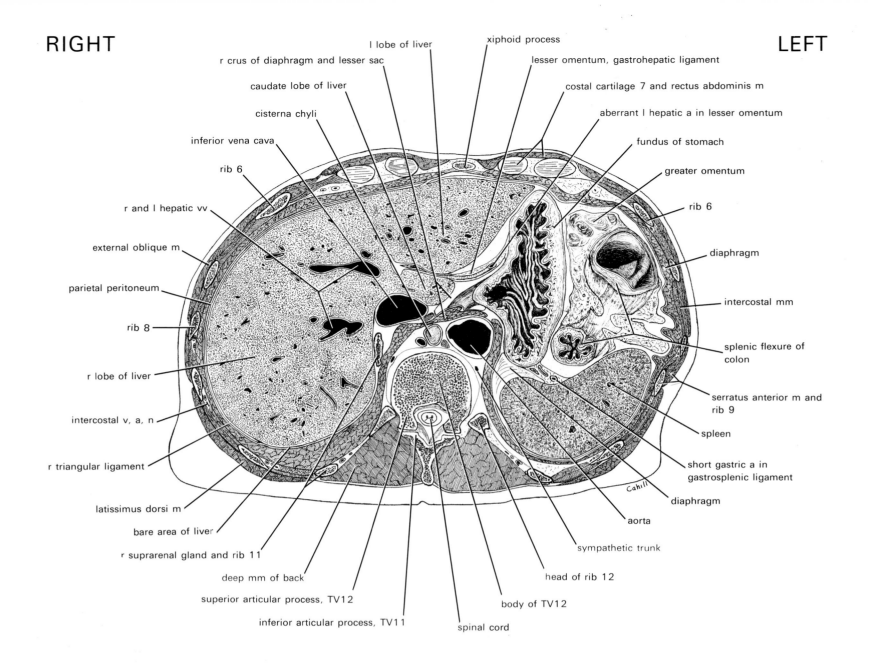

l lobe of liver
xiphoid process
r crus of diaphragm and lesser sac
lesser omentum, gastrohepatic ligament
caudate lobe of liver
costal cartilage 7 and rectus abdominis m
cisterna chyli
aberrant l hepatic a in lesser omentum
inferior vena cava
fundus of stomach
rib 6
greater omentum
r and l hepatic vv
rib 6
external oblique m
diaphragm
parietal peritoneum
intercostal mm
rib 8
splenic flexure of colon
r lobe of liver
serratus anterior m and rib 9
intercostal v, a, n
spleen
r triangular ligament
short gastric a in gastrosplenic ligament
latissimus dorsi m
diaphragm
bare area of liver
aorta
r suprarenal gland and rib 11
sympathetic trunk
deep mm of back
head of rib 12
superior articular process, TV12
body of TV12
inferior articular process, TV11
spinal cord

Cahill

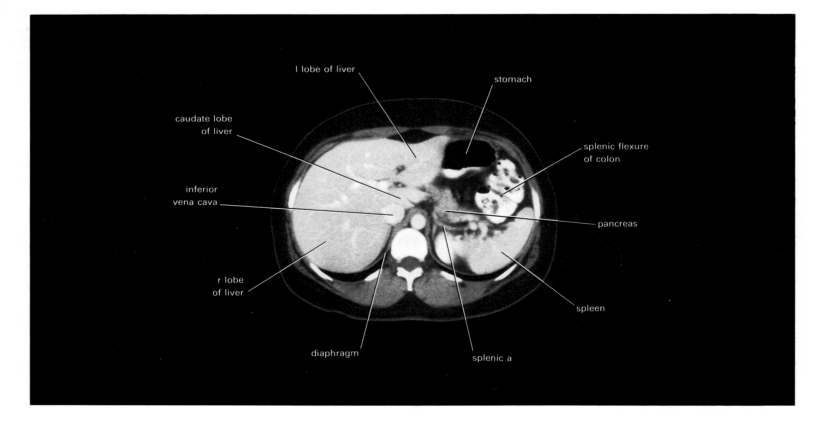

l lobe of liver
stomach
caudate lobe of liver
splenic flexure of colon
inferior vena cava
pancreas
r lobe of liver
spleen
diaphragm
splenic a

Section 16 from below.

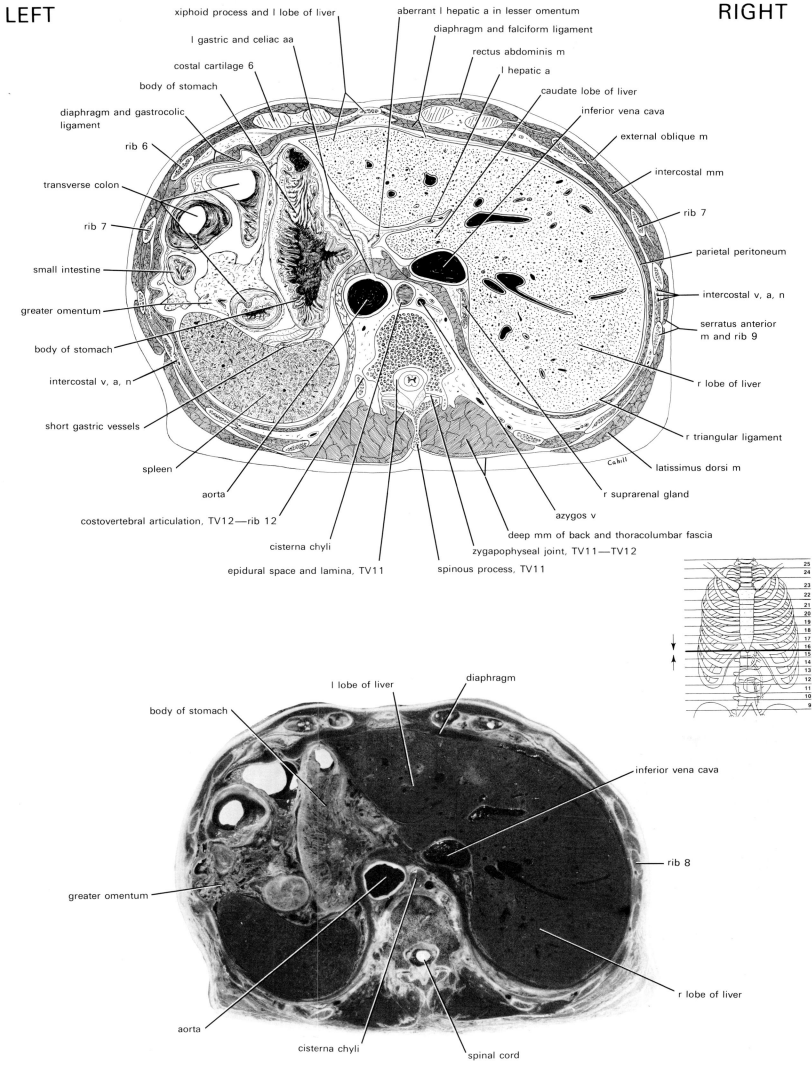

xiphoid process and l lobe of liver
aberrant l hepatic a in lesser omentum
diaphragm and falciform ligament
l gastric and celiac aa
rectus abdominis m
costal cartilage 6
l hepatic a
body of stomach
caudate lobe of liver
diaphragm and gastrocolic ligament
inferior vena cava
rib 6
external oblique m
transverse colon
intercostal mm
rib 7
rib 7
small intestine
parietal peritoneum
greater omentum
intercostal v, a, n
body of stomach
serratus anterior m and rib 9
intercostal v, a, n
r lobe of liver
short gastric vessels
r triangular ligament
spleen
latissimus dorsi m
aorta
r suprarenal gland
costovertebral articulation, TV12—rib 12
azygos v
deep mm of back and thoracolumbar fascia
cisterna chyli
zygapophyseal joint, TV11—TV12
epidural space and lamina, TV11
spinous process, TV11

Cahill

l lobe of liver
diaphragm
body of stomach
inferior vena cava
rib 8
greater omentum
r lobe of liver
aorta
cisterna chyli
spinal cord

Section 15 from above.

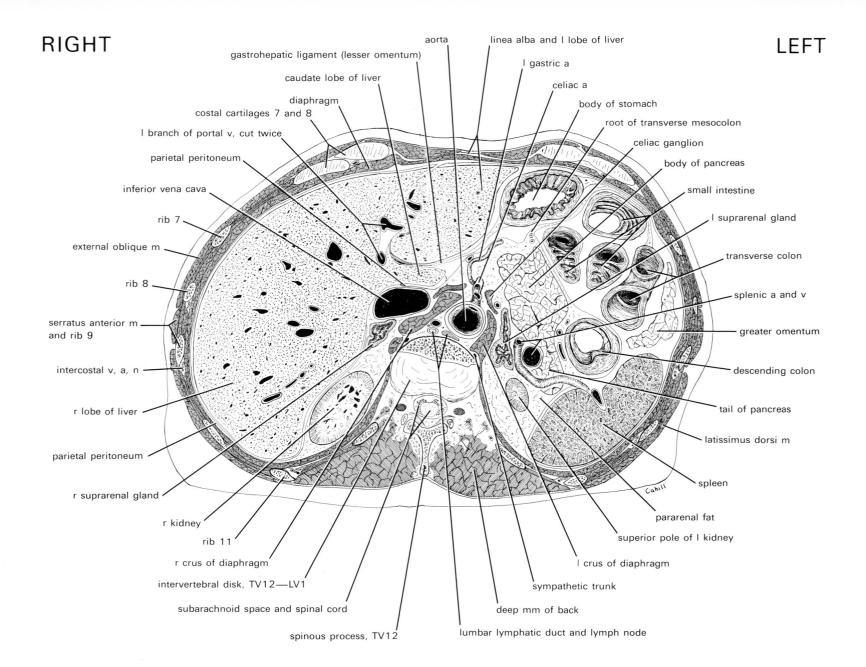

aorta
linea alba and l lobe of liver
gastrohepatic ligament (lesser omentum)
l gastric a
caudate lobe of liver
celiac a
diaphragm
body of stomach
costal cartilages 7 and 8
root of transverse mesocolon
l branch of portal v, cut twice
celiac ganglion
parietal peritoneum
body of pancreas
inferior vena cava
small intestine
rib 7
l suprarenal gland
external oblique m
transverse colon
rib 8
splenic a and v
serratus anterior m
and rib 9
greater omentum
intercostal v, a, n
descending colon
r lobe of liver
tail of pancreas
latissimus dorsi m
parietal peritoneum
spleen
r suprarenal gland
pararenal fat
r kidney
superior pole of l kidney
rib 11
l crus of diaphragm
r crus of diaphragm
intervertebral disk, TV12—LV1
sympathetic trunk
subarachnoid space and spinal cord
deep mm of back
spinous process, TV12
lumbar lymphatic duct and lymph node

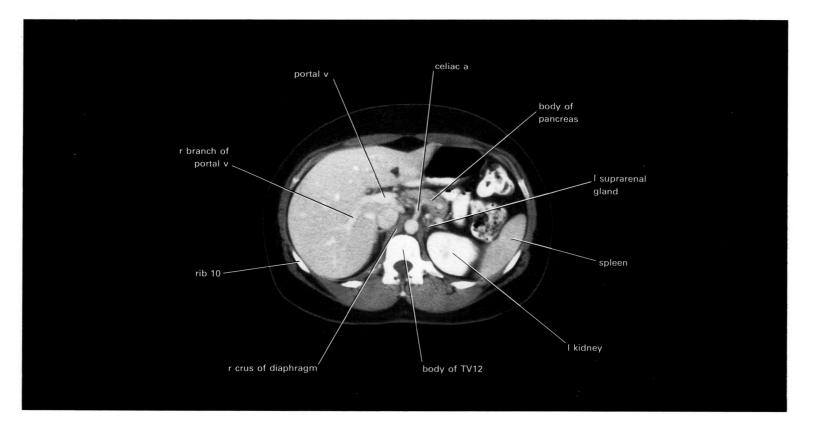

portal v
celiac a
body of
pancreas
r branch of
portal v
l suprarenal
gland
l kidney
rib 10
spleen
r crus of diaphragm
body of TV12

Section 15 from below.

falciform ligament and l lobe of liver

caudate lobe of liver

common hepatic a

round ligament of liver

aorta and celiac ganglion

rectus abdominis m

celiac a

inferior vena cava

body of stomach

costal cartilage 7

gastroepiploic vessels

parietal peritoneum

root of transverse mesocolon

diaphragm

body of pancreas

external oblique m

rib 7

rib 7

peritoneal cavity

intercostal mm

rib 8

splenic a

serratus anterior m

greater omentum

rib 9

rib 9

r lobe of liver

descending colon

parietal peritoneum

latissimus dorsi m

rib 10

spleen

r suprarenal gland

tail of pancreas

r kidney

splenic v

sympathetic trunk

l kidney

rib 12

l suprarenal gland and l crus of diaphragm

spinal n T12

deep mm of back

zygapophyseal joint, TV12—LV1

body, TV12

spinal cord   lamina, TV12

Cahill

body of stomach   aorta

inferior vena cava

diaphragm

greater omentum

r lobe of liver

tail of pancreas

r kidney

l suprarenal gland and l kidney

spinal cord

Section 14 from above.

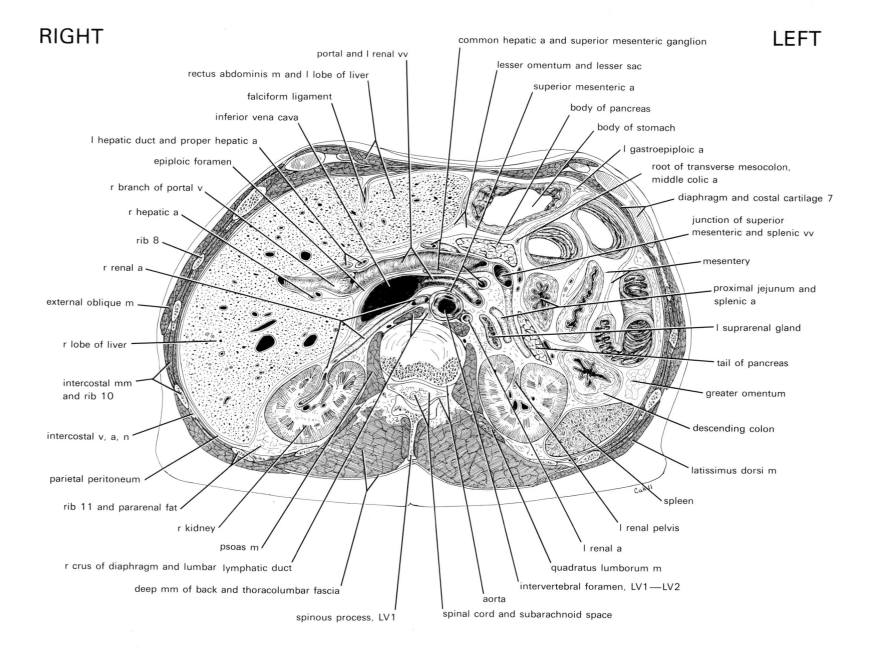

common hepatic a and superior mesenteric ganglion

portal and l renal vv

lesser omentum and lesser sac

rectus abdominis m and l lobe of liver

superior mesenteric a

falciform ligament

body of pancreas

inferior vena cava

body of stomach

l hepatic duct and proper hepatic a

l gastroepiploic a

epiploic foramen

root of transverse mesocolon, middle colic a

r branch of portal v

diaphragm and costal cartilage 7

r hepatic a

junction of superior mesenteric and splenic vv

rib 8

r renal a

mesentery

external oblique m

proximal jejunum and splenic a

r lobe of liver

l suprarenal gland

intercostal mm and rib 10

tail of pancreas

greater omentum

intercostal v, a, n

descending colon

parietal peritoneum

latissimus dorsi m

rib 11 and pararenal fat

spleen

r kidney

l renal pelvis

psoas m

l renal a

r crus of diaphragm and lumbar lymphatic duct

quadratus lumborum m

deep mm of back and thoracolumbar fascia

intervertebral foramen, LV1—LV2

aorta

spinous process, LV1

spinal cord and subarachnoid space

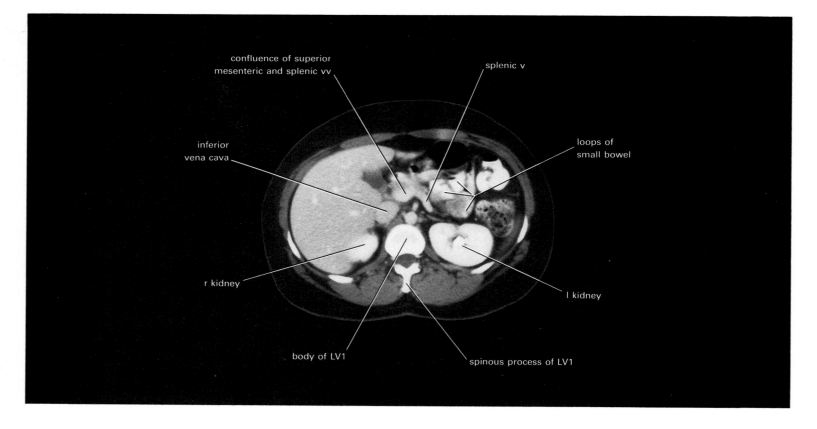

confluence of superior mesenteric and splenic vv

splenic v

inferior vena cava

loops of small bowel

r kidney

l kidney

body of LV1

spinous process of LV1

Section 14 from below.

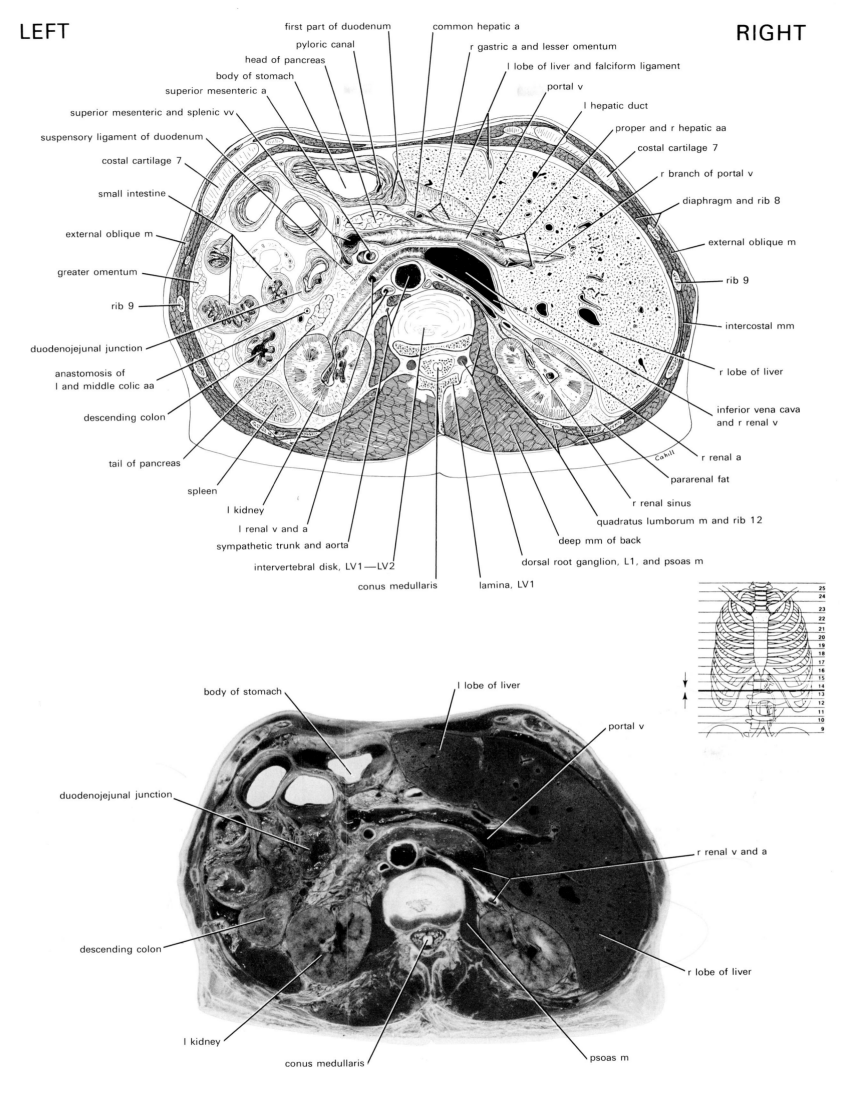

LEFT   RIGHT

first part of duodenum
common hepatic a
pyloric canal
r gastric a and lesser omentum
head of pancreas
l lobe of liver and falciform ligament
body of stomach
portal v
superior mesenteric a
l hepatic duct
superior mesenteric and splenic vv
proper and r hepatic aa
suspensory ligament of duodenum
costal cartilage 7
costal cartilage 7
r branch of portal v
small intestine
diaphragm and rib 8
external oblique m
external oblique m
greater omentum
rib 9
rib 9
intercostal mm
duodenojejunal junction
r lobe of liver
anastomosis of
l and middle colic aa
inferior vena cava
and r renal v
descending colon
r renal a
tail of pancreas
pararenal fat
spleen
r renal sinus
l kidney
quadratus lumborum m and rib 12
l renal v and a
deep mm of back
sympathetic trunk and aorta
dorsal root ganglion, L1, and psoas m
intervertebral disk, LV1—LV2
lamina, LV1
conus medullaris

body of stomach
l lobe of liver
duodenojejunal junction
portal v
r renal v and a
descending colon
r lobe of liver
l kidney
conus medullaris
psoas m

Section 13 from above.

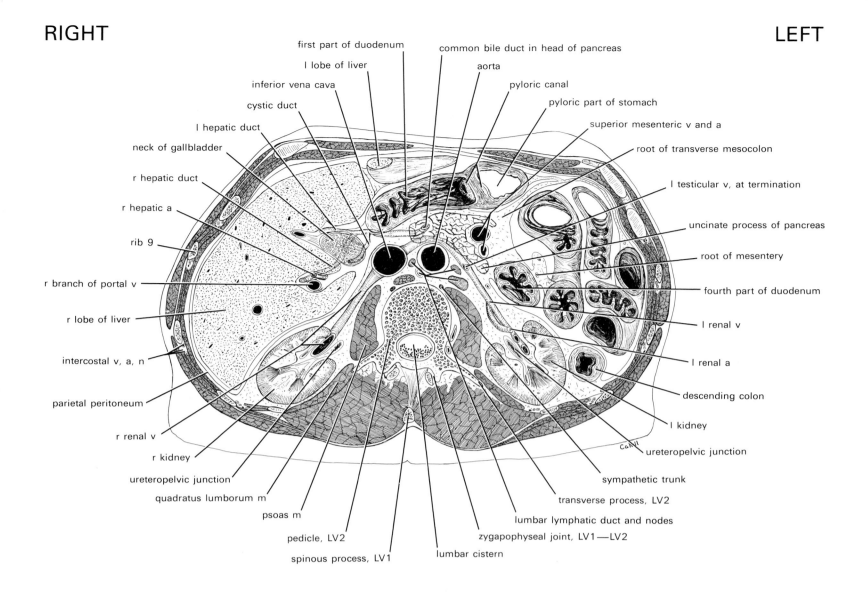

first part of duodenum
l lobe of liver
inferior vena cava
cystic duct
l hepatic duct
neck of gallbladder
r hepatic duct
r hepatic a
rib 9
r branch of portal v
r lobe of liver
intercostal v, a, n
parietal peritoneum
r renal v
r kidney
ureteropelvic junction
quadratus lumborum m
psoas m
pedicle, LV2
spinous process, LV1

common bile duct in head of pancreas
aorta
pyloric canal
pyloric part of stomach
superior mesenteric v and a
root of transverse mesocolon
l testicular v, at termination
uncinate process of pancreas
root of mesentery
fourth part of duodenum
l renal v
l renal a
descending colon
l kidney
ureteropelvic junction
sympathetic trunk
transverse process, LV2
lumbar lymphatic duct and nodes
zygapophyseal joint, LV1—LV2
lumbar cistern

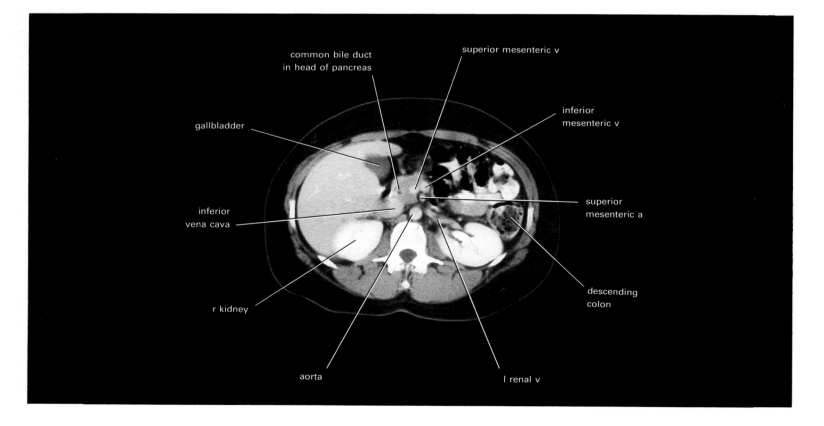

common bile duct
in head of pancreas
gallbladder
inferior
vena cava
r kidney
aorta

superior mesenteric v
inferior
mesenteric v
superior
mesenteric a
descending
colon
l renal v

Section 13 from above.

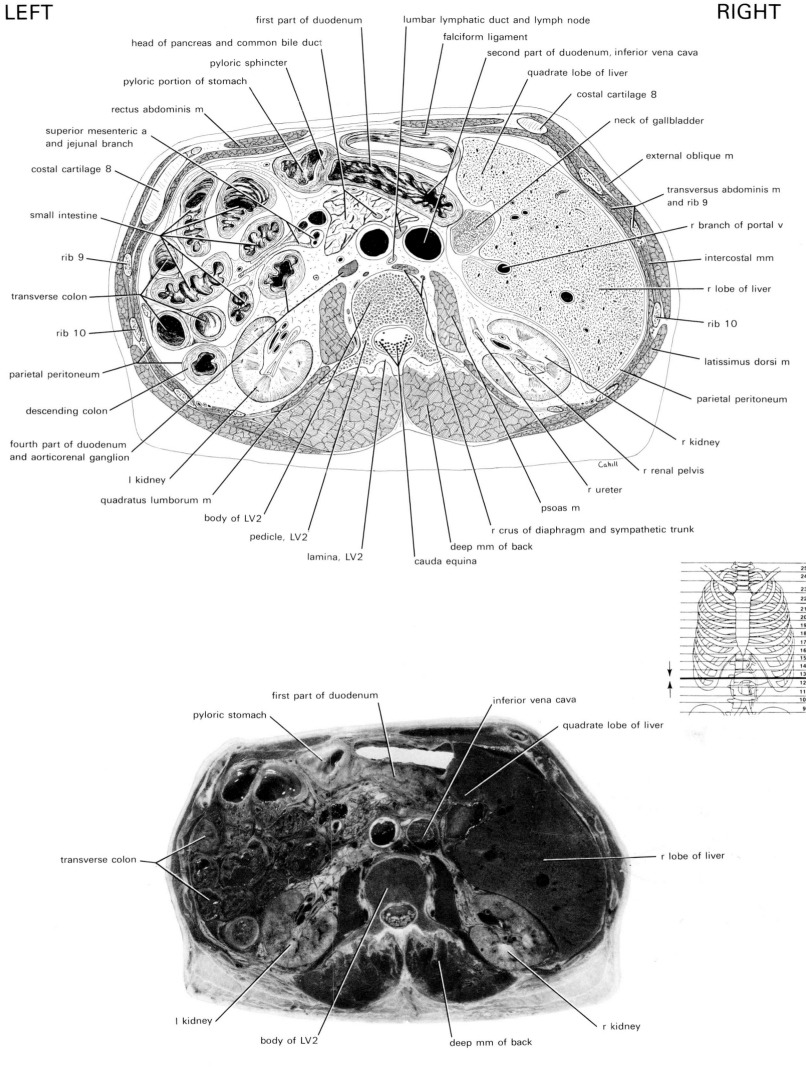

LEFT   RIGHT

first part of duodenum

lumbar lymphatic duct and lymph node

head of pancreas and common bile duct

falciform ligament

pyloric sphincter

second part of duodenum, inferior vena cava

pyloric portion of stomach

quadrate lobe of liver

costal cartilage 8

rectus abdominis m

neck of gallbladder

superior mesenteric a
and jejunal branch

external oblique m

costal cartilage 8

transversus abdominis m
and rib 9

small intestine

r branch of portal v

rib 9

intercostal mm

r lobe of liver

transverse colon

rib 10

rib 10

latissimus dorsi m

parietal peritoneum

parietal peritoneum

descending colon

r kidney

fourth part of duodenum
and aorticorenal ganglion

r renal pelvis

l kidney

r ureter

quadratus lumborum m

psoas m

body of LV2

r crus of diaphragm and sympathetic trunk

pedicle, LV2

deep mm of back

lamina, LV2

cauda equina

Cahill

first part of duodenum

inferior vena cava

pyloric stomach

quadrate lobe of liver

transverse colon

r lobe of liver

l kidney

r kidney

body of LV2

deep mm of back

Section 12 from above.

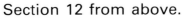

ATLAS OF HUMAN CROSS-SECTIONAL ANATOMY   31

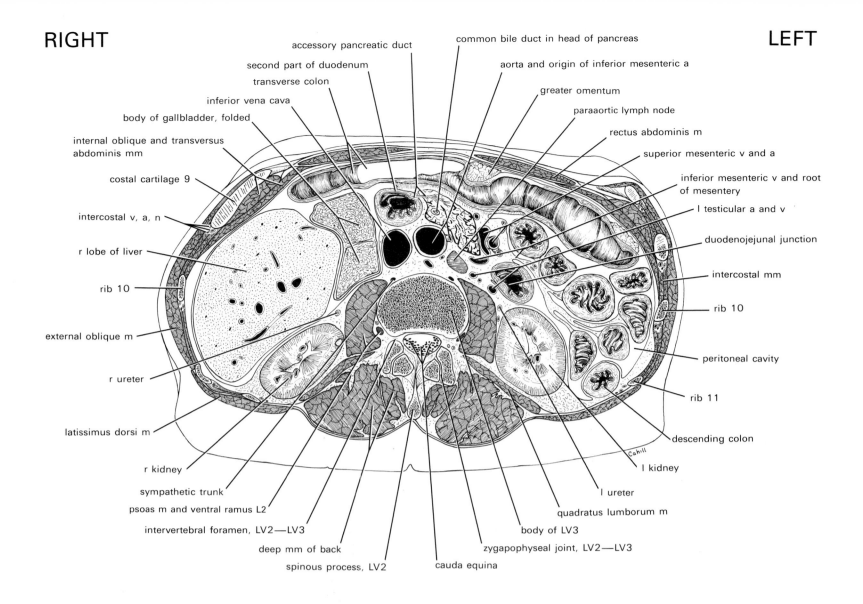

accessory pancreatic duct
second part of duodenum
transverse colon
inferior vena cava
body of gallbladder, folded
internal oblique and transversus
abdominis mm
costal cartilage 9
intercostal v, a, n
r lobe of liver
rib 10
external oblique m
r ureter
latissimus dorsi m
r kidney
sympathetic trunk
psoas m and ventral ramus L2
intervertebral foramen, LV2—LV3
deep mm of back
spinous process, LV2

common bile duct in head of pancreas
aorta and origin of inferior mesenteric a
greater omentum
paraaortic lymph node
rectus abdominis m
superior mesenteric v and a
inferior mesenteric v and root
of mesentery
l testicular a and v
duodenojejunal junction
intercostal mm
rib 10
peritoneal cavity
rib 11
descending colon
l kidney
l ureter
quadratus lumborum m
body of LV3
zygapophyseal joint, LV2—LV3
cauda equina

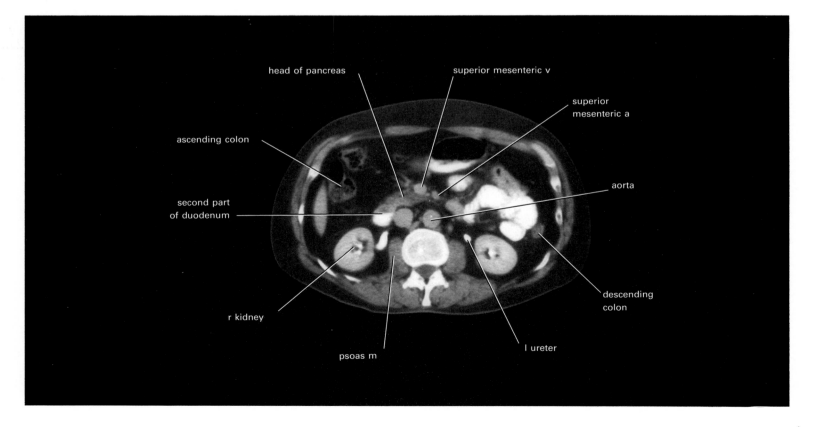

head of pancreas
superior mesenteric v
superior
mesenteric a
ascending colon
aorta
second part
of duodenum
r kidney
descending
colon
psoas m
l ureter

Section 12 from below.

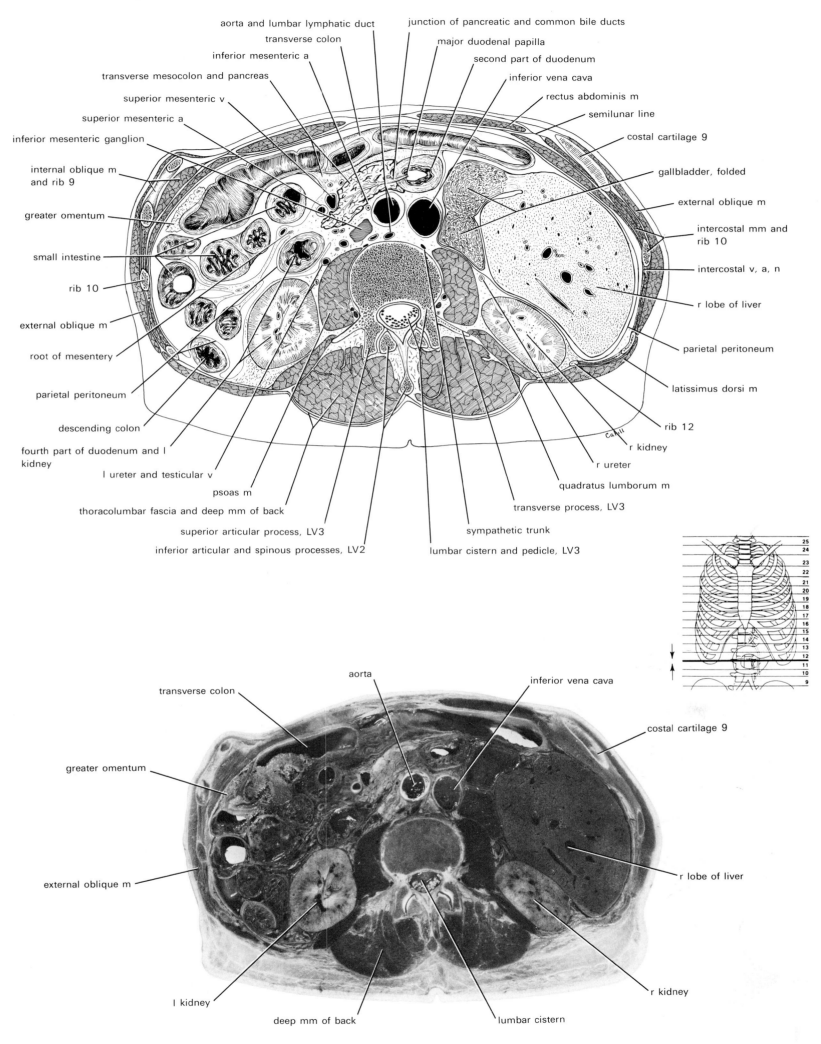

aorta and lumbar lymphatic duct

transverse colon

inferior mesenteric a

transverse mesocolon and pancreas

superior mesenteric v

superior mesenteric a

inferior mesenteric ganglion

internal oblique m and rib 9

greater omentum

small intestine

rib 10

external oblique m

root of mesentery

parietal peritoneum

descending colon

fourth part of duodenum and l kidney

l ureter and testicular v

psoas m

thoracolumbar fascia and deep mm of back

superior articular process, LV3

inferior articular and spinous processes, LV2

junction of pancreatic and common bile ducts

major duodenal papilla

second part of duodenum

inferior vena cava

rectus abdominis m

semilunar line

costal cartilage 9

gallbladder, folded

external oblique m

intercostal mm and rib 10

intercostal v, a, n

r lobe of liver

parietal peritoneum

latissimus dorsi m

rib 12

r kidney

r ureter

quadratus lumborum m

transverse process, LV3

sympathetic trunk

lumbar cistern and pedicle, LV3

transverse colon

aorta

inferior vena cava

greater omentum

costal cartilage 9

external oblique m

r lobe of liver

l kidney

r kidney

deep mm of back

lumbar cistern

Section 11 from above.

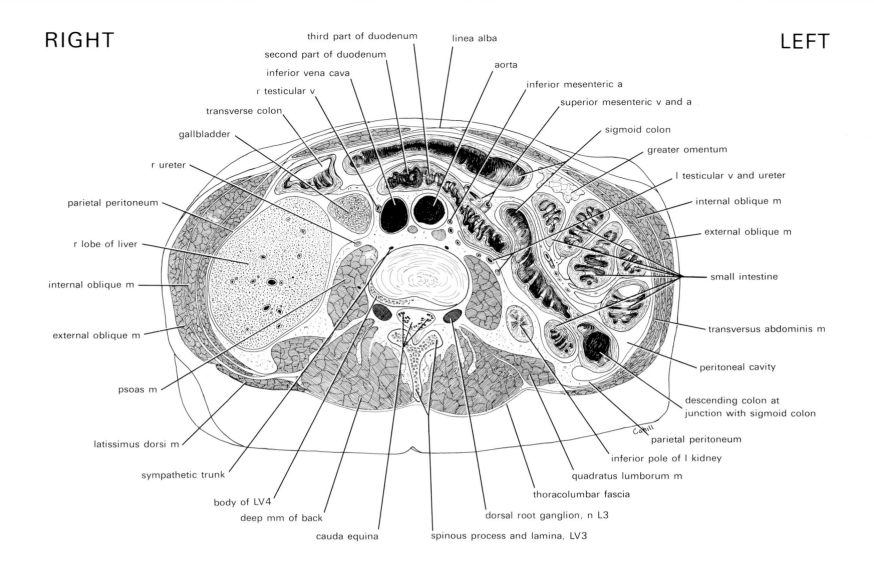

third part of duodenum
second part of duodenum
inferior vena cava
r testicular v
transverse colon
gallbladder
r ureter
parietal peritoneum
r lobe of liver
internal oblique m
external oblique m
psoas m
latissimus dorsi m
sympathetic trunk
body of LV4
deep mm of back
cauda equina

linea alba
aorta
inferior mesenteric a
superior mesenteric v and a
sigmoid colon
greater omentum
l testicular v and ureter
internal oblique m
external oblique m
small intestine
transversus abdominis m
peritoneal cavity
descending colon at junction with sigmoid colon
parietal peritoneum
inferior pole of l kidney
quadratus lumborum m
thoracolumbar fascia
dorsal root ganglion, n L3
spinous process and lamina, LV3

Cahill

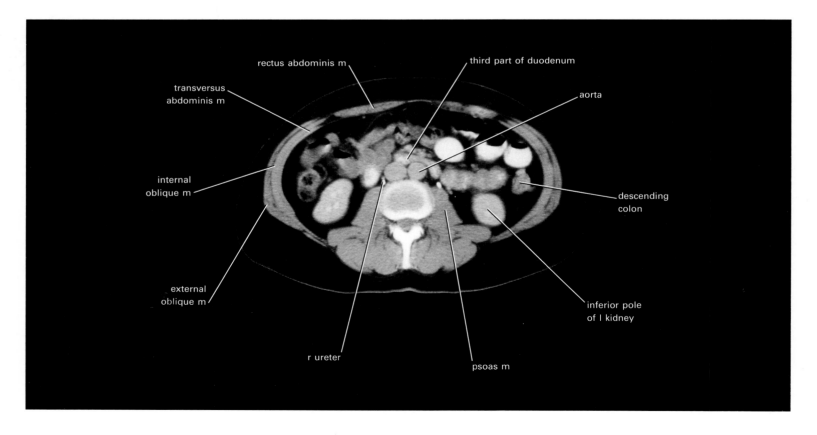

rectus abdominis m
transversus abdominis m
internal oblique m
external oblique m
r ureter

third part of duodenum
aorta
descending colon
inferior pole of l kidney
psoas m

Section 11 from below.

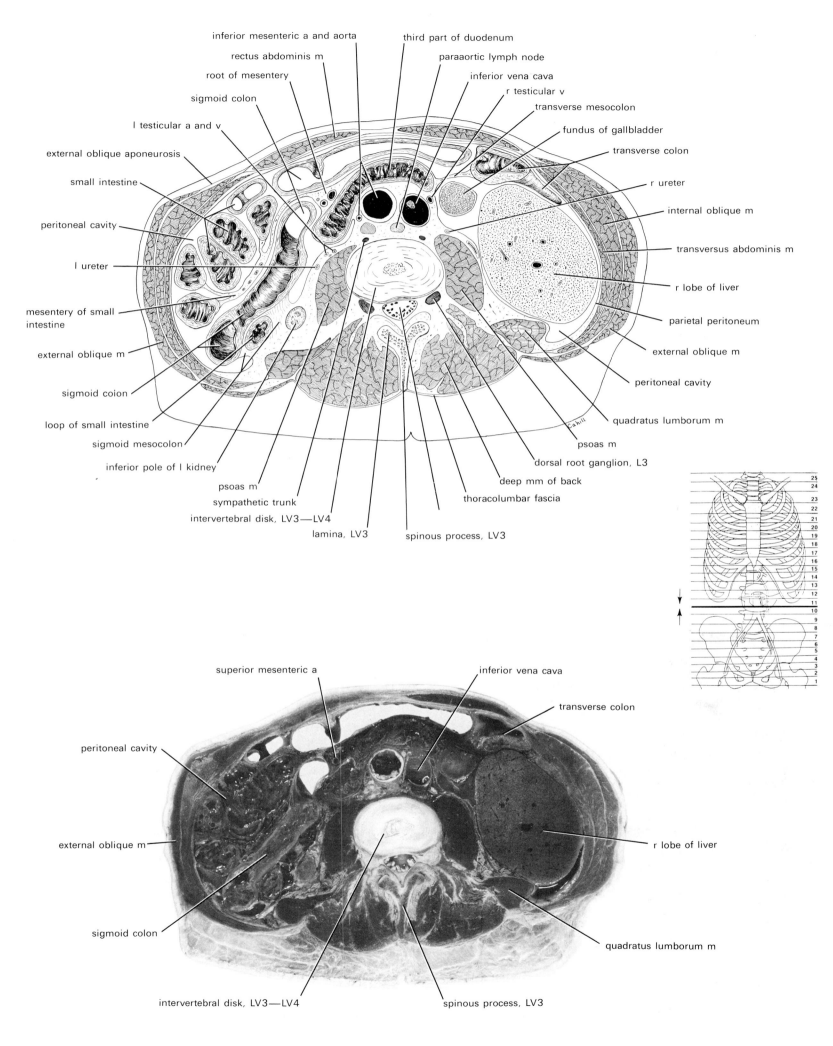

inferior mesenteric a and aorta
rectus abdominis m
root of mesentery
sigmoid colon
l testicular a and v
external oblique aponeurosis
small intestine
peritoneal cavity
l ureter
mesentery of small intestine
external oblique m
sigmoid coion
loop of small intestine
sigmoid mesocolon
inferior pole of l kidney
psoas m
sympathetic trunk
intervertebral disk, LV3—LV4
lamina, LV3

third part of duodenum
paraaortic lymph node
inferior vena cava
r testicular v
transverse mesocolon
fundus of gallbladder
transverse colon
r ureter
internal oblique m
transversus abdominis m
r lobe of liver
parietal peritoneum
external oblique m
peritoneal cavity
quadratus lumborum m
psoas m
dorsal root ganglion, L3
deep mm of back
thoracolumbar fascia
spinous process, LV3

superior mesenteric a
inferior vena cava
transverse colon
peritoneal cavity
r lobe of liver
external oblique m
sigmoid colon
quadratus lumborum m
intervertebral disk, LV3—LV4
spinous process, LV3

Section 10 from above.

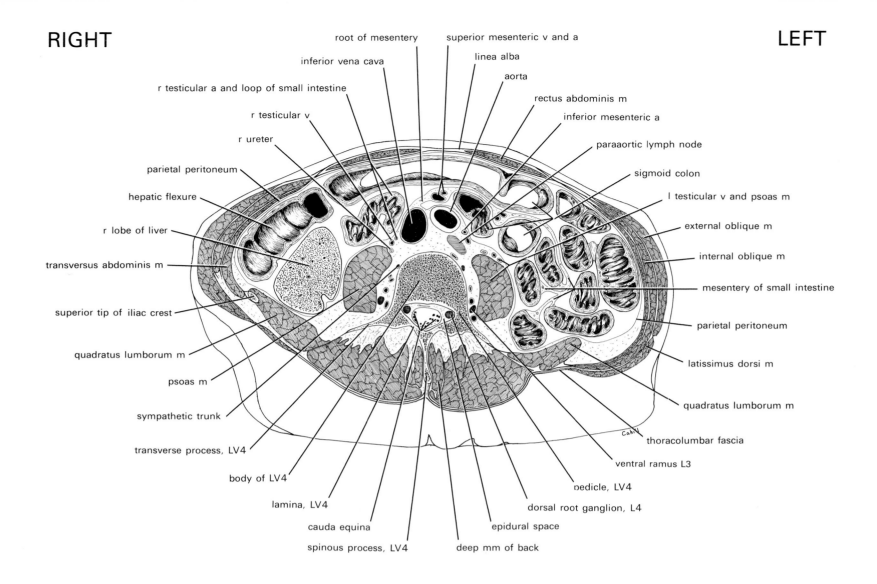

root of mesentery
superior mesenteric v and a
inferior vena cava
linea alba
aorta
r testicular a and loop of small intestine
rectus abdominis m
r testicular v
inferior mesenteric a
r ureter
paraaortic lymph node
parietal peritoneum
sigmoid colon
hepatic flexure
l testicular v and psoas m
r lobe of liver
external oblique m
transversus abdominis m
internal oblique m
mesentery of small intestine
superior tip of iliac crest
parietal peritoneum
quadratus lumborum m
latissimus dorsi m
psoas m
quadratus lumborum m
sympathetic trunk
thoracolumbar fascia
transverse process, LV4
ventral ramus L3
body of LV4
pedicle, LV4
lamina, LV4
dorsal root ganglion, L4
cauda equina
epidural space
spinous process, LV4
deep mm of back

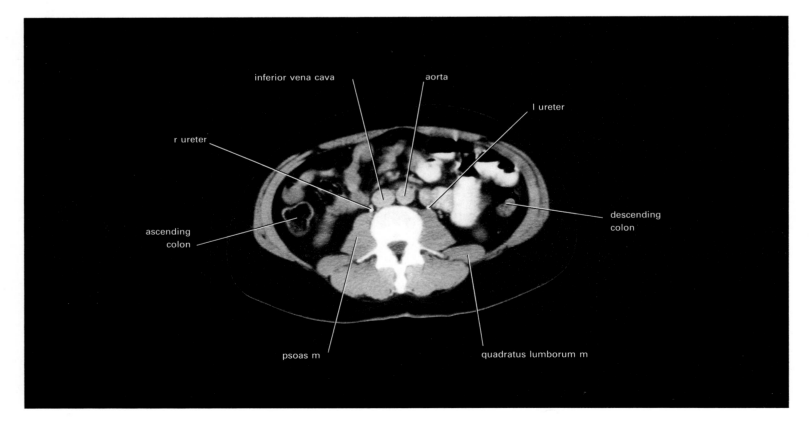

inferior vena cava
aorta
l ureter
r ureter
ascending colon
descending colon
psoas m
quadratus lumborum m

Section 10 from below.

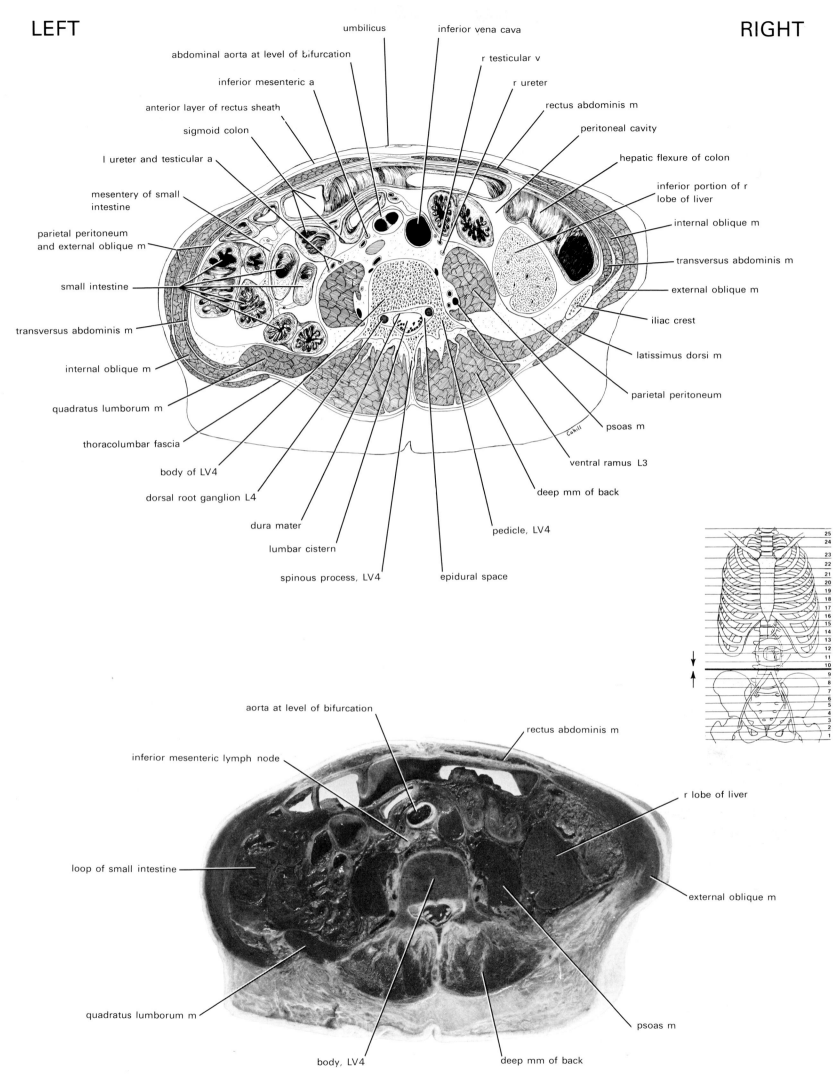

LEFT

RIGHT

umbilicus

abdominal aorta at level of bifurcation

inferior mesenteric a

anterior layer of rectus sheath

sigmoid colon

l ureter and testicular a

mesentery of small intestine

parietal peritoneum and external oblique m

small intestine

transversus abdominis m

internal oblique m

quadratus lumborum m

thoracolumbar fascia

body of LV4

dorsal root ganglion L4

dura mater

lumbar cistern

spinous process, LV4

inferior vena cava

r testicular v

r ureter

rectus abdominis m

peritoneal cavity

hepatic flexure of colon

inferior portion of r lobe of liver

internal oblique m

transversus abdominis m

external oblique m

iliac crest

latissimus dorsi m

parietal peritoneum

psoas m

ventral ramus L3

deep mm of back

pedicle, LV4

epidural space

aorta at level of bifurcation

inferior mesenteric lymph node

rectus abdominis m

r lobe of liver

loop of small intestine

external oblique m

quadratus lumborum m

psoas m

body, LV4

deep mm of back

Section 9 from above.

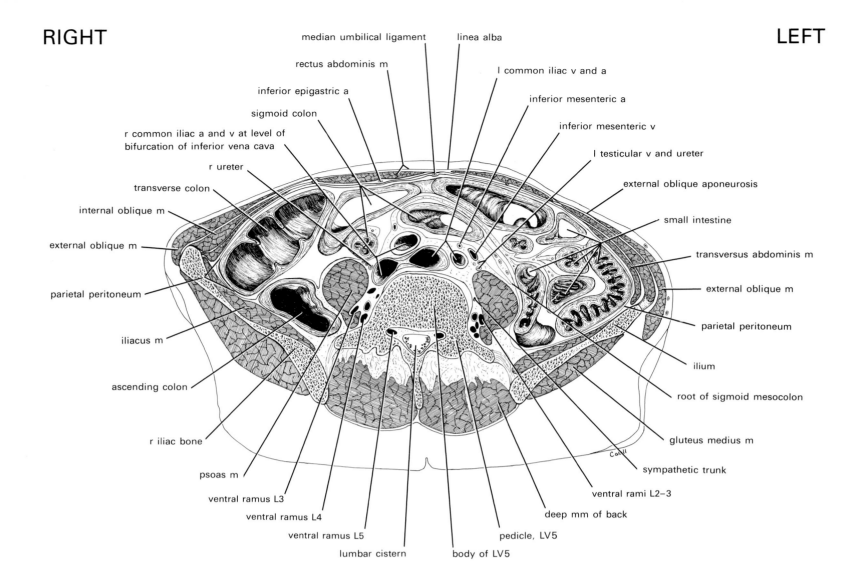

median umbilical ligament
linea alba
rectus abdominis m
l common iliac v and a
inferior epigastric a
inferior mesenteric a
sigmoid colon
inferior mesenteric v
r common iliac a and v at level of
bifurcation of inferior vena cava
l testicular v and ureter
r ureter
external oblique aponeurosis
transverse colon
small intestine
internal oblique m
transversus abdominis m
external oblique m
external oblique m
parietal peritoneum
parietal peritoneum
ilium
iliacus m
root of sigmoid mesocolon
ascending colon
gluteus medius m
r iliac bone
sympathetic trunk
psoas m
ventral rami L2–3
ventral ramus L3
deep mm of back
ventral ramus L4
ventral ramus L5
pedicle, LV5
lumbar cistern
body of LV5

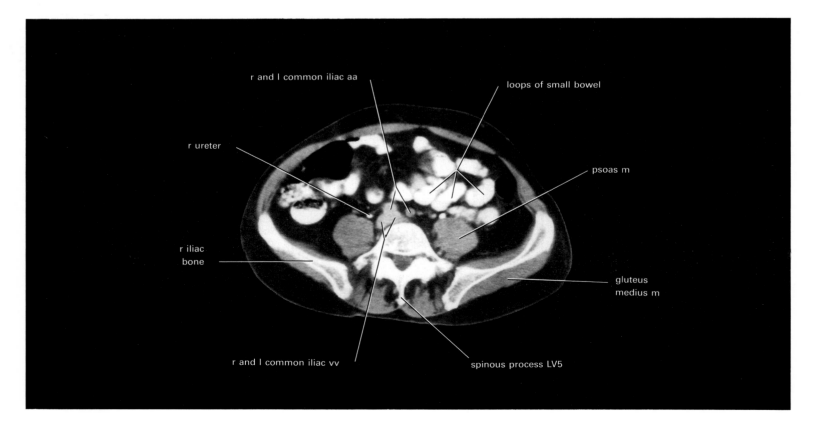

r and l common iliac aa
loops of small bowel
r ureter
psoas m
r iliac
bone
gluteus
medius m
r and l common iliac vv
spinous process LV5

Section 9 from below.

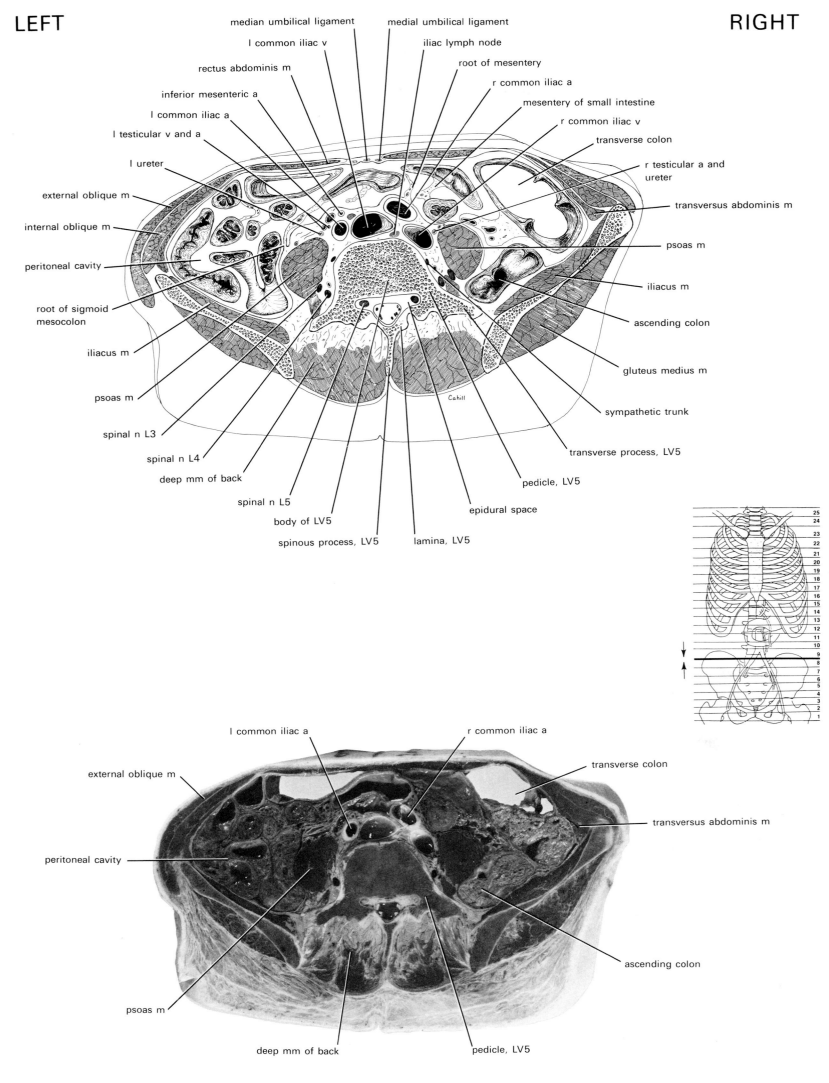

LEFT   RIGHT

median umbilical ligament
l common iliac v
rectus abdominis m
inferior mesenteric a
l common iliac a
l testicular v and a
l ureter
external oblique m
internal oblique m
peritoneal cavity
root of sigmoid
mesocolon
iliacus m
psoas m
spinal n L3
spinal n L4
deep mm of back
spinal n L5
body of LV5
spinous process, LV5
lamina, LV5

medial umbilical ligament
iliac lymph node
root of mesentery
r common iliac a
mesentery of small intestine
r common iliac v
transverse colon
r testicular a and
ureter
transversus abdominis m
psoas m
iliacus m
ascending colon
gluteus medius m
sympathetic trunk
transverse process, LV5
pedicle, LV5
epidural space

Cahill

l common iliac a
r common iliac a
external oblique m
transverse colon
transversus abdominis m
peritoneal cavity
ascending colon
psoas m
deep mm of back
pedicle, LV5

25
24
23
22
21
20
19
18
17
16
15
14
13
12
11
10
9
8
7
6
5
4
3
2
1

Section 8 from above.

Upper Abdomen Supplement

The next four sections, drawn from a second cadaver, illustrate a common variation of upper abdominal anatomy referenced by levels 16–13.

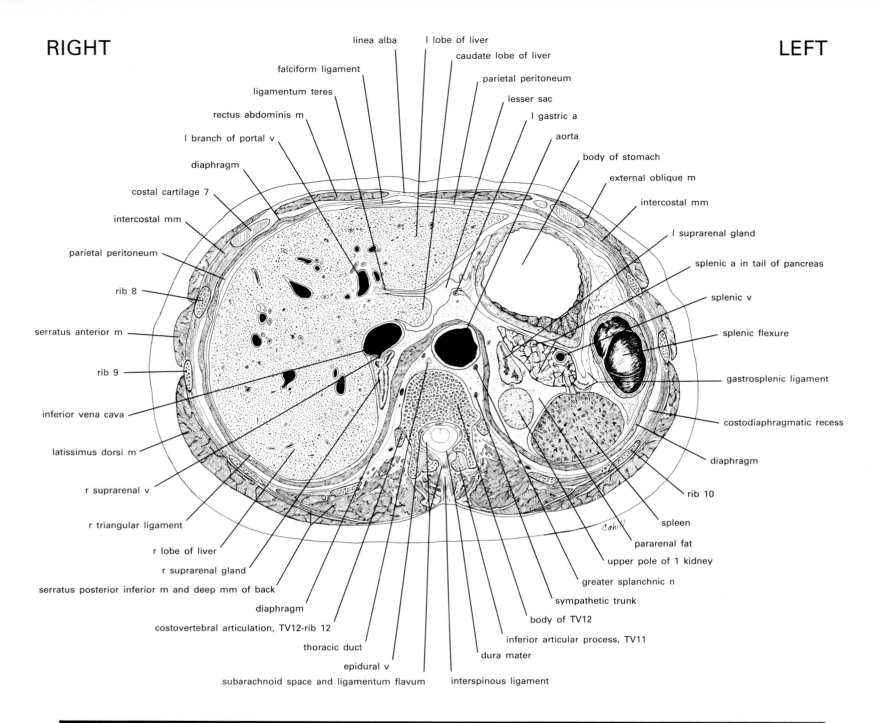

linea alba
l lobe of liver
falciform ligament
caudate lobe of liver
ligamentum teres
parietal peritoneum
rectus abdominis m
lesser sac
l gastric a
l branch of portal v
aorta
diaphragm
body of stomach
costal cartilage 7
external oblique m
intercostal mm
intercostal mm
l suprarenal gland
parietal peritoneum
splenic a in tail of pancreas
rib 8
splenic v
serratus anterior m
splenic flexure
rib 9
gastrosplenic ligament
inferior vena cava
costodiaphragmatic recess
latissimus dorsi m
diaphragm
r suprarenal v
rib 10
r triangular ligament
spleen
r lobe of liver
pararenal fat
r suprarenal gland
upper pole of 1 kidney
serratus posterior inferior m and deep mm of back
greater splanchnic n
diaphragm
sympathetic trunk
costovertebral articulation, TV12-rib 12
body of TV12
thoracic duct
inferior articular process, TV11
epidural v
dura mater
subarachnoid space and ligamentum flavum
interspinous ligament

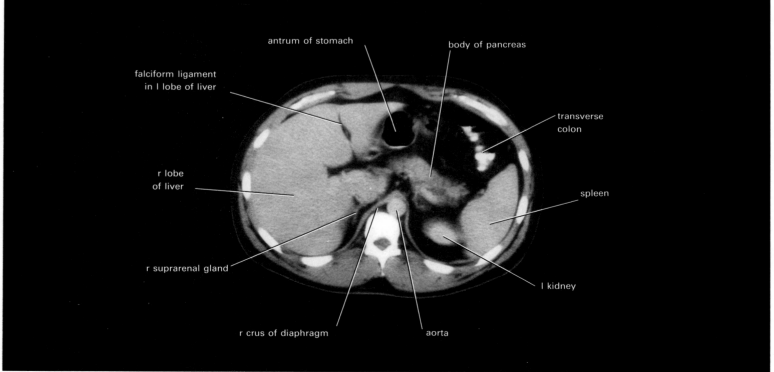

antrum of stomach
body of pancreas
falciform ligament
in l lobe of liver
transverse
colon
r lobe
of liver
spleen
r suprarenal gland
l kidney
r crus of diaphragm
aorta

Supplemental level 16.

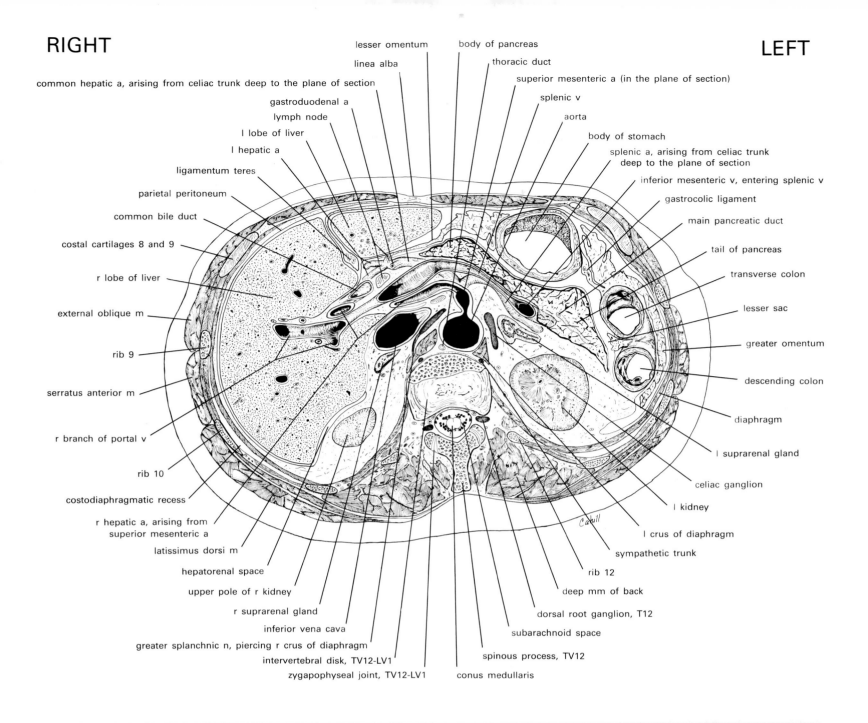

lesser omentum
body of pancreas
linea alba
thoracic duct
common hepatic a, arising from celiac trunk deep to the plane of section
superior mesenteric a (in the plane of section)
gastroduodenal a
splenic v
lymph node
aorta
l lobe of liver
body of stomach
l hepatic a
splenic a, arising from celiac trunk
deep to the plane of section
ligamentum teres
inferior mesenteric v, entering splenic v
parietal peritoneum
gastrocolic ligament
common bile duct
main pancreatic duct
costal cartilages 8 and 9
tail of pancreas
r lobe of liver
transverse colon
external oblique m
lesser sac
rib 9
greater omentum
serratus anterior m
descending colon
r branch of portal v
diaphragm
rib 10
l suprarenal gland
costodiaphragmatic recess
celiac ganglion
r hepatic a, arising from
superior mesenteric a
l kidney
latissimus dorsi m
l crus of diaphragm
hepatorenal space
sympathetic trunk
upper pole of r kidney
rib 12
r suprarenal gland
deep mm of back
inferior vena cava
dorsal root ganglion, T12
greater splanchnic n, piercing r crus of diaphragm
subarachnoid space
intervertebral disk, TV12-LV1
spinous process, TV12
zygapophyseal joint, TV12-LV1
conus medullaris

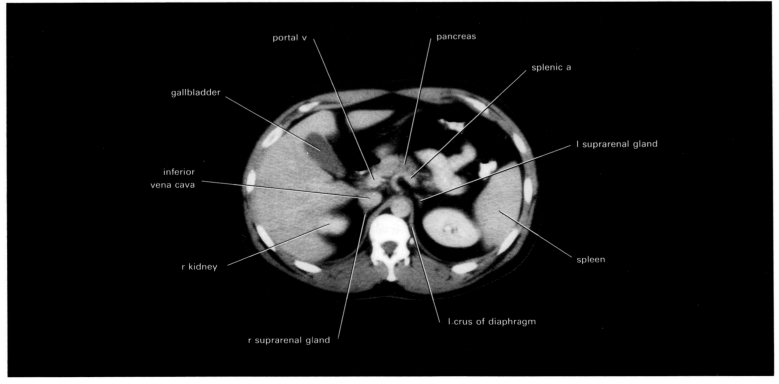

portal v
pancreas
splenic a
gallbladder
l suprarenal gland
inferior
vena cava
spleen
r kidney
l crus of diaphragm
r suprarenal gland

Supplemental level 15.

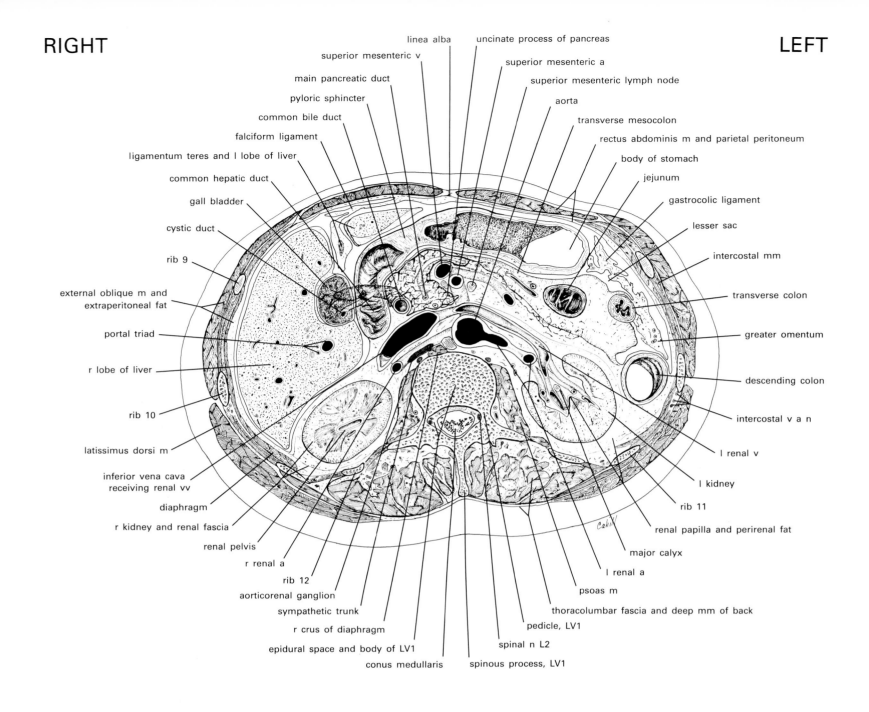

linea alba
superior mesenteric v
main pancreatic duct
pyloric sphincter
common bile duct
falciform ligament
ligamentum teres and l lobe of liver
common hepatic duct
gall bladder
cystic duct
rib 9
external oblique m and extraperitoneal fat
portal triad
r lobe of liver
rib 10
latissimus dorsi m
inferior vena cava receiving renal vv
diaphragm
r kidney and renal fascia
renal pelvis
r renal a
rib 12
aorticorenal ganglion
sympathetic trunk
r crus of diaphragm
epidural space and body of LV1
conus medullaris

uncinate process of pancreas
superior mesenteric a
superior mesenteric lymph node
aorta
transverse mesocolon
rectus abdominis m and parietal peritoneum
body of stomach
jejunum
gastrocolic ligament
lesser sac
intercostal mm
transverse colon
greater omentum
descending colon
intercostal v a n
l renal v
l kidney
rib 11
renal papilla and perirenal fat
major calyx
l renal a
psoas m
thoracolumbar fascia and deep mm of back
pedicle, LV1
spinal n L2
spinous process, LV1

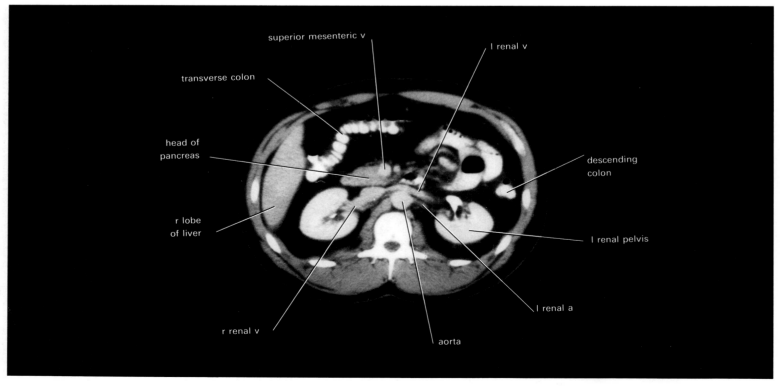

superior mesenteric v
l renal v
transverse colon
head of pancreas
descending colon
r lobe of liver
l renal pelvis
r renal v
aorta
l renal a

Supplemental level 14.

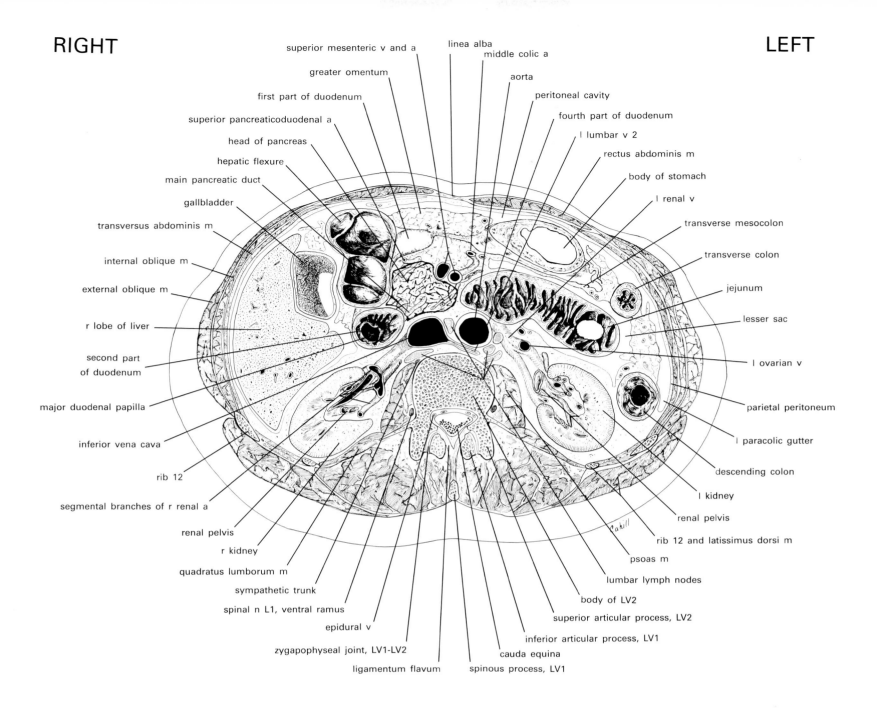

superior mesenteric v and a
linea alba
middle colic a
greater omentum
aorta
first part of duodenum
peritoneal cavity
superior pancreaticoduodenal a
fourth part of duodenum
head of pancreas
l lumbar v 2
hepatic flexure
rectus abdominis m
main pancreatic duct
body of stomach
gallbladder
l renal v
transversus abdominis m
transverse mesocolon
internal oblique m
transverse colon
external oblique m
jejunum
r lobe of liver
lesser sac
second part
of duodenum
l ovarian v
major duodenal papilla
parietal peritoneum
inferior vena cava
l paracolic gutter
rib 12
descending colon
segmental branches of r renal a
l kidney
renal pelvis
renal pelvis
r kidney
rib 12 and latissimus dorsi m
quadratus lumborum m
psoas m
sympathetic trunk
lumbar lymph nodes
spinal n L1, ventral ramus
body of LV2
epidural v
superior articular process, LV2
zygapophyseal joint, LV1-LV2
inferior articular process, LV1
ligamentum flavum
cauda equina
spinous process, LV1

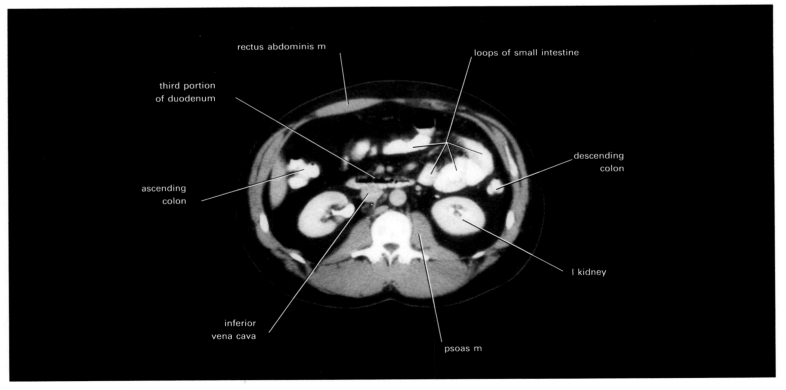

rectus abdominis m
loops of small intestine
third portion
of duodenum
descending
colon
ascending
colon
l kidney
inferior
vena cava
psoas m

Supplemental level 13.

Upper Abdominal Variations, CT

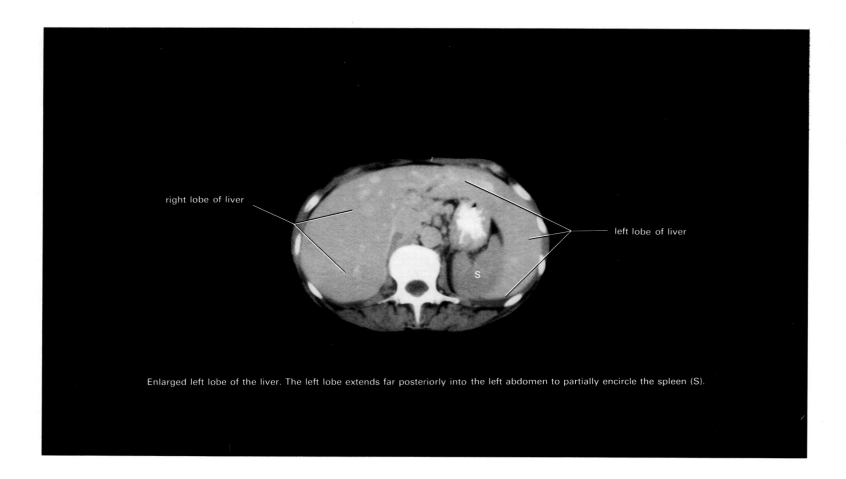

Enlarged left lobe of the liver. The left lobe extends far posteriorly into the left abdomen to partially encircle the spleen (S).

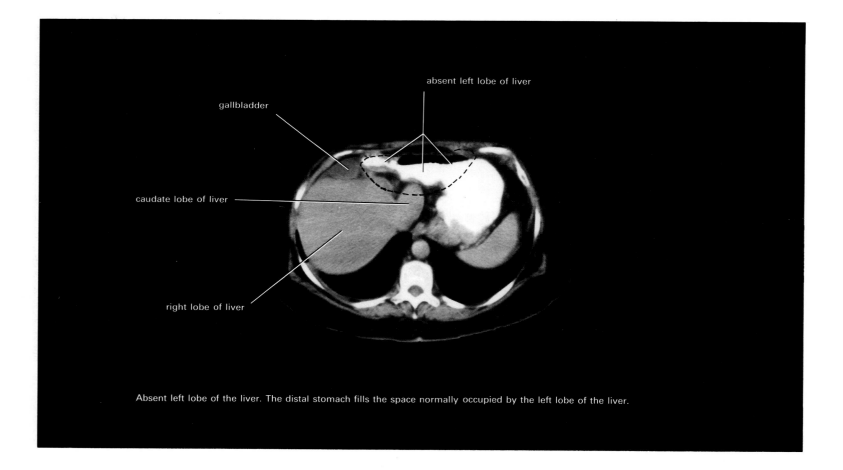

Absent left lobe of the liver. The distal stomach fills the space normally occupied by the left lobe of the liver.

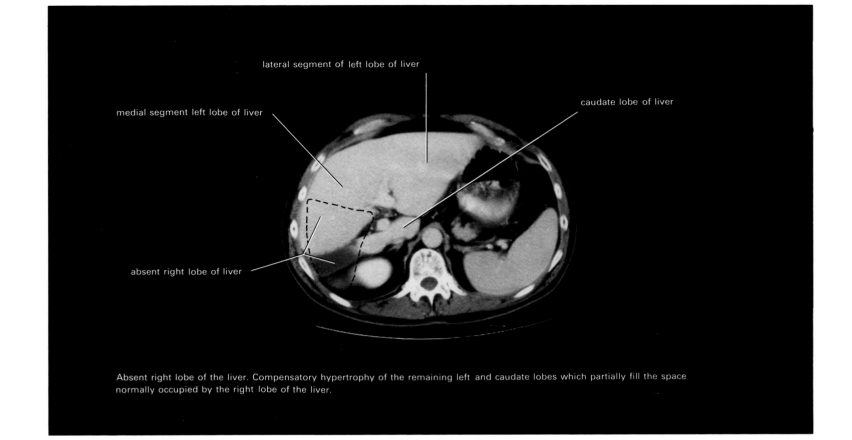

lateral segment of left lobe of liver

medial segment left lobe of liver

caudate lobe of liver

absent right lobe of liver

Absent right lobe of the liver. Compensatory hypertrophy of the remaining left and caudate lobes which partially fill the space normally occupied by the right lobe of the liver.

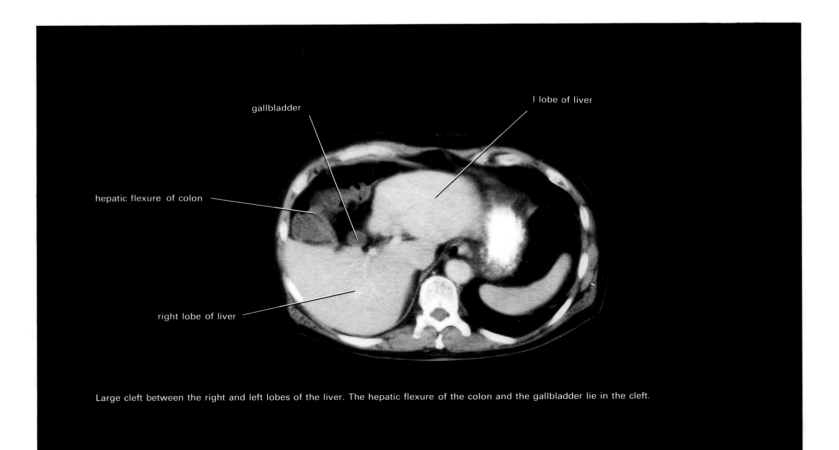

gallbladder

l lobe of liver

hepatic flexure of colon

right lobe of liver

Large cleft between the right and left lobes of the liver. The hepatic flexure of the colon and the gallbladder lie in the cleft.

The Male Pelvis

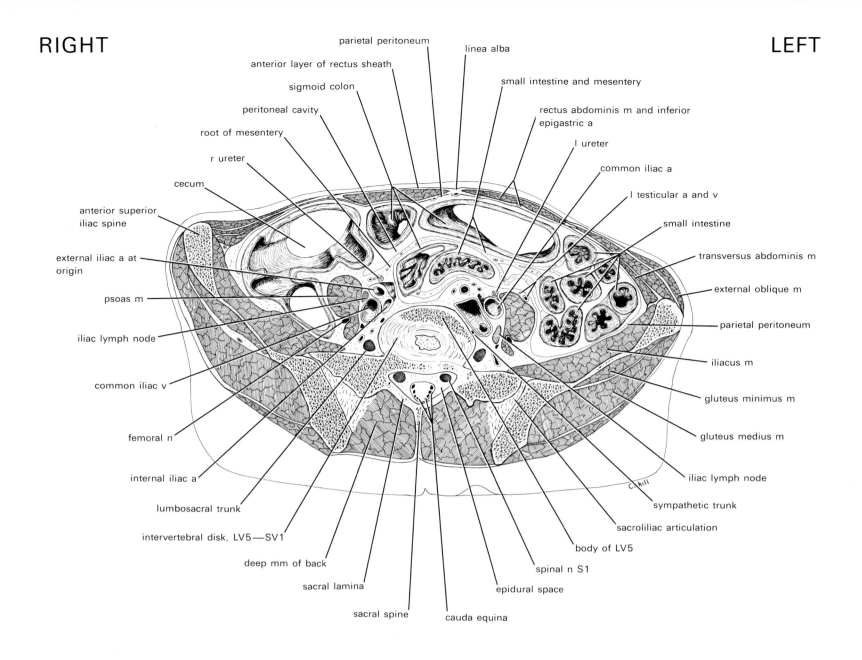

parietal peritoneum
anterior layer of rectus sheath
sigmoid colon
peritoneal cavity
root of mesentery
r ureter
cecum
anterior superior iliac spine
external iliac a at origin
psoas m
iliac lymph node
common iliac v
femoral n
internal iliac a
lumbosacral trunk
intervertebral disk, LV5—SV1
deep mm of back
sacral lamina
sacral spine
linea alba
small intestine and mesentery
rectus abdominis m and inferior epigastric a
l ureter
common iliac a
l testicular a and v
small intestine
transversus abdominis m
external oblique m
parietal peritoneum
iliacus m
gluteus minimus m
gluteus medius m
iliac lymph node
sympathetic trunk
sacroliliac articulation
body of LV5
spinal n S1
epidural space
cauda equina

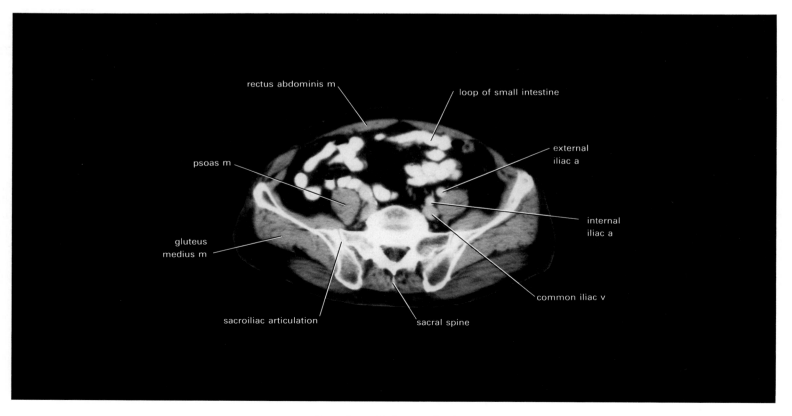

rectus abdominis m
loop of small intestine
psoas m
external iliac a
gluteus medius m
internal iliac a
sacroiliac articulation
sacral spine
common iliac v

Section 8 from below.

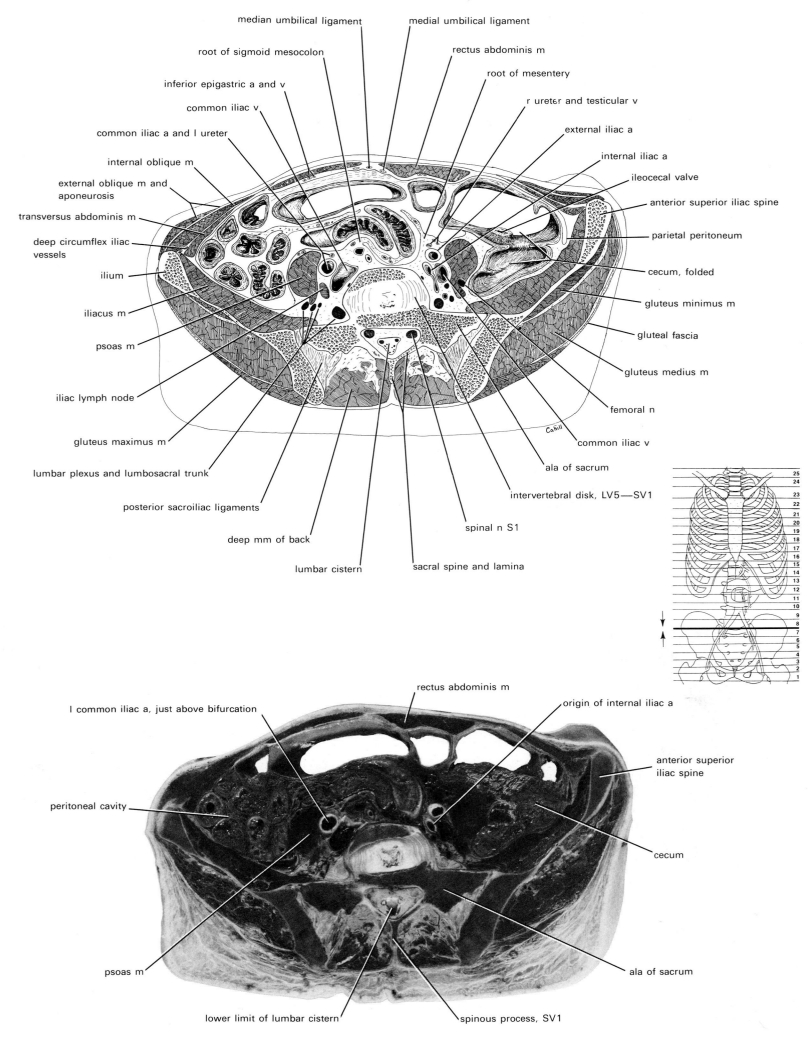

median umbilical ligament

medial umbilical ligament

root of sigmoid mesocolon

rectus abdominis m

inferior epigastric a and v

root of mesentery

common iliac v

r ureter and testicular v

common iliac a and l ureter

external iliac a

internal oblique m

internal iliac a

external oblique m and
aponeurosis

ileocecal valve

transversus abdominis m

anterior superior iliac spine

deep circumflex iliac
vessels

parietal peritoneum

ilium

cecum, folded

iliacus m

gluteus minimus m

psoas m

gluteal fascia

iliac lymph node

gluteus medius m

gluteus maximus m

femoral n

lumbar plexus and lumbosacral trunk

common iliac v

posterior sacroiliac ligaments

ala of sacrum

deep mm of back

intervertebral disk, LV5—SV1

spinal n S1

lumbar cistern

sacral spine and lamina

Cahill

rectus abdominis m

l common iliac a, just above bifurcation

origin of internal iliac a

anterior superior
iliac spine

peritoneal cavity

cecum

psoas m

ala of sacrum

lower limit of lumbar cistern

spinous process, SV1

Section 7 from above.

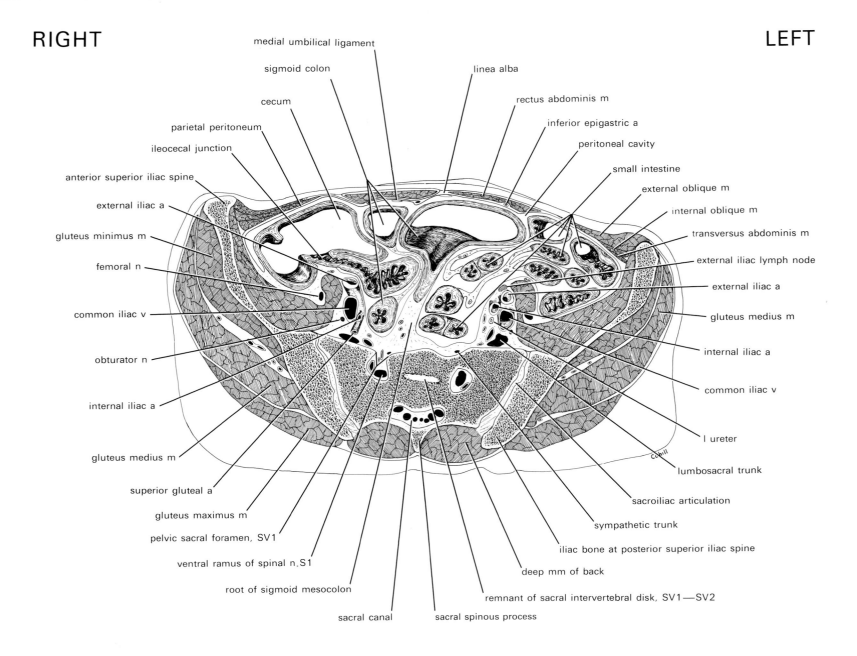

medial umbilical ligament
sigmoid colon
cecum
parietal peritoneum
ileocecal junction
anterior superior iliac spine
external iliac a
gluteus minimus m
femoral n
common iliac v
obturator n
internal iliac a
gluteus medius m
superior gluteal a
gluteus maximus m
pelvic sacral foramen, SV1
ventral ramus of spinal n. S1
root of sigmoid mesocolon
sacral canal
sacral spinous process

linea alba
rectus abdominis m
inferior epigastric a
peritoneal cavity
small intestine
external oblique m
internal oblique m
transversus abdominis m
external iliac lymph node
external iliac a
gluteus medius m
internal iliac a
common iliac v
l ureter
lumbosacral trunk
sacroiliac articulation
sympathetic trunk
iliac bone at posterior superior iliac spine
deep mm of back
remnant of sacral intervertebral disk, SV1—SV2

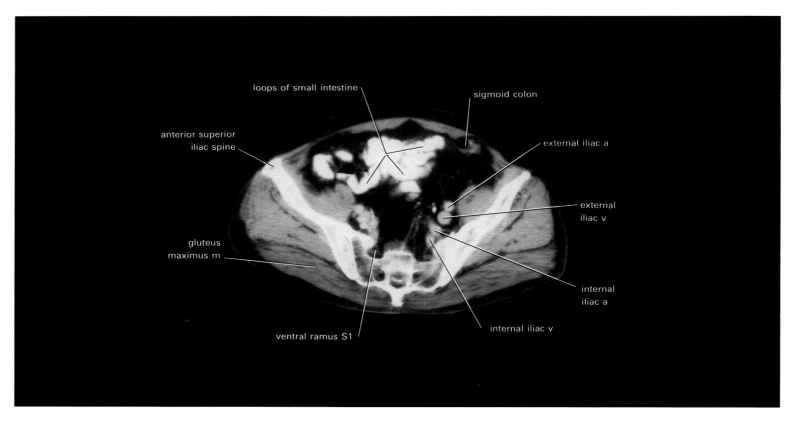

loops of small intestine
sigmoid colon
anterior superior iliac spine
external iliac a
external iliac v
gluteus maximus m
internal iliac a
ventral ramus S1
internal iliac v

Section 7 from below.

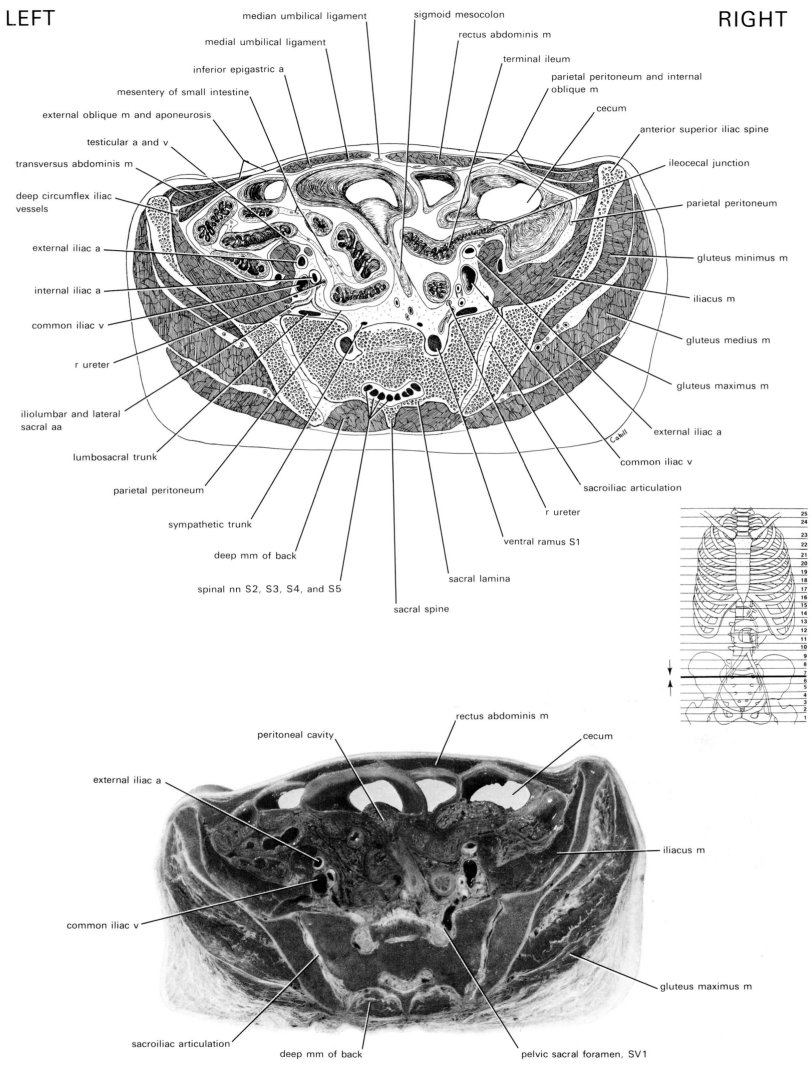

median umbilical ligament
medial umbilical ligament
inferior epigastric a
mesentery of small intestine
external oblique m and aponeurosis
testicular a and v
transversus abdominis m
deep circumflex iliac vessels
external iliac a
internal iliac a
common iliac v
r ureter
iliolumbar and lateral sacral aa
lumbosacral trunk
parietal peritoneum
sympathetic trunk
deep mm of back
spinal nn S2, S3, S4, and S5
sacral spine

sigmoid mesocolon
rectus abdominis m
terminal ileum
parietal peritoneum and internal oblique m
cecum
anterior superior iliac spine
ileocecal junction
parietal peritoneum
gluteus minimus m
iliacus m
gluteus medius m
gluteus maximus m
external iliac a
common iliac v
sacroiliac articulation
r ureter
ventral ramus S1
sacral lamina

Cahill

peritoneal cavity
external iliac a
common iliac v
sacroiliac articulation
deep mm of back

rectus abdominis m
cecum
iliacus m
gluteus maximus m
pelvic sacral foramen, SV1

Section 6 from above.

25
24
23
22
21
20
19
18
17
16
15
14
13
12
11
10
9
8
7
6
5
4
3
2
1

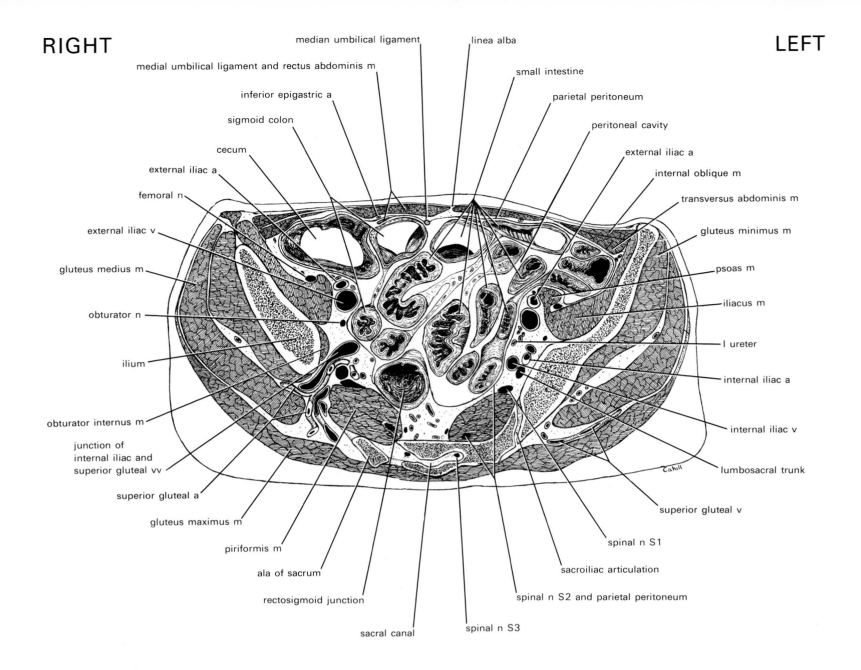

median umbilical ligament

linea alba

medial umbilical ligament and rectus abdominis m

small intestine

inferior epigastric a

parietal peritoneum

sigmoid colon

peritoneal cavity

cecum

external iliac a

external iliac a

internal oblique m

femoral n

transversus abdominis m

external iliac v

gluteus minimus m

gluteus medius m

psoas m

obturator n

iliacus m

l ureter

ilium

internal iliac a

obturator internus m

internal iliac v

junction of internal iliac and superior gluteal vv

lumbosacral trunk

superior gluteal a

superior gluteal v

gluteus maximus m

spinal n S1

piriformis m

sacroiliac articulation

ala of sacrum

spinal n S2 and parietal peritoneum

rectosigmoid junction

sacral canal

spinal n S3

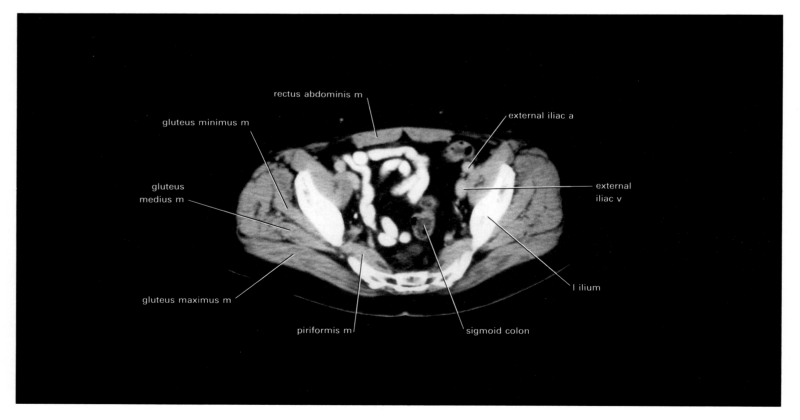

rectus abdominis m

gluteus minimus m

external iliac a

gluteus medius m

external iliac v

gluteus maximus m

l ilium

piriformis m

sigmoid colon

Section 6 from below.

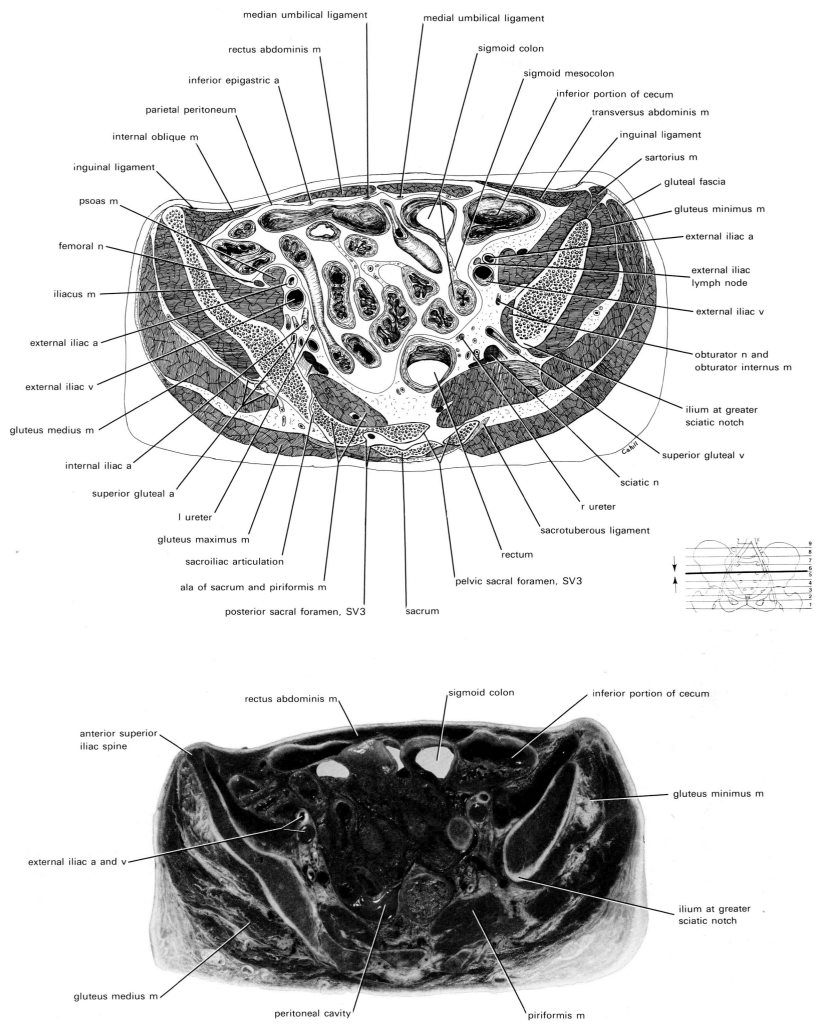

median umbilical ligament

medial umbilical ligament

rectus abdominis m

sigmoid colon

inferior epigastric a

sigmoid mesocolon

parietal peritoneum

inferior portion of cecum

internal oblique m

transversus abdominis m

inguinal ligament

inguinal ligament

sartorius m

psoas m

gluteal fascia

gluteus minimus m

femoral n

external iliac a

iliacus m

external iliac
lymph node

external iliac a

external iliac v

external iliac v

obturator n and
obturator internus m

gluteus medius m

ilium at greater
sciatic notch

internal iliac a

superior gluteal v

superior gluteal a

sciatic n

l ureter

r ureter

gluteus maximus m

sacrotuberous ligament

sacroiliac articulation

rectum

ala of sacrum and piriformis m

pelvic sacral foramen, SV3

posterior sacral foramen, SV3

sacrum

Cahill

rectus abdominis m

sigmoid colon

inferior portion of cecum

anterior superior
iliac spine

gluteus minimus m

external iliac a and v

ilium at greater
sciatic notch

gluteus medius m

peritoneal cavity

piriformis m

Section 5 from above.

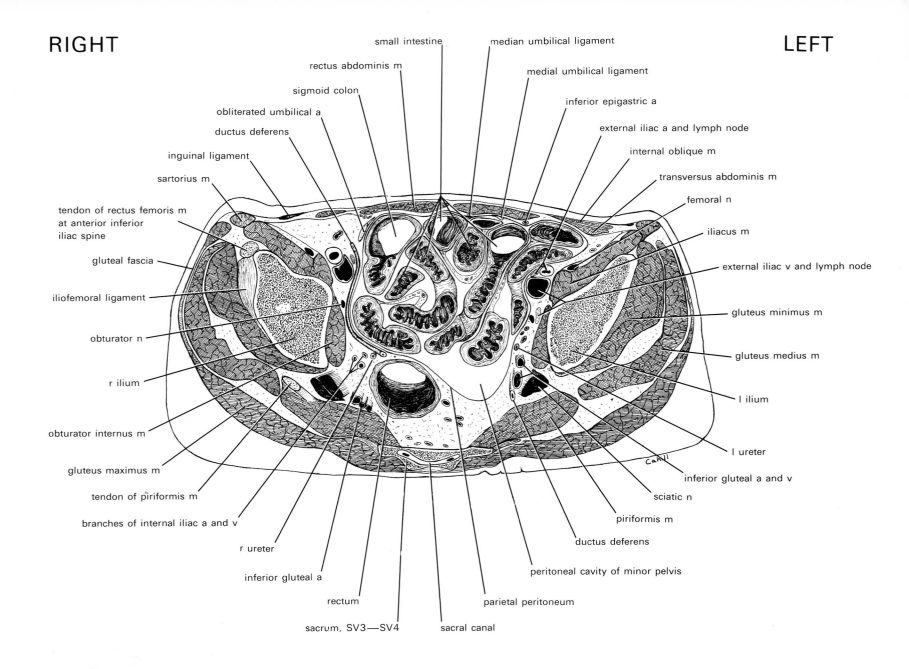

small intestine
median umbilical ligament
rectus abdominis m
medial umbilical ligament
sigmoid colon
inferior epigastric a
obliterated umbilical a
external iliac a and lymph node
ductus deferens
internal oblique m
inguinal ligament
transversus abdominis m
sartorius m
femoral n
tendon of rectus femoris m at anterior inferior iliac spine
iliacus m
gluteal fascia
external iliac v and lymph node
iliofemoral ligament
gluteus minimus m
obturator n
gluteus medius m
r ilium
l ilium
obturator internus m
l ureter
gluteus maximus m
inferior gluteal a and v
tendon of piriformis m
sciatic n
branches of internal iliac a and v
piriformis m
r ureter
ductus deferens
inferior gluteal a
peritoneal cavity of minor pelvis
rectum
parietal peritoneum
sacrum, SV3—SV4
sacral canal

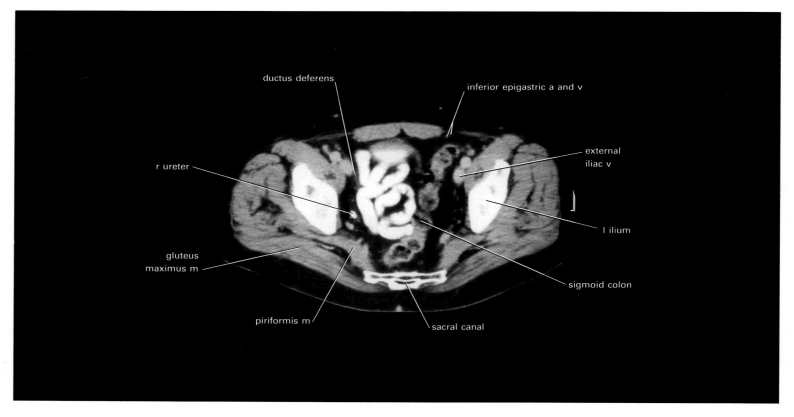

ductus deferens
inferior epigastric a and v
r ureter
external iliac v
gluteus maximus m
l ilium
piriformis m
sigmoid colon
sacral canal

Section 5 from below.

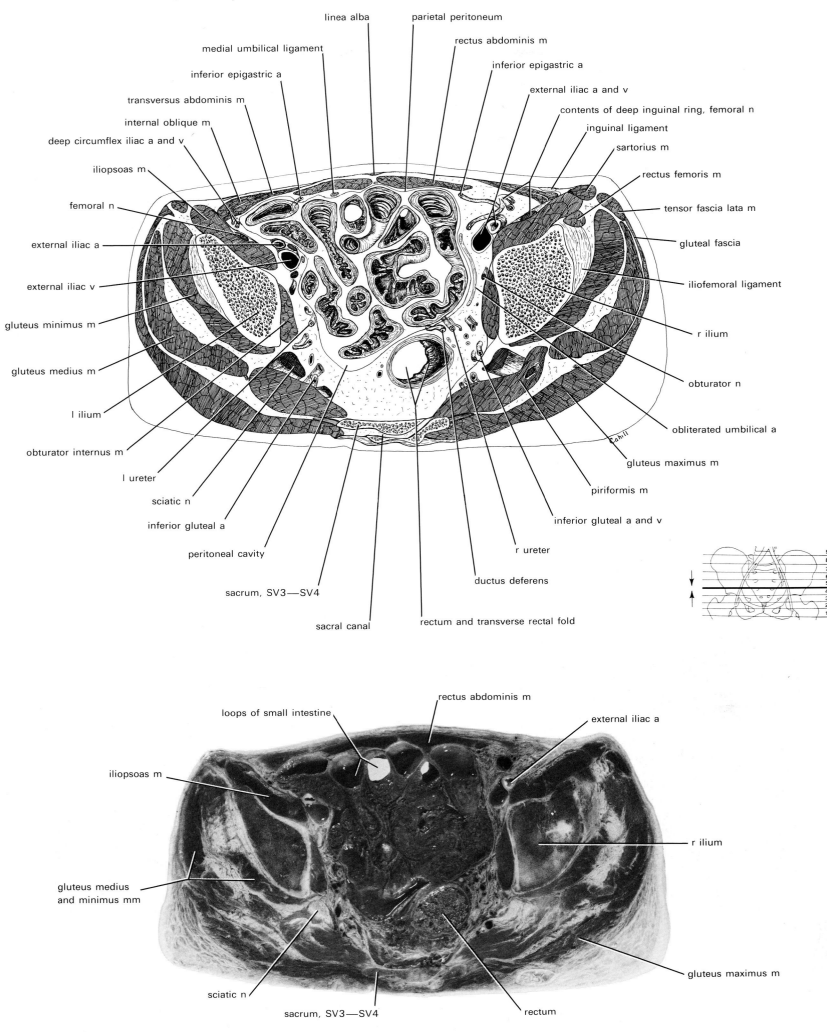

linea alba

parietal peritoneum

medial umbilical ligament

rectus abdominis m

inferior epigastric a

inferior epigastric a

transversus abdominis m

external iliac a and v

internal oblique m

contents of deep inguinal ring, femoral n

deep circumflex iliac a and v

inguinal ligament

iliopsoas m

sartorius m

femoral n

rectus femoris m

external iliac a

tensor fascia lata m

external iliac v

gluteal fascia

gluteus minimus m

iliofemoral ligament

gluteus medius m

r ilium

l ilium

obturator n

obturator internus m

obliterated umbilical a

l ureter

gluteus maximus m

sciatic n

piriformis m

inferior gluteal a

inferior gluteal a and v

peritoneal cavity

r ureter

sacrum, SV3—SV4

ductus deferens

sacral canal

rectum and transverse rectal fold

loops of small intestine

rectus abdominis m

external iliac a

iliopsoas m

r ilium

gluteus medius
and minimus mm

gluteus maximus m

sciatic n

rectum

sacrum, SV3—SV4

Section 4 from above.

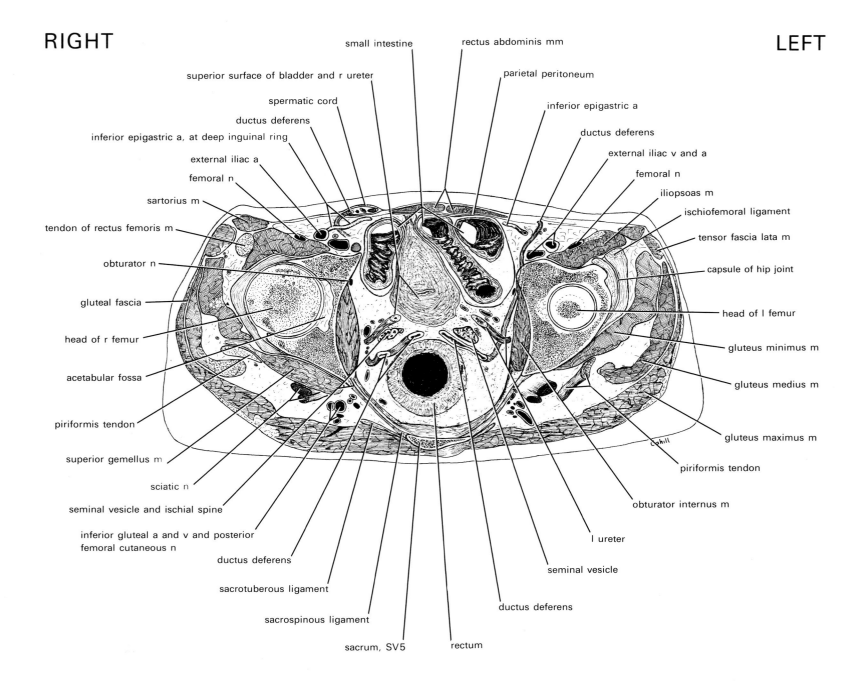

small intestine

rectus abdominis mm

superior surface of bladder and r ureter

parietal peritoneum

spermatic cord

inferior epigastric a

ductus deferens

ductus deferens

inferior epigastric a, at deep inguinal ring

external iliac v and a

external iliac a

femoral n

femoral n

iliopsoas m

sartorius m

ischiofemoral ligament

tendon of rectus femoris m

tensor fascia lata m

obturator n

capsule of hip joint

gluteal fascia

head of l femur

head of r femur

gluteus minimus m

acetabular fossa

gluteus medius m

piriformis tendon

gluteus maximus m

superior gemellus m

piriformis tendon

sciatic n

obturator internus m

seminal vesicle and ischial spine

l ureter

inferior gluteal a and v and posterior
femoral cutaneous n

seminal vesicle

ductus deferens

ductus deferens

sacrotuberous ligament

sacrospinous ligament

sacrum, SV5

rectum

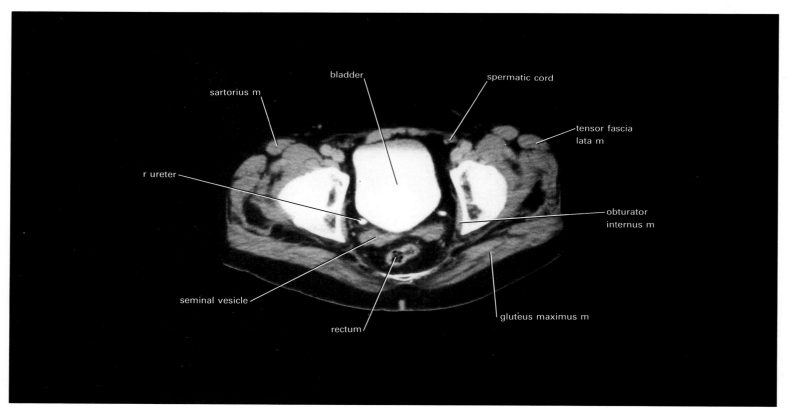

bladder

spermatic cord

sartorius m

tensor fascia
lata m

r ureter

obturator
internus m

seminal vesicle

gluteus maximus m

rectum

Section 4 from below.

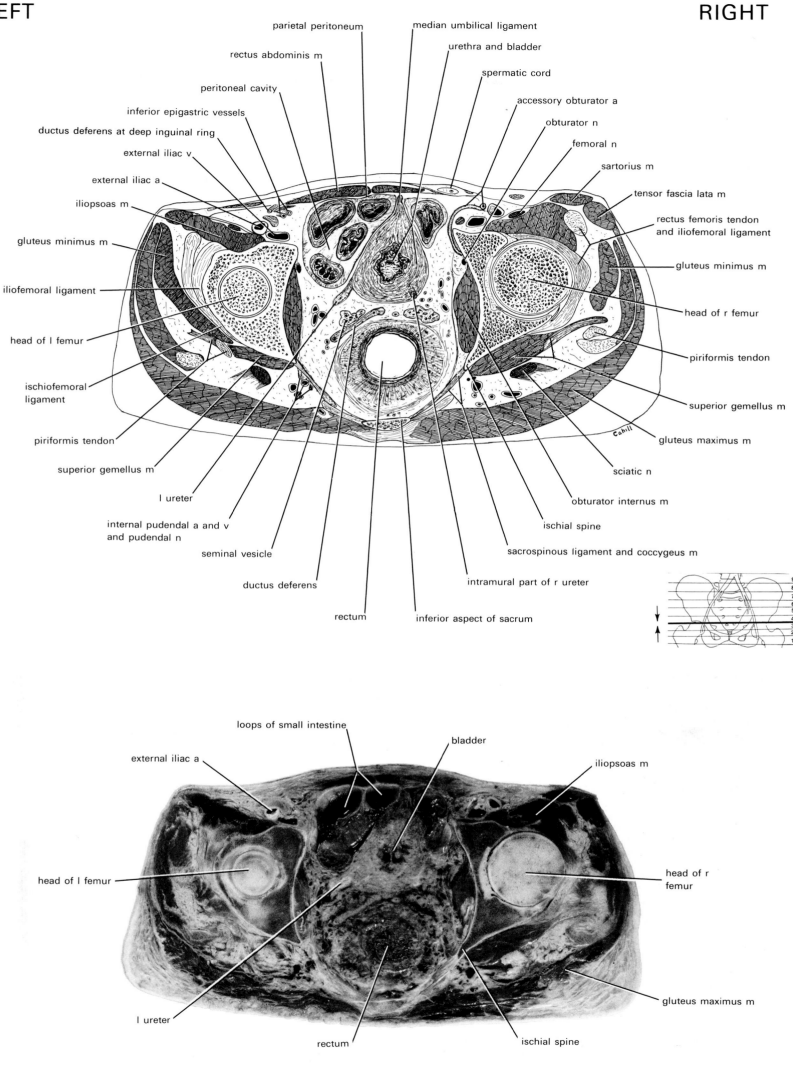

parietal peritoneum

median umbilical ligament

rectus abdominis m

urethra and bladder

spermatic cord

peritoneal cavity

accessory obturator a

inferior epigastric vessels

obturator n

ductus deferens at deep inguinal ring

femoral n

external iliac v

sartorius m

external iliac a

tensor fascia lata m

iliopsoas m

rectus femoris tendon
and iliofemoral ligament

gluteus minimus m

gluteus minimus m

iliofemoral ligament

head of r femur

head of l femur

piriformis tendon

ischiofemoral
ligament

superior gemellus m

piriformis tendon

gluteus maximus m

superior gemellus m

sciatic n

l ureter

obturator internus m

internal pudendal a and v
and pudendal n

ischial spine

seminal vesicle

sacrospinous ligament and coccygeus m

ductus deferens

intramural part of r ureter

rectum

inferior aspect of sacrum

loops of small intestine

bladder

external iliac a

iliopsoas m

head of l femur

head of r
femur

l ureter

gluteus maximus m

rectum

ischial spine

Section 3 from above.

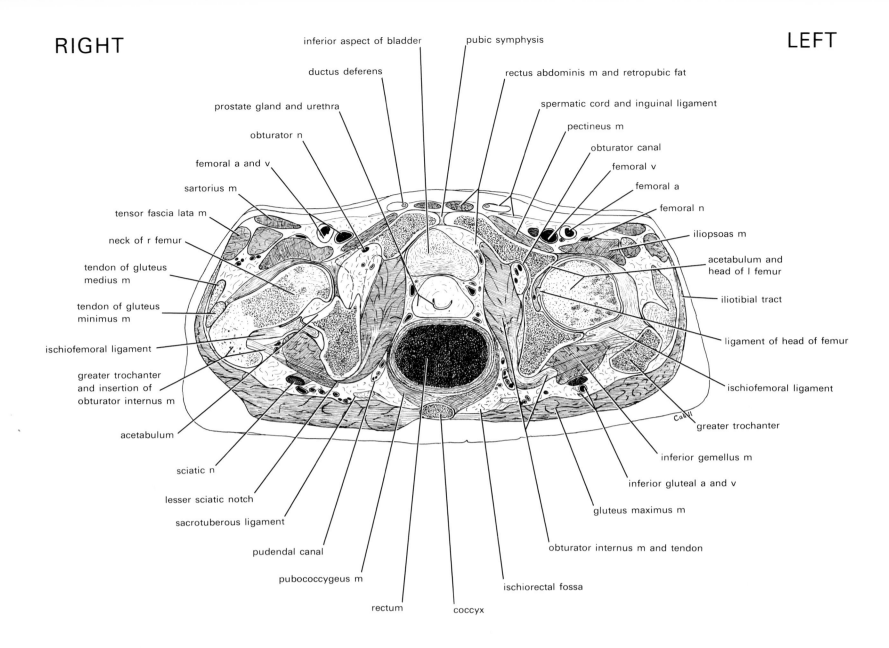

inferior aspect of bladder
pubic symphysis
ductus deferens
rectus abdominis m and retropubic fat
prostate gland and urethra
spermatic cord and inguinal ligament
obturator n
pectineus m
femoral a and v
obturator canal
sartorius m
femoral v
tensor fascia lata m
femoral a
neck of r femur
femoral n
tendon of gluteus medius m
iliopsoas m
tendon of gluteus minimus m
acetabulum and head of l femur
ischiofemoral ligament
iliotibial tract
greater trochanter and insertion of obturator internus m
ligament of head of femur
acetabulum
ischiofemoral ligament
sciatic n
greater trochanter
lesser sciatic notch
inferior gemellus m
sacrotuberous ligament
inferior gluteal a and v
pudendal canal
gluteus maximus m
pubococcygeus m
obturator internus m and tendon
rectum
ischiorectal fossa
coccyx

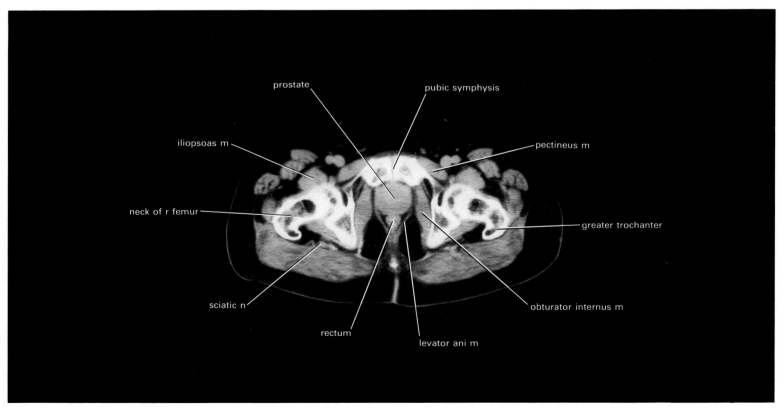

prostate
pubic symphysis
iliopsoas m
pectineus m
neck of r femur
greater trochanter
sciatic n
obturator internus m
rectum
levator ani m

Section 3 from below.

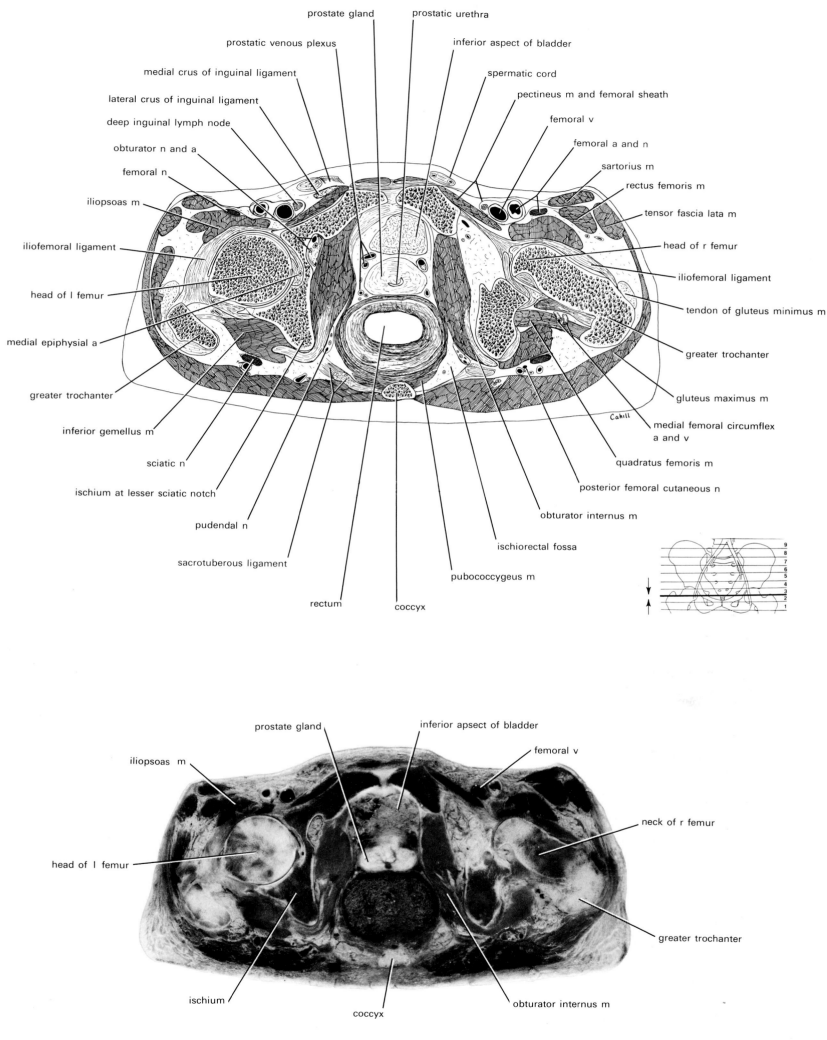

Section 2 from above.

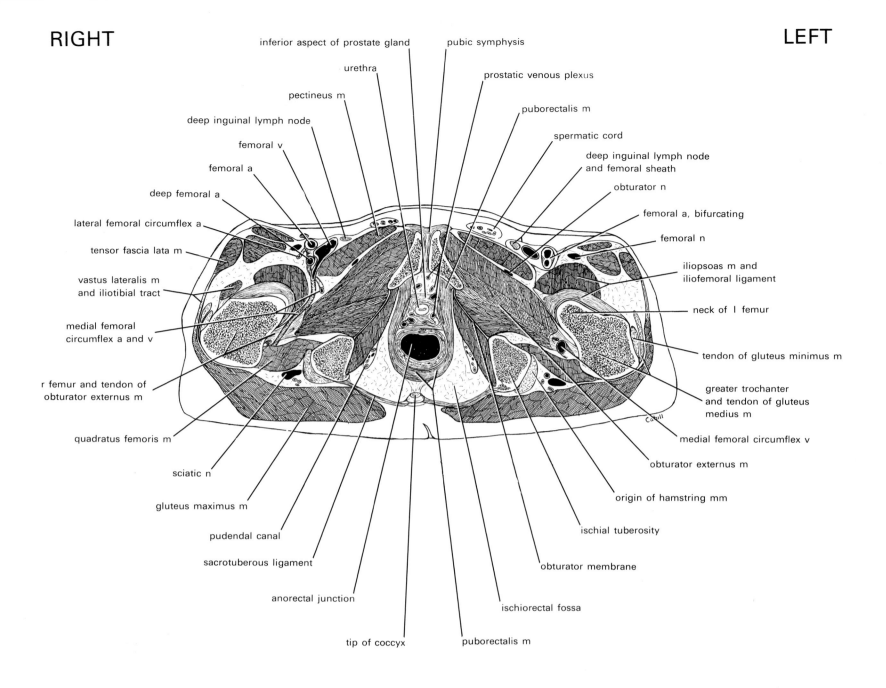

inferior aspect of prostate gland
pubic symphysis
urethra
prostatic venous plexus
pectineus m
puborectalis m
deep inguinal lymph node
spermatic cord
femoral v
deep inguinal lymph node
and femoral sheath
femoral a
obturator n
deep femoral a
femoral a, bifurcating
lateral femoral circumflex a
femoral n
tensor fascia lata m
iliopsoas m and
iliofemoral ligament
vastus lateralis m
and iliotibial tract
neck of l femur
medial femoral
circumflex a and v
tendon of gluteus minimus m
r femur and tendon of
obturator externus m
greater trochanter
and tendon of gluteus
medius m
quadratus femoris m
medial femoral circumflex v
sciatic n
obturator externus m
gluteus maximus m
origin of hamstring mm
pudendal canal
ischial tuberosity
sacrotuberous ligament
obturator membrane
anorectal junction
ischiorectal fossa
tip of coccyx
puborectalis m

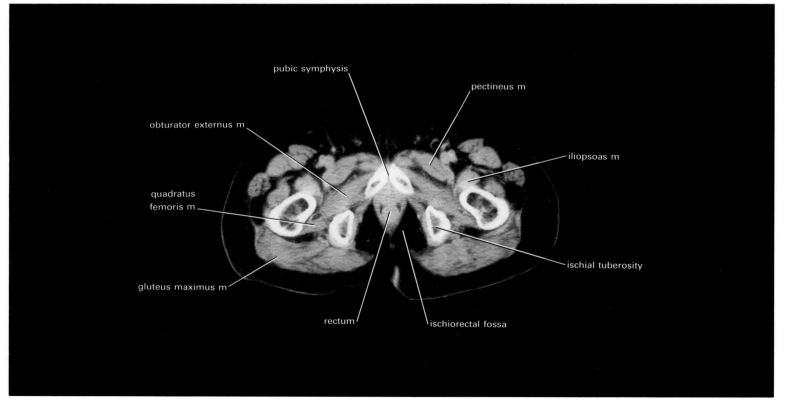

pubic symphysis
pectineus m
obturator externus m
iliopsoas m
quadratus
femoris m
gluteus maximus m
ischial tuberosity
rectum
ischiorectal fossa

Section 2 from below.

LEFT   RIGHT

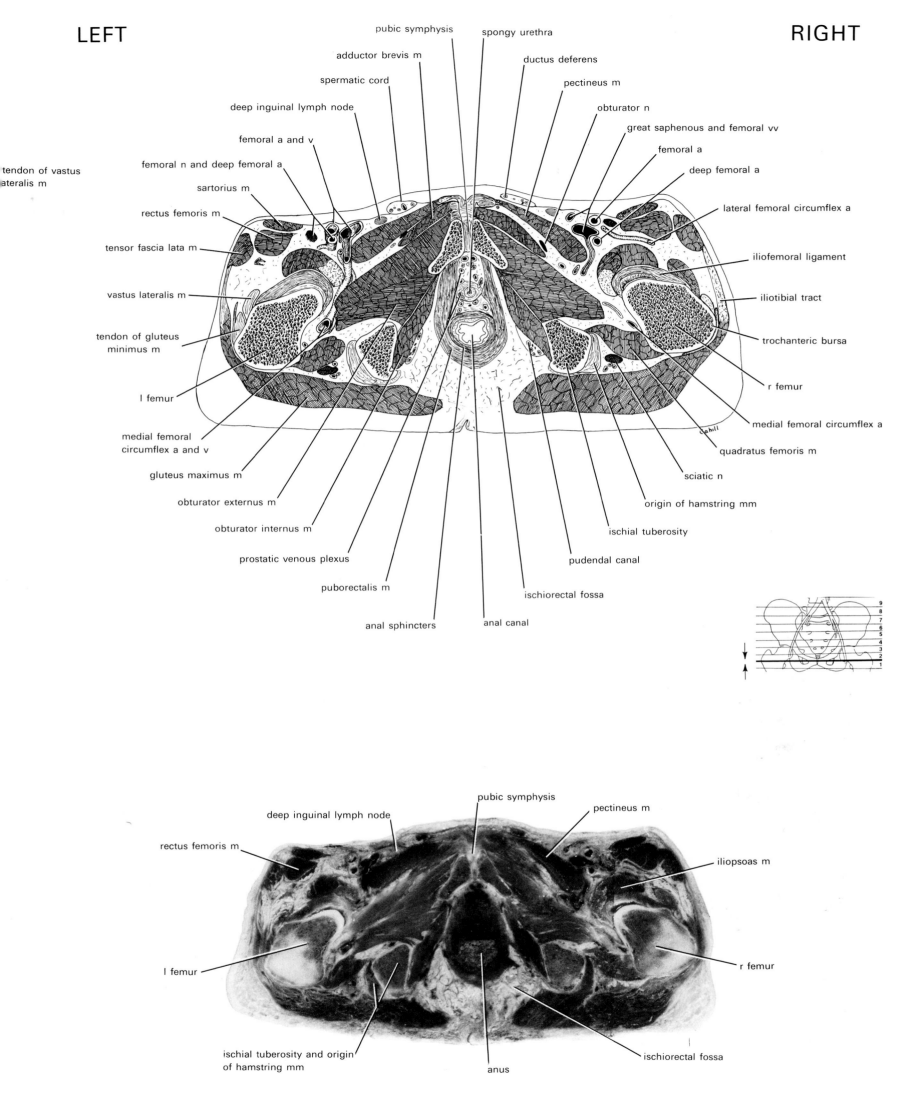

pubic symphysis
adductor brevis m
spermatic cord
deep inguinal lymph node
femoral a and v
femoral n and deep femoral a
sartorius m
rectus femoris m
tensor fascia lata m
vastus lateralis m
tendon of gluteus minimus m
l femur
medial femoral circumflex a and v
gluteus maximus m
obturator externus m
obturator internus m
prostatic venous plexus
puborectalis m
anal sphincters

tendon of vastus lateralis m

spongy urethra
ductus deferens
pectineus m
obturator n
great saphenous and femoral vv
femoral a
deep femoral a
lateral femoral circumflex a
iliofemoral ligament
iliotibial tract
trochanteric bursa
r femur
medial femoral circumflex a
quadratus femoris m
sciatic n
origin of hamstring mm
ischial tuberosity
pudendal canal
ischiorectal fossa
anal canal

pubic symphysis
deep inguinal lymph node
rectus femoris m
l femur
ischial tuberosity and origin of hamstring mm
anus

pectineus m
iliopsoas m
r femur
ischiorectal fossa

Section 1 from above.

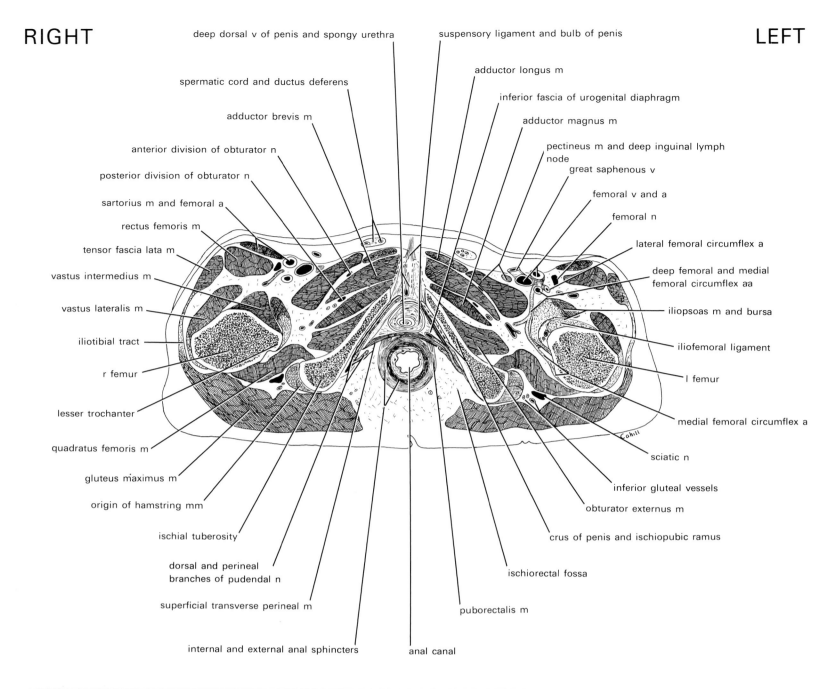

deep dorsal v of penis and spongy urethra

suspensory ligament and bulb of penis

spermatic cord and ductus deferens

adductor longus m

adductor brevis m

inferior fascia of urogenital diaphragm

anterior division of obturator n

adductor magnus m

posterior division of obturator n

pectineus m and deep inguinal lymph node

sartorius m and femoral a

great saphenous v

rectus femoris m

femoral v and a

tensor fascia lata m

femoral n

vastus intermedius m

lateral femoral circumflex a

vastus lateralis m

deep femoral and medial femoral circumflex aa

iliotibial tract

iliopsoas m and bursa

r femur

iliofemoral ligament

lesser trochanter

l femur

quadratus femoris m

medial femoral circumflex a

gluteus maximus m

sciatic n

origin of hamstring mm

inferior gluteal vessels

ischial tuberosity

obturator externus m

dorsal and perineal branches of pudendal n

crus of penis and ischiopubic ramus

superficial transverse perineal m

ischiorectal fossa

puborectalis m

internal and external anal sphincters

anal canal

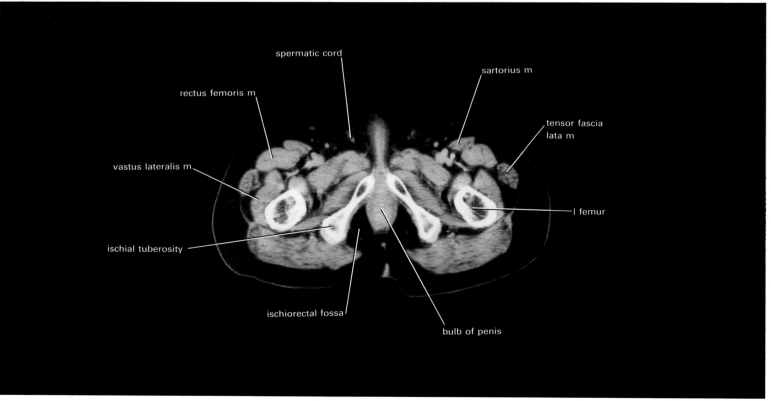

spermatic cord

sartorius m

rectus femoris m

tensor fascia lata m

vastus lateralis m

l femur

ischial tuberosity

ischiorectal fossa

bulb of penis

Section 1 from below.

The Female Pelvis

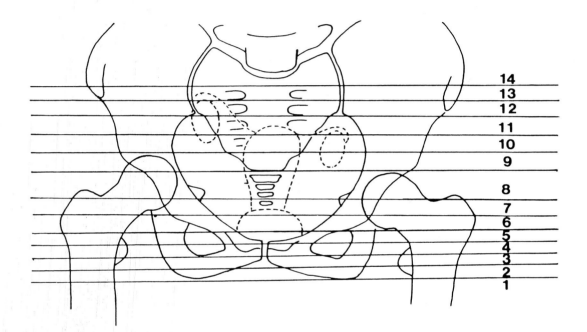

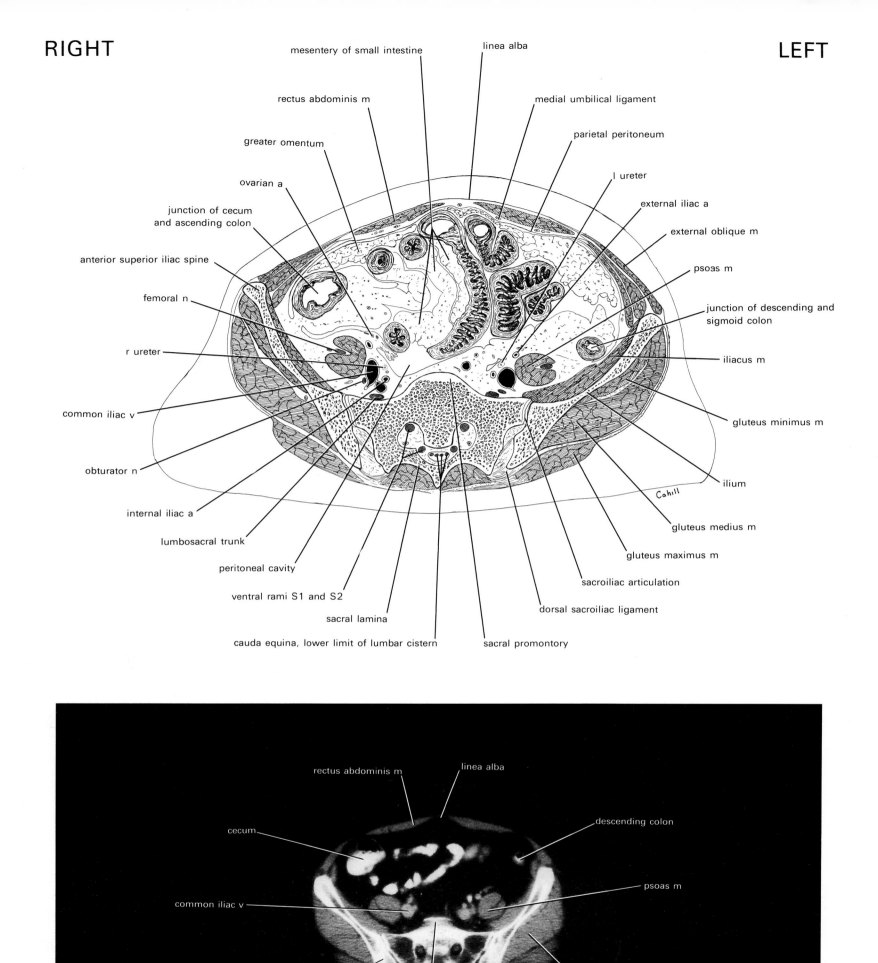

RIGHT   LEFT

mesentery of small intestine   linea alba

rectus abdominis m   medial umbilical ligament

greater omentum   parietal peritoneum

ovarian a   l ureter

junction of cecum   external iliac a
and ascending colon
                                                        external oblique m
anterior superior iliac spine
                                                        psoas m
femoral n
                                                        junction of descending and
                                                        sigmoid colon
r ureter
                                                        iliacus m
common iliac v
                                                        gluteus minimus m

obturator n
                                                        ilium
internal iliac a
                                                        gluteus medius m
lumbosacral trunk
                                                        gluteus maximus m
peritoneal cavity
                                                        sacroiliac articulation
ventral rami S1 and S2
                                                        dorsal sacroiliac ligament
sacral lamina

cauda equina, lower limit of lumbar cistern   sacral promontory

Cahill

rectus abdominis m   linea alba

cecum   descending colon

                                               psoas m
common iliac v

                                               gluteus medius m
ilium

sacral promontory

Section 14 from below.

68 ATLAS OF HUMAN CROSS-SECTIONAL ANATOMY

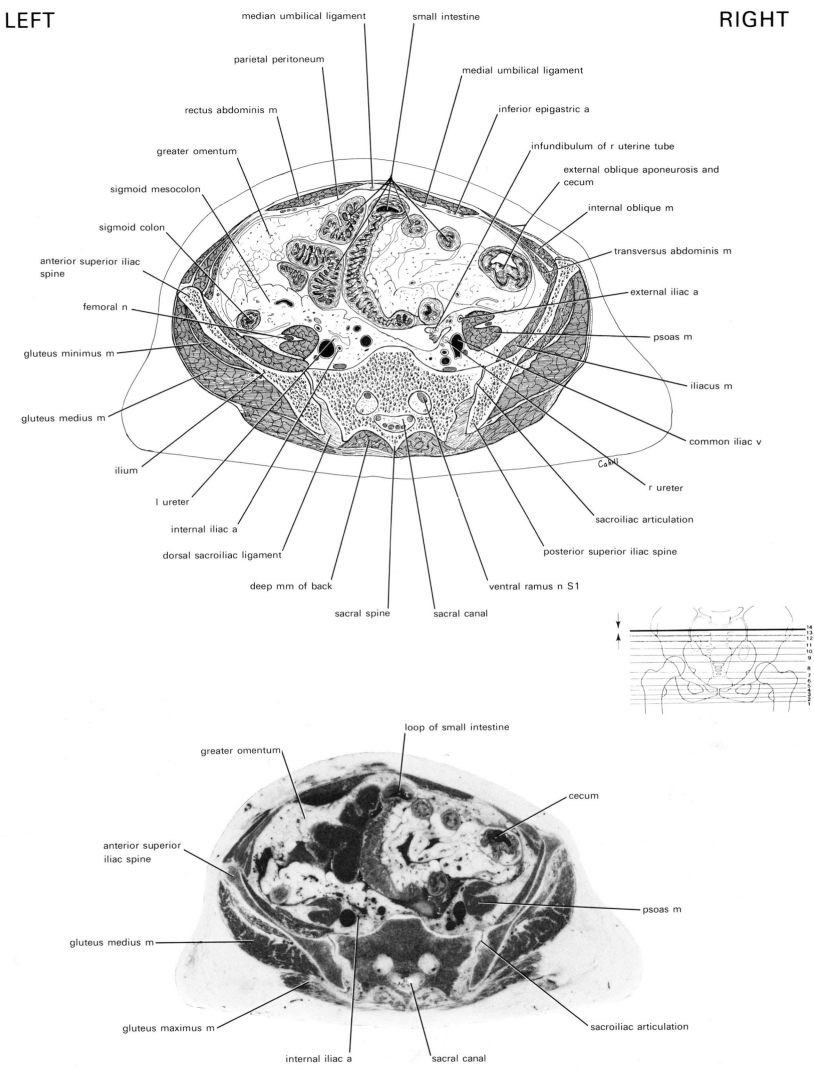

LEFT

RIGHT

median umbilical ligament

small intestine

parietal peritoneum

medial umbilical ligament

rectus abdominis m

inferior epigastric a

greater omentum

infundibulum of r uterine tube

sigmoid mesocolon

external oblique aponeurosis and cecum

sigmoid colon

internal oblique m

anterior superior iliac spine

transversus abdominis m

femoral n

external iliac a

gluteus minimus m

psoas m

gluteus medius m

iliacus m

ilium

common iliac v

l ureter

r ureter

internal iliac a

sacroiliac articulation

dorsal sacroiliac ligament

posterior superior iliac spine

deep mm of back

ventral ramus n S1

sacral spine

sacral canal

loop of small intestine

greater omentum

cecum

anterior superior iliac spine

psoas m

gluteus medius m

gluteus maximus m

sacroiliac articulation

internal iliac a

sacral canal

Section 13 from above.

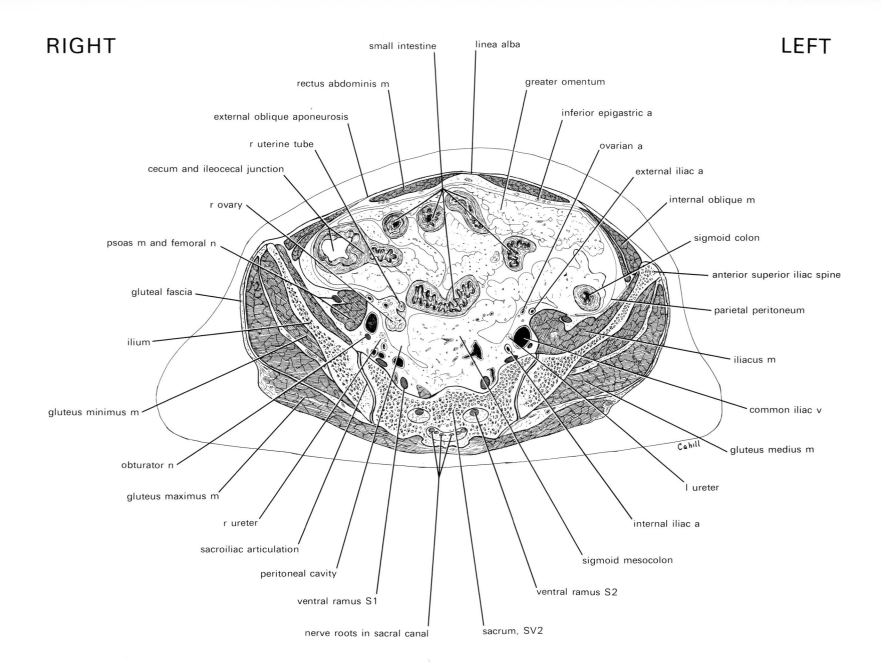

small intestine

linea alba

rectus abdominis m

greater omentum

external oblique aponeurosis

inferior epigastric a

r uterine tube

ovarian a

cecum and ileocecal junction

external iliac a

r ovary

internal oblique m

psoas m and femoral n

sigmoid colon

anterior superior iliac spine

gluteal fascia

parietal peritoneum

ilium

iliacus m

gluteus minimus m

common iliac v

Cahill

gluteus medius m

obturator n

l ureter

gluteus maximus m

internal iliac a

r ureter

sigmoid mesocolon

sacroiliac articulation

peritoneal cavity

ventral ramus S2

ventral ramus S1

nerve roots in sacral canal

sacrum, SV2

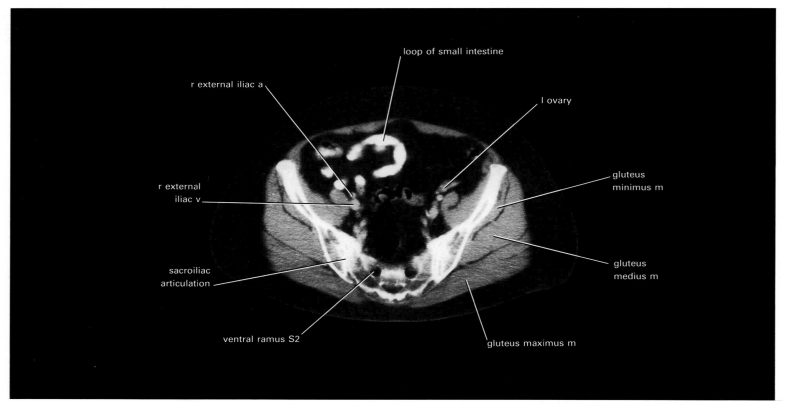

loop of small intestine

r external iliac a

l ovary

r external
iliac v

gluteus
minimus m

gluteus
medius m

sacroiliac
articulation

ventral ramus S2

gluteus maximus m

Section 13 from below.

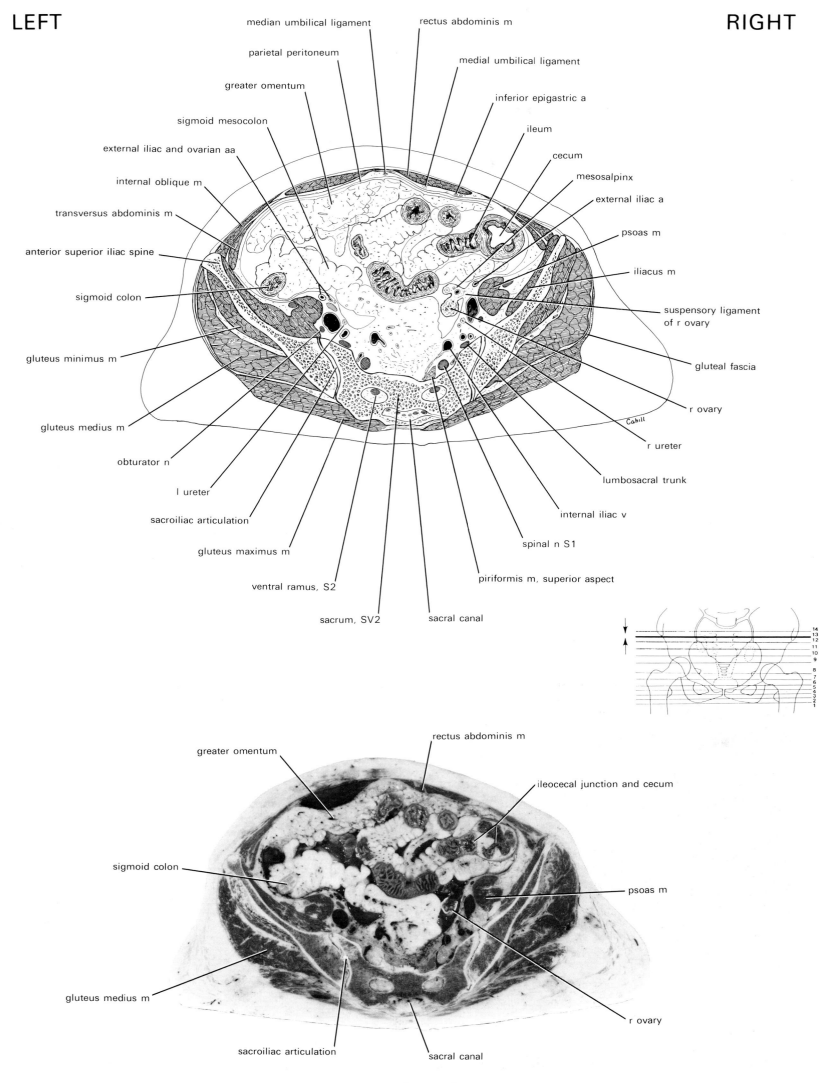

LEFT   RIGHT

median umbilical ligament

parietal peritoneum

greater omentum

sigmoid mesocolon

external iliac and ovarian aa

internal oblique m

transversus abdominis m

anterior superior iliac spine

sigmoid colon

gluteus minimus m

gluteus medius m

obturator n

l ureter

sacroiliac articulation

gluteus maximus m

ventral ramus, S2

sacrum, SV2

rectus abdominis m

medial umbilical ligament

inferior epigastric a

ileum

cecum

mesosalpinx

external iliac a

psoas m

iliacus m

suspensory ligament
of r ovary

gluteal fascia

r ovary

r ureter

lumbosacral trunk

internal iliac v

spinal n S1

piriformis m, superior aspect

sacral canal

greater omentum

sigmoid colon

gluteus medius m

sacroiliac articulation

rectus abdominis m

ileocecal junction and cecum

psoas m

r ovary

sacral canal

Section 12 from above.

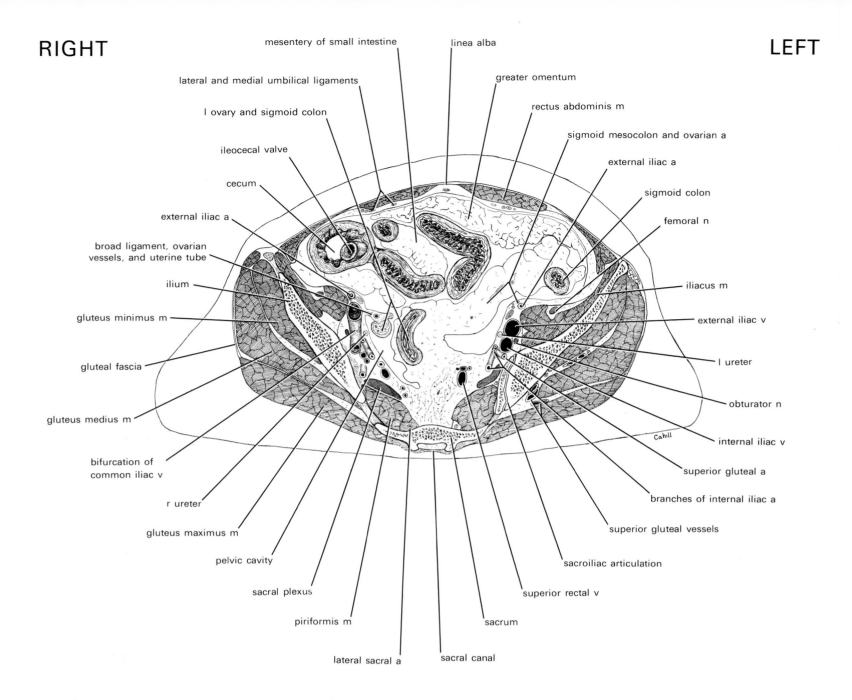

mesentery of small intestine

linea alba

lateral and medial umbilical ligaments

greater omentum

l ovary and sigmoid colon

rectus abdominis m

ileocecal valve

sigmoid mesocolon and ovarian a

cecum

external iliac a

external iliac a

sigmoid colon

broad ligament, ovarian vessels, and uterine tube

femoral n

ilium

iliacus m

gluteus minimus m

external iliac v

gluteal fascia

l ureter

gluteus medius m

obturator n

bifurcation of common iliac v

internal iliac v

r ureter

superior gluteal a

gluteus maximus m

branches of internal iliac a

pelvic cavity

superior gluteal vessels

sacral plexus

sacroiliac articulation

piriformis m

superior rectal v

lateral sacral a

sacrum

sacral canal

Cahill

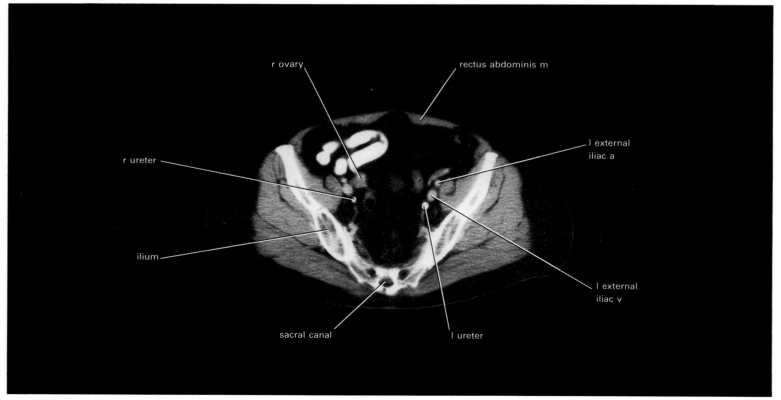

r ovary

rectus abdominis m

r ureter

l external iliac a

ilium

l external iliac v

sacral canal

l ureter

Section 12 from below.

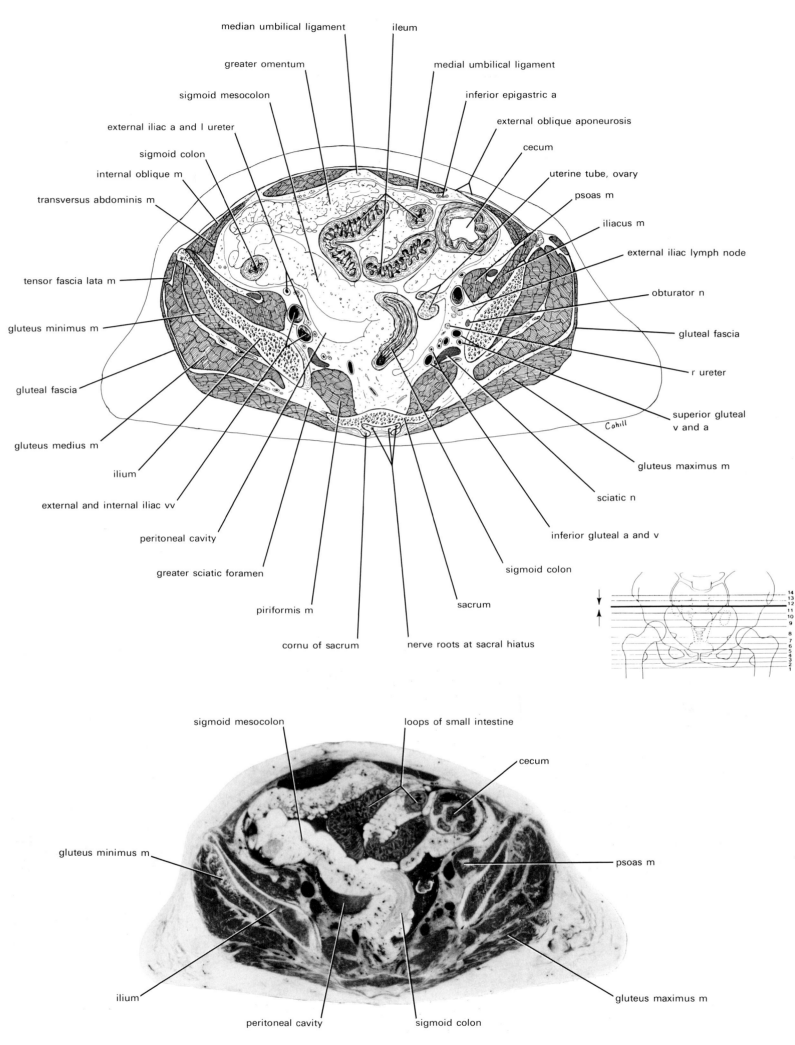

median umbilical ligament
ileum
greater omentum
medial umbilical ligament
sigmoid mesocolon
inferior epigastric a
external iliac a and l ureter
external oblique aponeurosis
sigmoid colon
cecum
internal oblique m
uterine tube, ovary
transversus abdominis m
psoas m
iliacus m
external iliac lymph node
tensor fascia lata m
obturator n
gluteus minimus m
gluteal fascia
gluteal fascia
r ureter
gluteus medius m
superior gluteal v and a
ilium
gluteus maximus m
external and internal iliac vv
sciatic n
peritoneal cavity
inferior gluteal a and v
greater sciatic foramen
sigmoid colon
piriformis m
sacrum
cornu of sacrum
nerve roots at sacral hiatus

Cahill

sigmoid mesocolon
loops of small intestine

cecum

gluteus minimus m

psoas m

ilium

gluteus maximus m

peritoneal cavity
sigmoid colon

Section 11 from above.

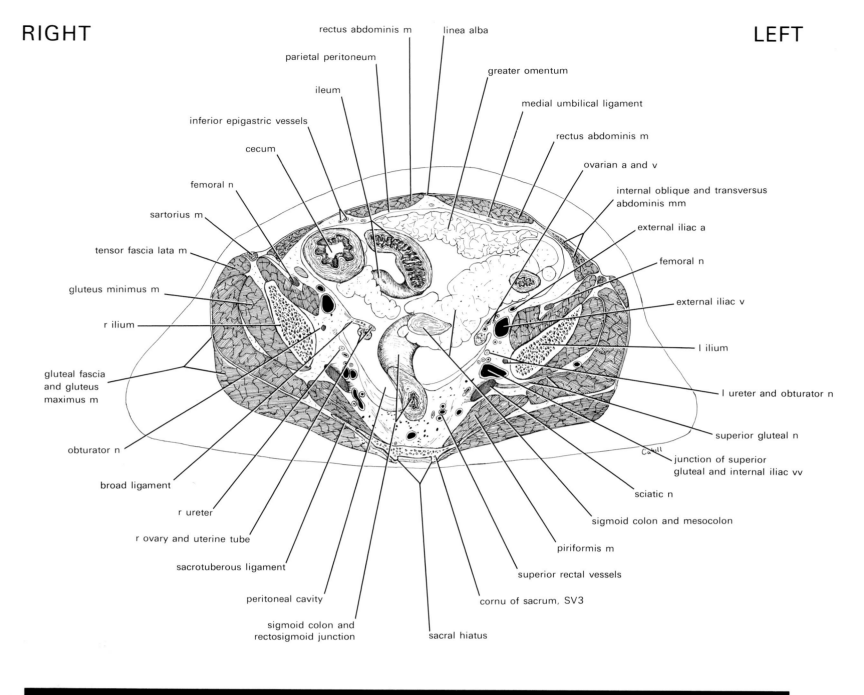

rectus abdominis m — linea alba

parietal peritoneum

greater omentum

ileum

medial umbilical ligament

inferior epigastric vessels

rectus abdominis m

cecum

ovarian a and v

femoral n

internal oblique and transversus abdominis mm

sartorius m

external iliac a

tensor fascia lata m

femoral n

gluteus minimus m

external iliac v

r ilium

l ilium

gluteal fascia and gluteus maximus m

l ureter and obturator n

superior gluteal n

obturator n

junction of superior gluteal and internal iliac vv

broad ligament

sciatic n

r ureter

sigmoid colon and mesocolon

r ovary and uterine tube

piriformis m

sacrotuberous ligament

superior rectal vessels

peritoneal cavity

cornu of sacrum, SV3

sigmoid colon and rectosigmoid junction

sacral hiatus

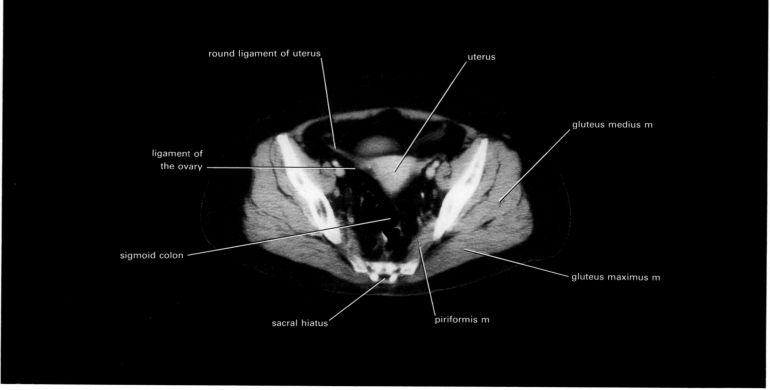

round ligament of uterus

uterus

gluteus medius m

ligament of the ovary

sigmoid colon

gluteus maximus m

sacral hiatus

piriformis m

Section 11 from below.

LEFT   RIGHT

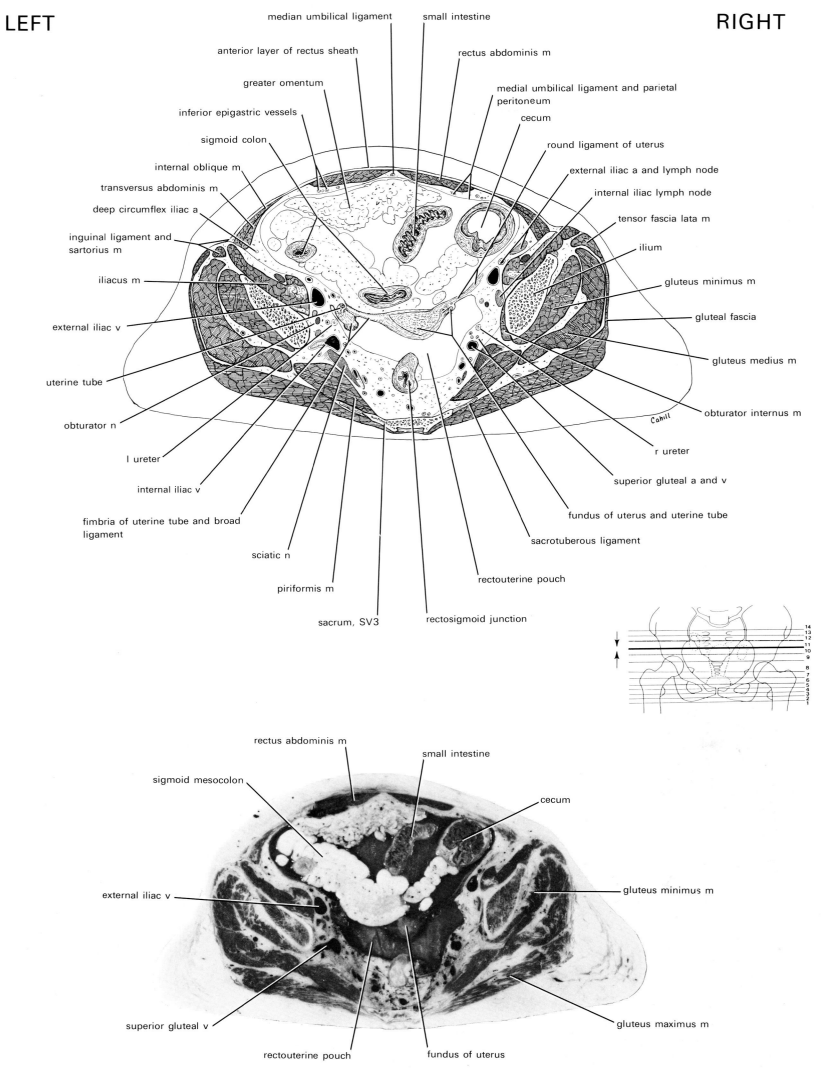

median umbilical ligament
small intestine
anterior layer of rectus sheath
rectus abdominis m
greater omentum
medial umbilical ligament and parietal peritoneum
inferior epigastric vessels
cecum
sigmoid colon
round ligament of uterus
internal oblique m
external iliac a and lymph node
transversus abdominis m
internal iliac lymph node
deep circumflex iliac a
tensor fascia lata m
inguinal ligament and sartorius m
ilium
iliacus m
gluteus minimus m
external iliac v
gluteal fascia
uterine tube
gluteus medius m
obturator n
obturator internus m
l ureter
r ureter
internal iliac v
superior gluteal a and v
fimbria of uterine tube and broad ligament
fundus of uterus and uterine tube
sciatic n
sacrotuberous ligament
piriformis m
rectouterine pouch
sacrum, SV3
rectosigmoid junction

rectus abdominis m
small intestine
sigmoid mesocolon
cecum
external iliac v
gluteus minimus m
superior gluteal v
gluteus maximus m
rectouterine pouch
fundus of uterus

Section 10 from above.

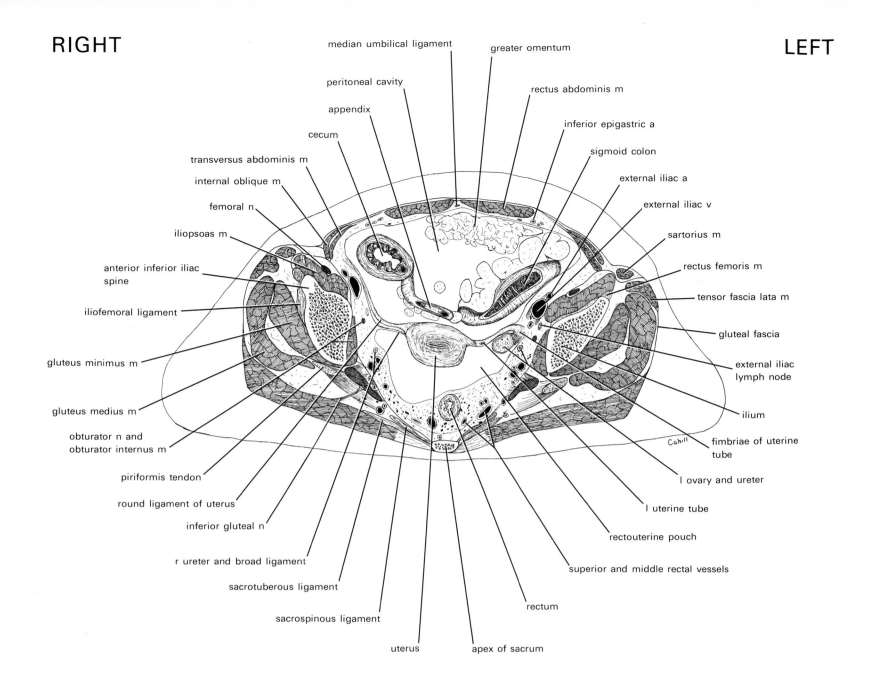

median umbilical ligament

greater omentum

peritoneal cavity

rectus abdominis m

appendix

inferior epigastric a

cecum

sigmoid colon

transversus abdominis m

external iliac a

internal oblique m

external iliac v

femoral n

sartorius m

iliopsoas m

rectus femoris m

anterior inferior iliac spine

tensor fascia lata m

iliofemoral ligament

gluteal fascia

gluteus minimus m

external iliac lymph node

gluteus medius m

ilium

obturator n and obturator internus m

fimbriae of uterine tube

piriformis tendon

l ovary and ureter

round ligament of uterus

l uterine tube

inferior gluteal n

rectouterine pouch

r ureter and broad ligament

superior and middle rectal vessels

sacrotuberous ligament

rectum

sacrospinous ligament

uterus

apex of sacrum

Cahill

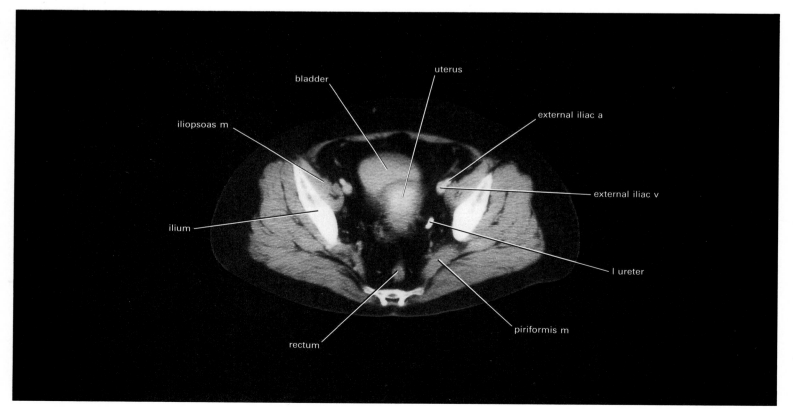

bladder

uterus

iliopsoas m

external iliac a

external iliac v

ilium

l ureter

piriformis m

rectum

Section 10 from below.

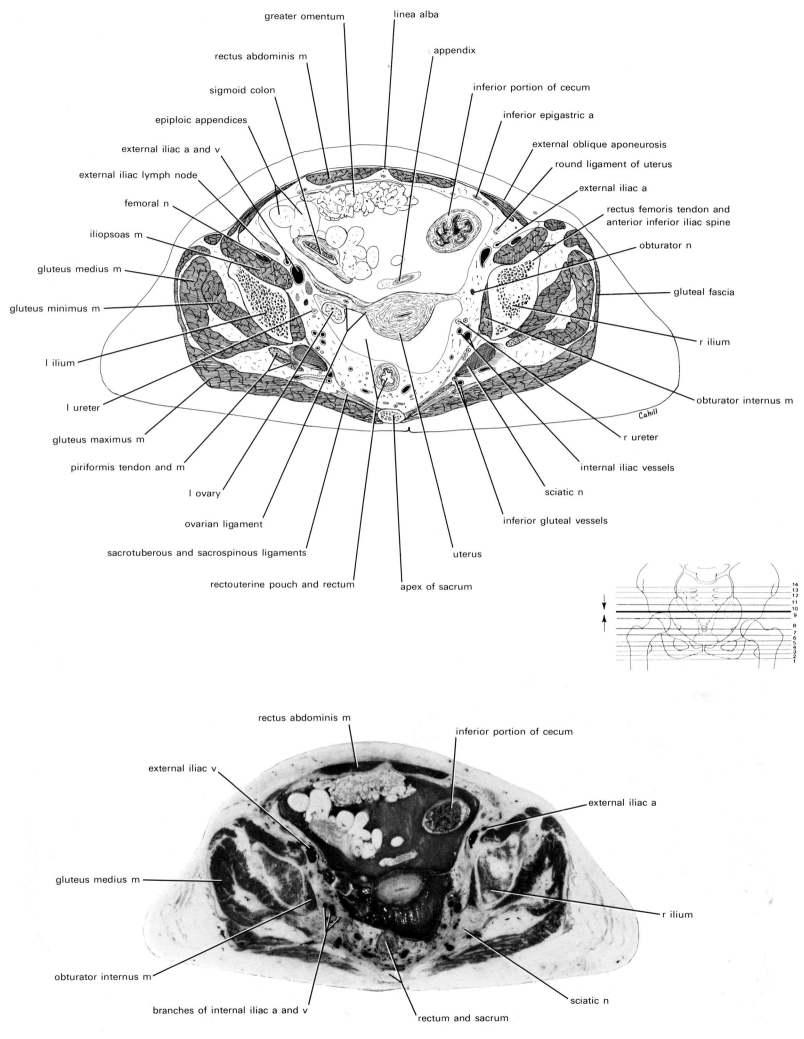

greater omentum

linea alba

rectus abdominis m

appendix

sigmoid colon

inferior portion of cecum

epiploic appendices

inferior epigastric a

external iliac a and v

external oblique aponeurosis

external iliac lymph node

round ligament of uterus

femoral n

external iliac a

iliopsoas m

rectus femoris tendon and
anterior inferior iliac spine

gluteus medius m

obturator n

gluteus minimus m

gluteal fascia

l ilium

r ilium

l ureter

obturator internus m

gluteus maximus m

r ureter

piriformis tendon and m

internal iliac vessels

l ovary

sciatic n

ovarian ligament

inferior gluteal vessels

sacrotuberous and sacrospinous ligaments

uterus

rectouterine pouch and rectum

apex of sacrum

Cahill

rectus abdominis m

inferior portion of cecum

external iliac v

external iliac a

gluteus medius m

r ilium

obturator internus m

sciatic n

branches of internal iliac a and v

rectum and sacrum

Section 9 from above.

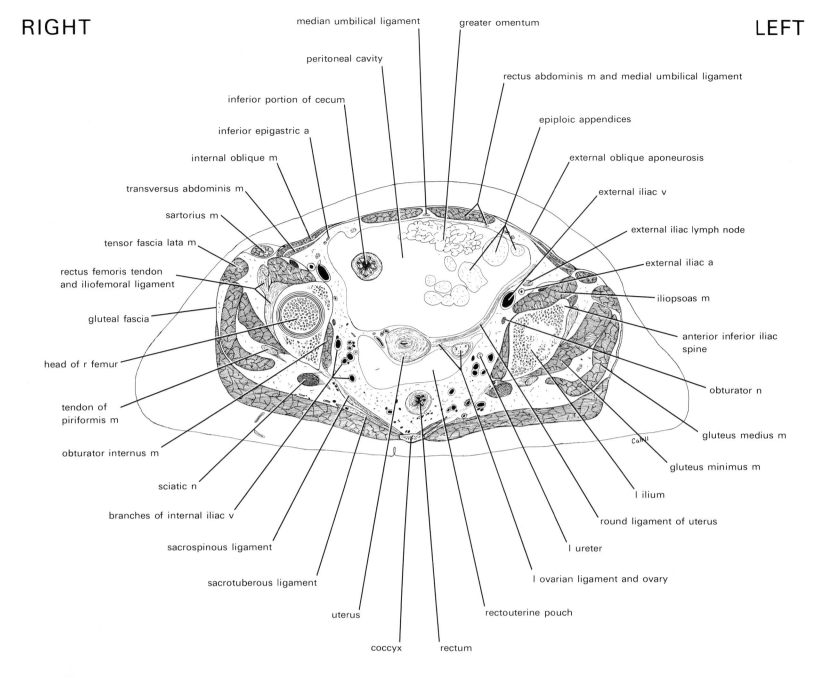

median umbilical ligament
greater omentum
peritoneal cavity
rectus abdominis m and medial umbilical ligament
inferior portion of cecum
epiploic appendices
inferior epigastric a
external oblique aponeurosis
internal oblique m
external iliac v
transversus abdominis m
external iliac lymph node
sartorius m
external iliac a
tensor fascia lata m
iliopsoas m
rectus femoris tendon and iliofemoral ligament
anterior inferior iliac spine
gluteal fascia
head of r femur
obturator n
tendon of piriformis m
gluteus medius m
obturator internus m
gluteus minimus m
sciatic n
l ilium
branches of internal iliac v
round ligament of uterus
sacrospinous ligament
l ureter
sacrotuberous ligament
l ovarian ligament and ovary
uterus
rectouterine pouch
coccyx   rectum

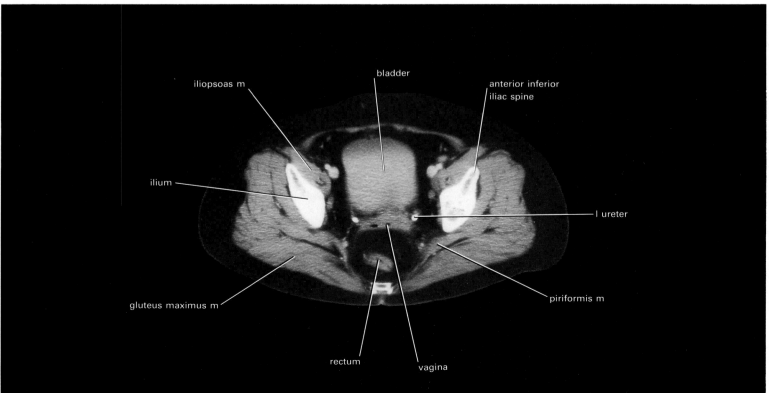

iliopsoas m
bladder
anterior inferior iliac spine
ilium
l ureter
gluteus maximus m
piriformis m
rectum   vagina

Section 9 from below.

LEFT   RIGHT

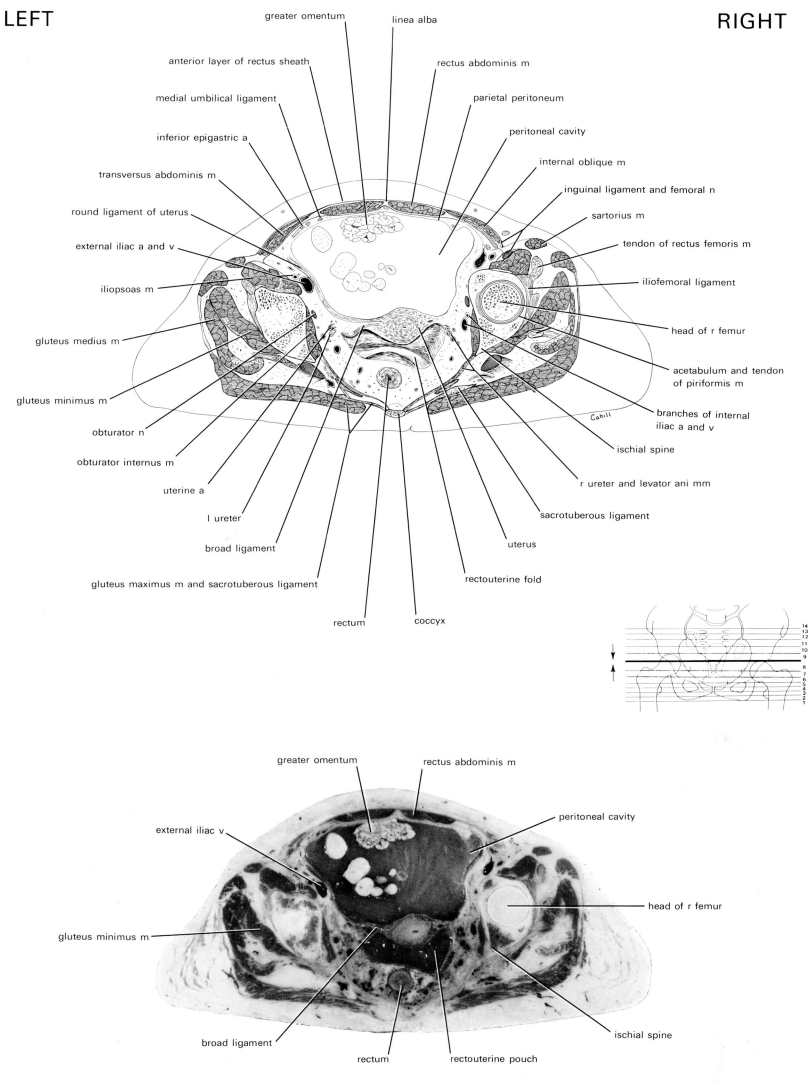

greater omentum   linea alba

anterior layer of rectus sheath   rectus abdominis m

medial umbilical ligament   parietal peritoneum

inferior epigastric a   peritoneal cavity

transversus abdominis m   internal oblique m

round ligament of uterus   inguinal ligament and femoral n

external iliac a and v   sartorius m

iliopsoas m   tendon of rectus femoris m

gluteus medius m   iliofemoral ligament

gluteus minimus m   head of r femur

obturator n   acetabulum and tendon
of piriformis m

obturator internus m   branches of internal
iliac a and v

uterine a   ischial spine

l ureter   r ureter and levator ani mm

broad ligament   sacrotuberous ligament

gluteus maximus m and sacrotuberous ligament   uterus

rectum   coccyx   rectouterine fold

Cahill

greater omentum   rectus abdominis m

external iliac v   peritoneal cavity

gluteus minimus m   head of r femur

broad ligament   ischial spine

rectum   rectouterine pouch

Section 8 from above.

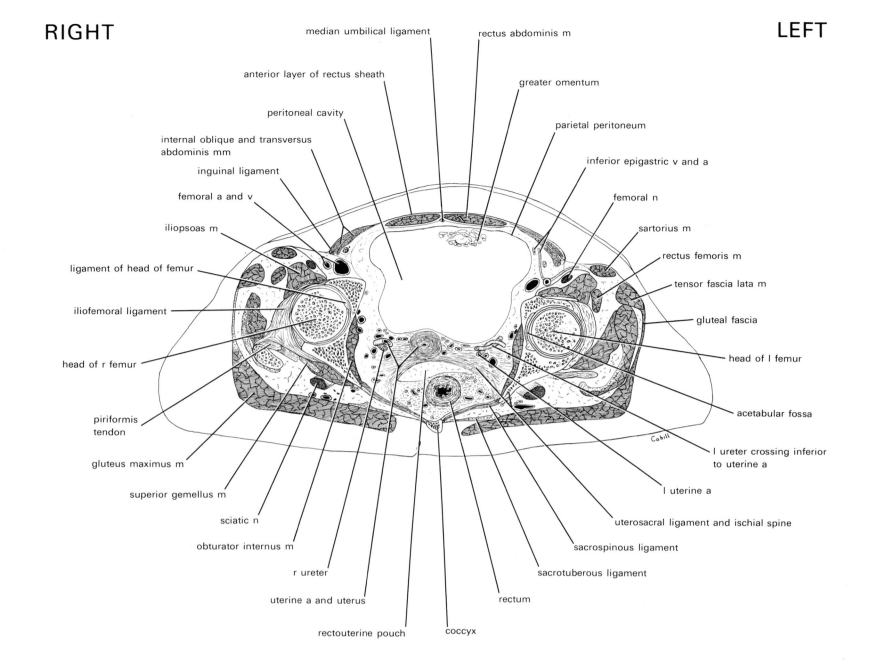

median umbilical ligament

rectus abdominis m

anterior layer of rectus sheath

greater omentum

peritoneal cavity

parietal peritoneum

internal oblique and transversus
abdominis mm

inferior epigastric v and a

inguinal ligament

femoral n

femoral a and v

sartorius m

iliopsoas m

rectus femoris m

ligament of head of femur

tensor fascia lata m

iliofemoral ligament

gluteal fascia

head of r femur

head of l femur

acetabular fossa

piriformis
tendon

l ureter crossing inferior
to uterine a

gluteus maximus m

l uterine a

superior gemellus m

uterosacral ligament and ischial spine

sciatic n

sacrospinous ligament

obturator internus m

sacrotuberous ligament

r ureter

rectum

uterine a and uterus

rectouterine pouch

coccyx

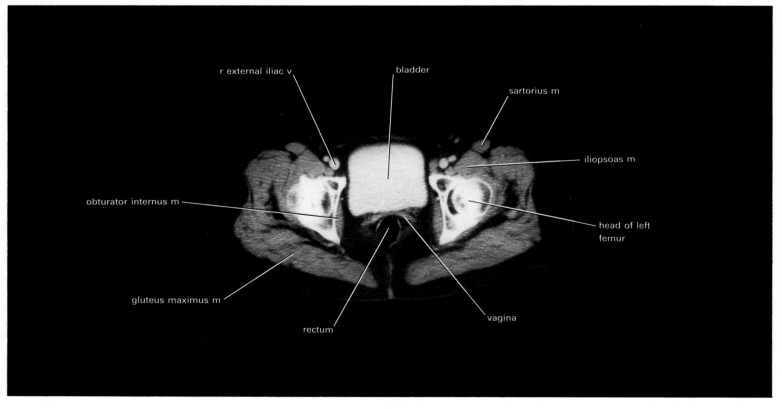

r external iliac v

bladder

sartorius m

iliopsoas m

obturator internus m

head of left
femur

gluteus maximus m

rectum

vagina

Section 8 from below.

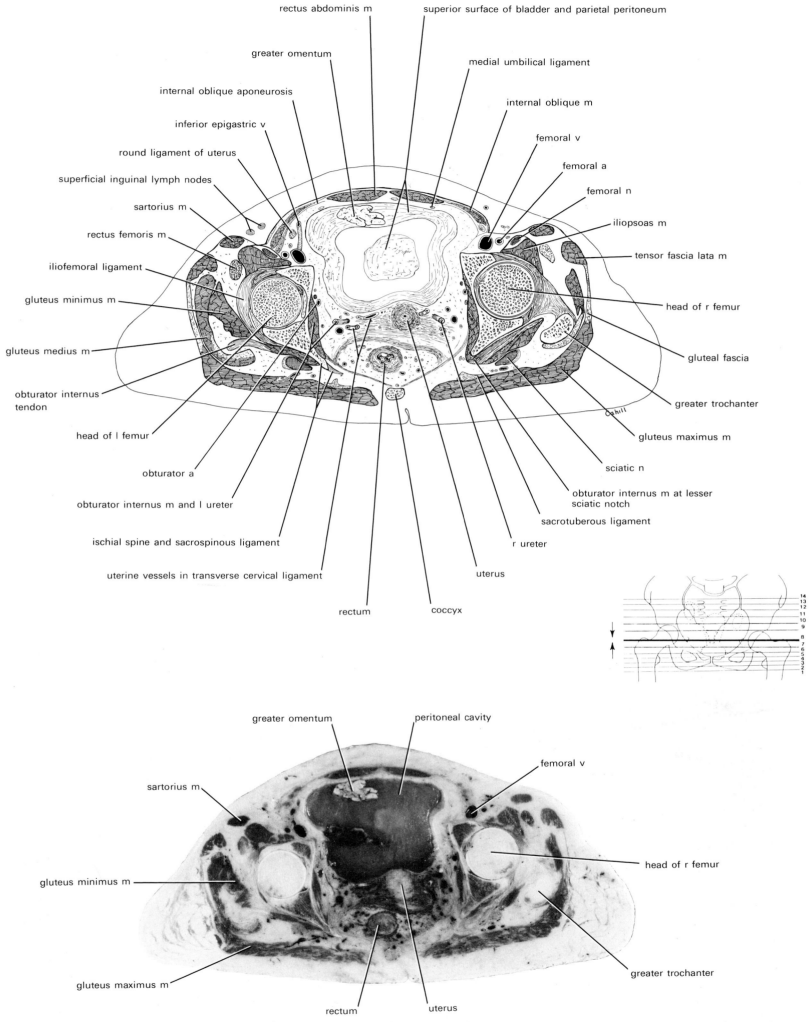

rectus abdominis m

superior surface of bladder and parietal peritoneum

greater omentum

medial umbilical ligament

internal oblique aponeurosis

internal oblique m

inferior epigastric v

femoral v

round ligament of uterus

femoral a

superficial inguinal lymph nodes

femoral n

sartorius m

iliopsoas m

rectus femoris m

tensor fascia lata m

iliofemoral ligament

gluteus minimus m

head of r femur

gluteus medius m

gluteal fascia

obturator internus tendon

greater trochanter

head of l femur

gluteus maximus m

obturator a

sciatic n

obturator internus m and l ureter

obturator internus m at lesser sciatic notch

ischial spine and sacrospinous ligament

sacrotuberous ligament

r ureter

uterine vessels in transverse cervical ligament

uterus

rectum

coccyx

greater omentum

peritoneal cavity

femoral v

sartorius m

head of r femur

gluteus minimus m

greater trochanter

gluteus maximus m

rectum

uterus

Section 7 from above.

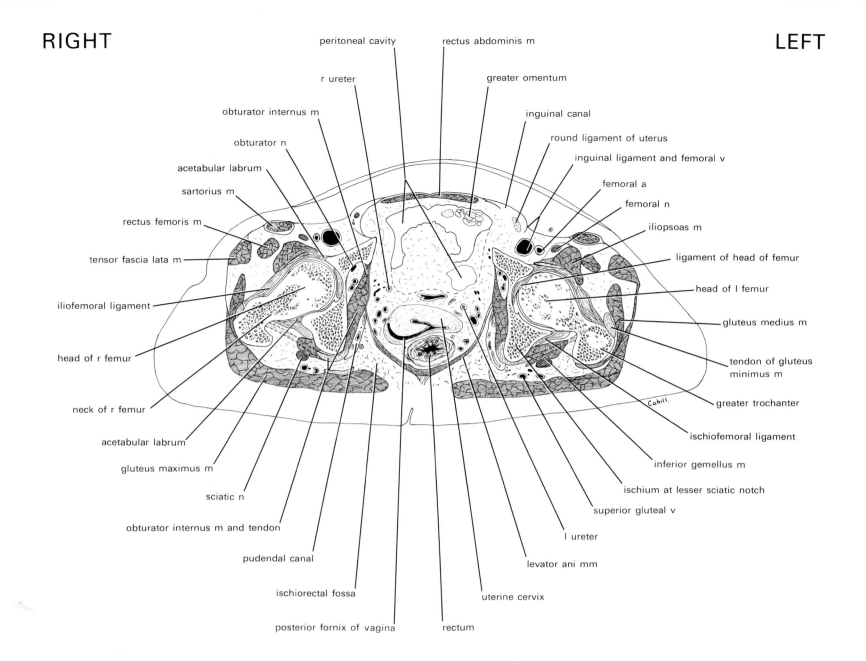

peritoneal cavity
rectus abdominis m
r ureter
greater omentum
obturator internus m
inguinal canal
obturator n
round ligament of uterus
acetabular labrum
inguinal ligament and femoral v
sartorius m
femoral a
rectus femoris m
femoral n
iliopsoas m
tensor fascia lata m
ligament of head of femur
head of l femur
iliofemoral ligament
gluteus medius m
head of r femur
tendon of gluteus minimus m
neck of r femur
greater trochanter
acetabular labrum
ischiofemoral ligament
gluteus maximus m
inferior gemellus m
sciatic n
ischium at lesser sciatic notch
obturator internus m and tendon
superior gluteal v
pudendal canal
l ureter
ischiorectal fossa
levator ani mm
posterior fornix of vagina
rectum
uterine cervix

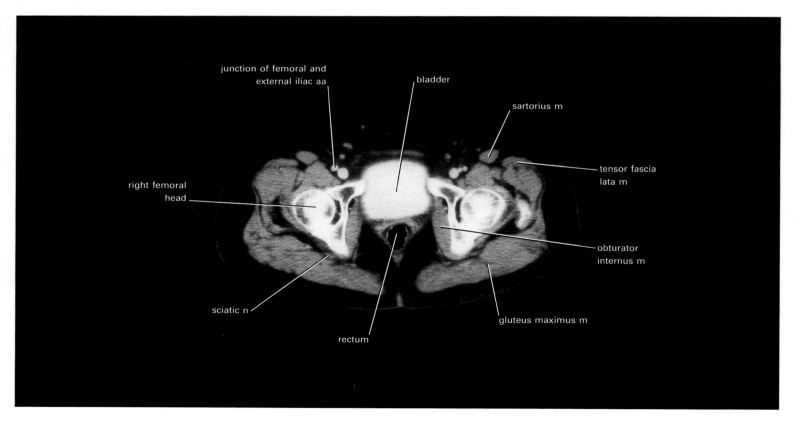

junction of femoral and
external iliac aa
bladder
sartorius m
right femoral
head
tensor fascia
lata m
obturator
internus m
sciatic n
gluteus maximus m
rectum

Section 7 from below.

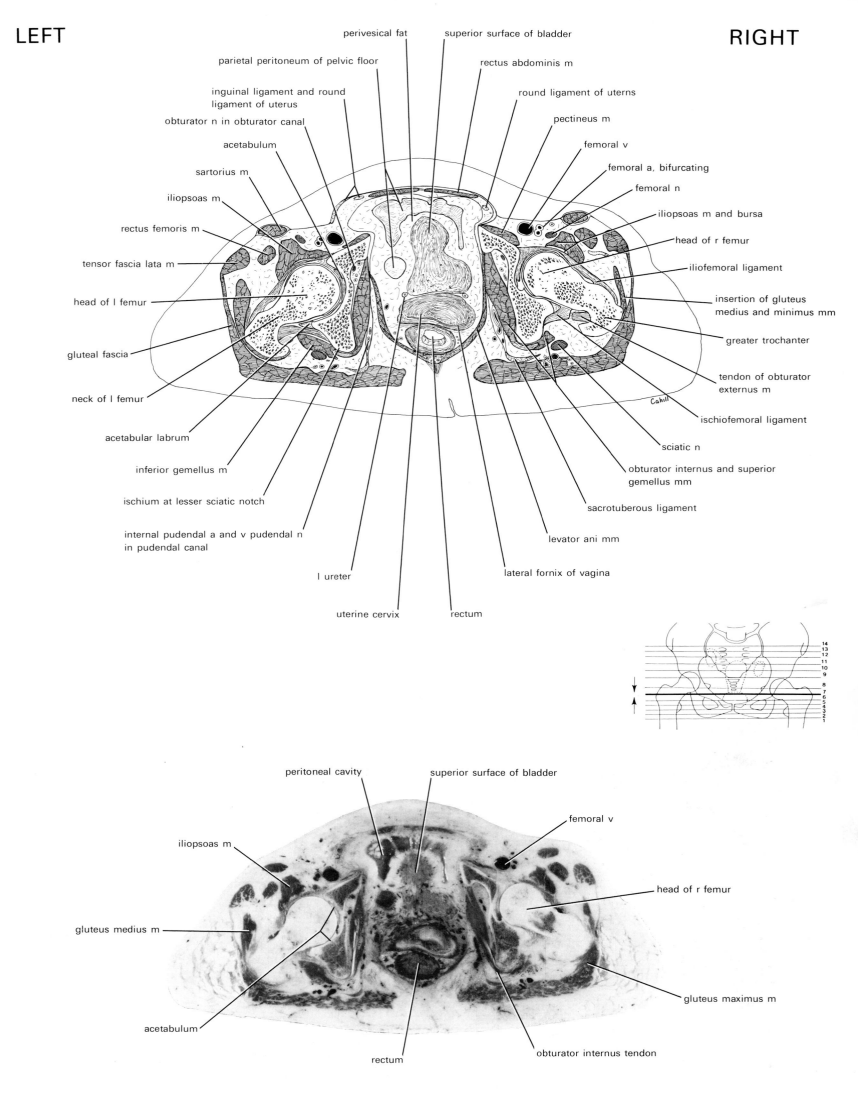

LEFT   RIGHT

perivesical fat

superior surface of bladder

parietal peritoneum of pelvic floor

rectus abdominis m

inguinal ligament and round ligament of uterus

round ligament of uterns

obturator n in obturator canal

pectineus m

acetabulum

femoral v

sartorius m

femoral a, bifurcating

iliopsoas m

femoral n

rectus femoris m

iliopsoas m and bursa

tensor fascia lata m

head of r femur

head of l femur

iliofemoral ligament

insertion of gluteus medius and minimus mm

gluteal fascia

greater trochanter

neck of l femur

tendon of obturator externus m

acetabular labrum

ischiofemoral ligament

inferior gemellus m

sciatic n

ischium at lesser sciatic notch

obturator internus and superior gemellus mm

internal pudendal a and v pudendal n in pudendal canal

sacrotuberous ligament

levator ani mm

l ureter

lateral fornix of vagina

uterine cervix

rectum

peritoneal cavity

superior surface of bladder

iliopsoas m

femoral v

gluteus medius m

head of r femur

acetabulum

gluteus maximus m

rectum

obturator internus tendon

Section 6 from above.

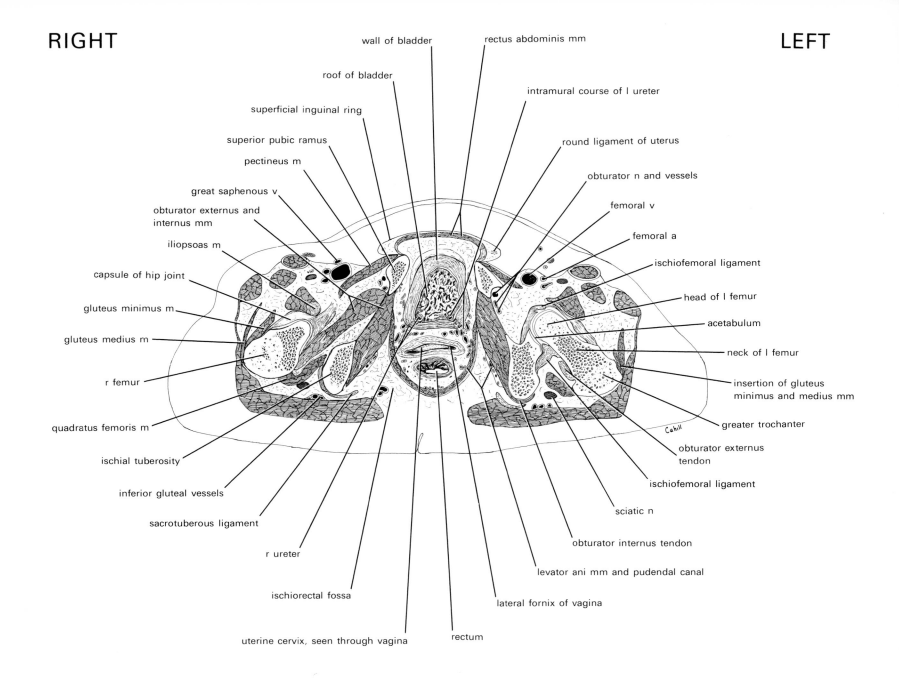

wall of bladder

rectus abdominis mm

roof of bladder

intramural course of l ureter

superficial inguinal ring

round ligament of uterus

superior pubic ramus

obturator n and vessels

pectineus m

femoral v

great saphenous v

femoral a

obturator externus and
internus mm

ischiofemoral ligament

iliopsoas m

head of l femur

capsule of hip joint

acetabulum

gluteus minimus m

neck of l femur

gluteus medius m

insertion of gluteus
minimus and medius mm

r femur

quadratus femoris m

greater trochanter

ischial tuberosity

obturator externus
tendon

inferior gluteal vessels

ischiofemoral ligament

sacrotuberous ligament

sciatic n

r ureter

obturator internus tendon

ischiorectal fossa

levator ani mm and pudendal canal

lateral fornix of vagina

uterine cervix, seen through vagina

rectum

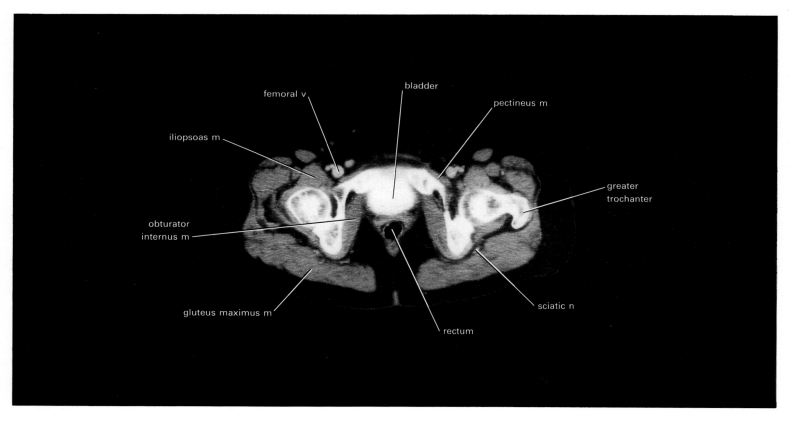

femoral v

bladder

pectineus m

iliopsoas m

greater
trochanter

obturator
internus m

gluteus maximus m

sciatic n

rectum

Section 6 from below.

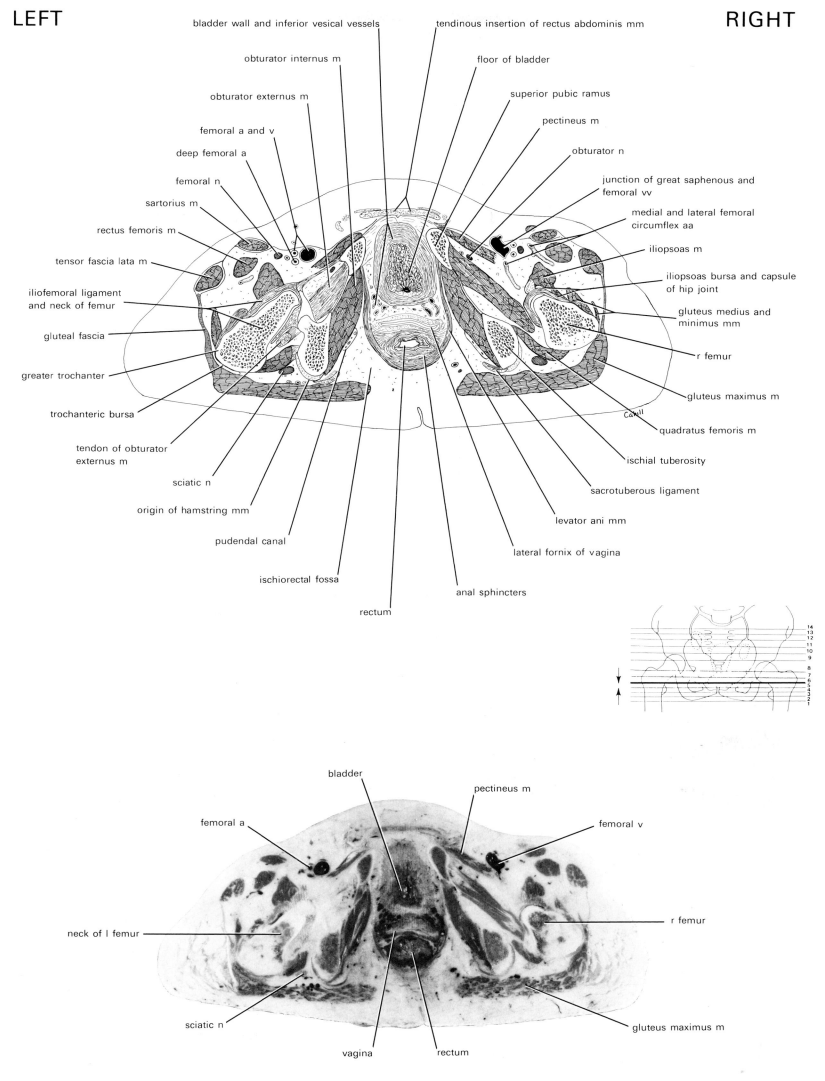

LEFT   RIGHT

bladder wall and inferior vesical vessels

tendinous insertion of rectus abdominis mm

obturator internus m

floor of bladder

obturator externus m

superior pubic ramus

femoral a and v

pectineus m

deep femoral a

obturator n

femoral n

junction of great saphenous and
femoral vv

sartorius m

medial and lateral femoral
circumflex aa

rectus femoris m

iliopsoas m

tensor fascia lata m

iliopsoas bursa and capsule
of hip joint

iliofemoral ligament
and neck of femur

gluteus medius and
minimus mm

gluteal fascia

r femur

greater trochanter

gluteus maximus m

trochanteric bursa

quadratus femoris m

tendon of obturator
externus m

ischial tuberosity

sciatic n

sacrotuberous ligament

origin of hamstring mm

levator ani mm

pudendal canal

lateral fornix of vagina

ischiorectal fossa

anal sphincters

rectum

Cahll

bladder

pectineus m

femoral a

femoral v

neck of l femur

r femur

sciatic n

gluteus maximus m

vagina

rectum

Section 5 from above.

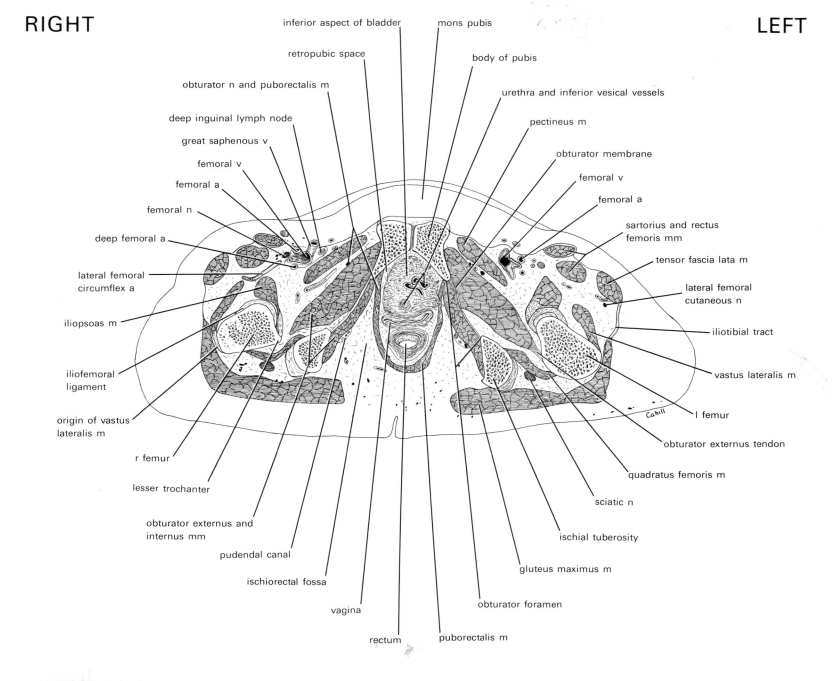

inferior aspect of bladder

mons pubis

retropubic space

body of pubis

obturator n and puborectalis m

urethra and inferior vesical vessels

deep inguinal lymph node

pectineus m

great saphenous v

obturator membrane

femoral v

femoral v

femoral a

femoral a

femoral n

sartorius and rectus femoris mm

deep femoral a

tensor fascia lata m

lateral femoral circumflex a

lateral femoral cutaneous n

iliopsoas m

iliotibial tract

iliofemoral ligament

vastus lateralis m

origin of vastus lateralis m

l femur

r femur

obturator externus tendon

lesser trochanter

quadratus femoris m

obturator externus and internus mm

sciatic n

pudendal canal

ischial tuberosity

ischiorectal fossa

gluteus maximus m

vagina

obturator foramen

rectum

puborectalis m

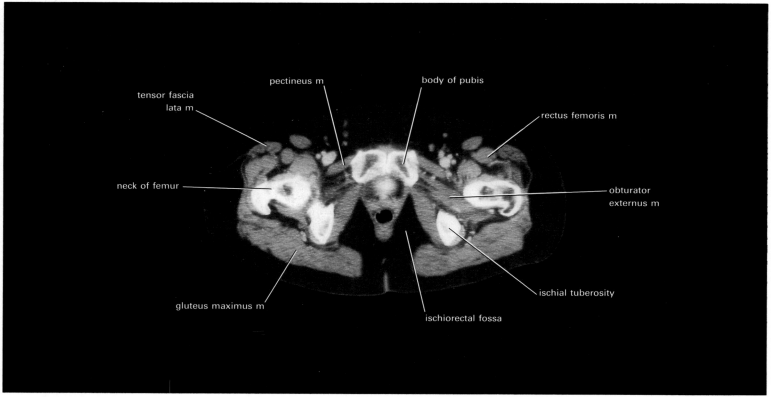

tensor fascia lata m

pectineus m

body of pubis

rectus femoris m

neck of femur

obturator externus m

gluteus maximus m

ischial tuberosity

ischiorectal fossa

Section 5 from below.

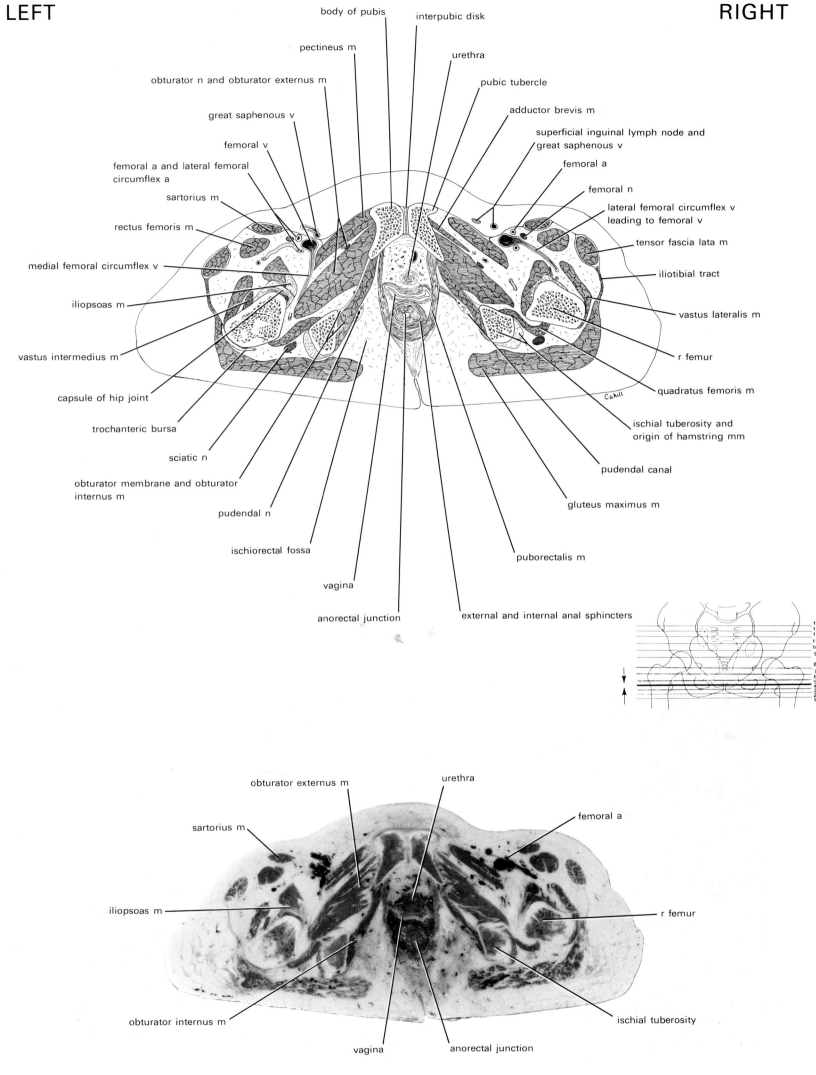

body of pubis

interpubic disk

pectineus m

urethra

obturator n and obturator externus m

pubic tubercle

great saphenous v

adductor brevis m

superficial inguinal lymph node and
great saphenous v

femoral v

femoral a

femoral a and lateral femoral
circumflex a

femoral n

sartorius m

lateral femoral circumflex v
leading to femoral v

rectus femoris m

tensor fascia lata m

medial femoral circumflex v

iliotibial tract

iliopsoas m

vastus lateralis m

vastus intermedius m

r femur

capsule of hip joint

quadratus femoris m

trochanteric bursa

ischial tuberosity and
origin of hamstring mm

sciatic n

pudendal canal

obturator membrane and obturator
internus m

gluteus maximus m

pudendal n

puborectalis m

ischiorectal fossa

vagina

external and internal anal sphincters

anorectal junction

obturator externus m

urethra

sartorius m

femoral a

iliopsoas m

r femur

obturator internus m

ischial tuberosity

vagina

anorectal junction

Section 4 from above.

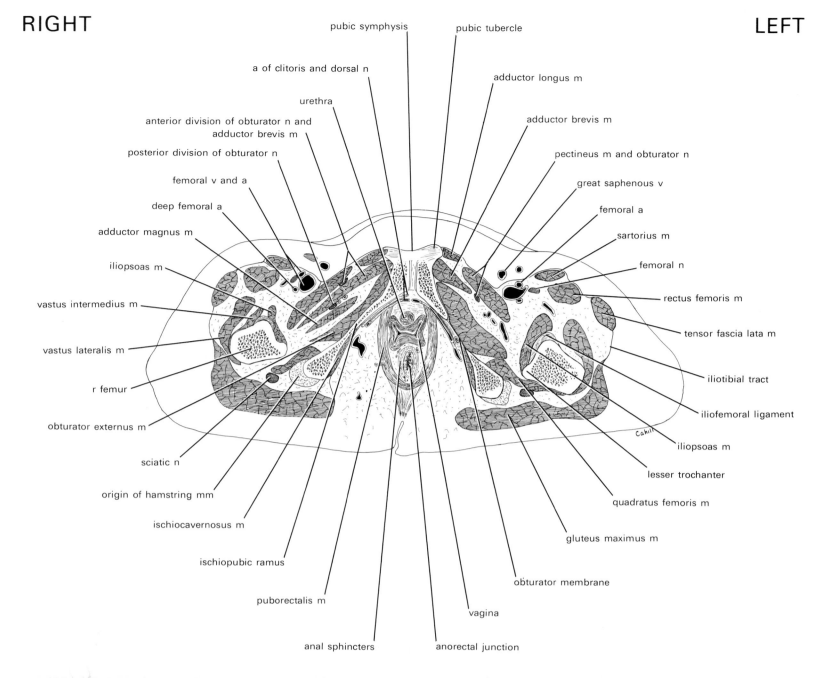

pubic symphysis

pubic tubercle

a of clitoris and dorsal n

adductor longus m

urethra

anterior division of obturator n and
adductor brevis m

adductor brevis m

posterior division of obturator n

pectineus m and obturator n

femoral v and a

great saphenous v

deep femoral a

femoral a

adductor magnus m

sartorius m

iliopsoas m

femoral n

vastus intermedius m

rectus femoris m

vastus lateralis m

tensor fascia lata m

r femur

iliotibial tract

obturator externus m

iliofemoral ligament

sciatic n

iliopsoas m

origin of hamstring mm

lesser trochanter

ischiocavernosus m

quadratus femoris m

ischiopubic ramus

gluteus maximus m

puborectalis m

obturator membrane

vagina

anal sphincters

anorectal junction

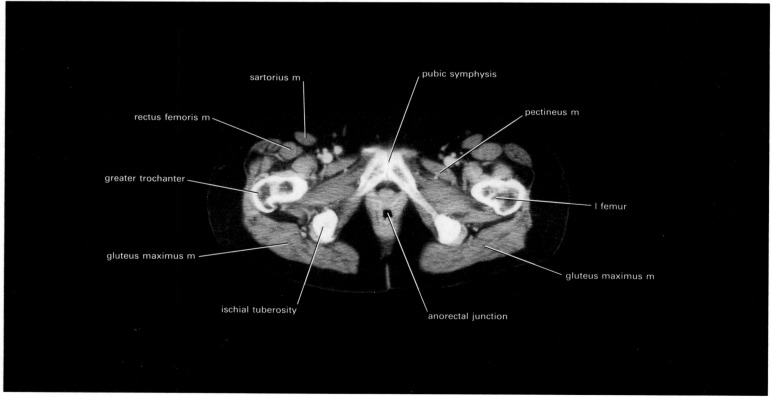

sartorius m

pubic symphysis

rectus femoris m

pectineus m

greater trochanter

l femur

gluteus maximus m

gluteus maximus m

ischial tuberosity

anorectal junction

Section 4 from below.

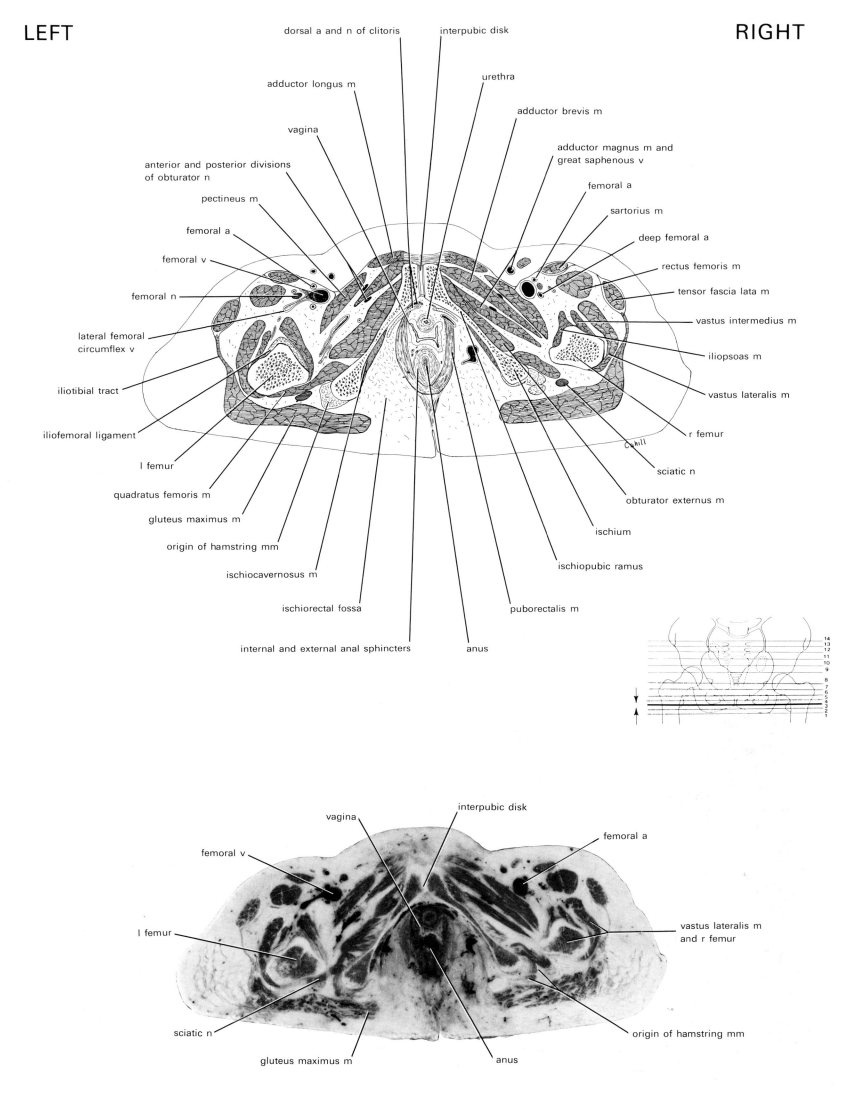

LEFT   RIGHT

dorsal a and n of clitoris   interpubic disk

adductor longus m   urethra

vagina   adductor brevis m

anterior and posterior divisions   adductor magnus m and
of obturator n   great saphenous v

pectineus m   femoral a

femoral a   sartorius m

femoral v   deep femoral a

femoral n   rectus femoris m

lateral femoral   tensor fascia lata m
circumflex v
                                                                     vastus intermedius m

iliotibial tract   iliopsoas m

iliofemoral ligament   vastus lateralis m

l femur   r femur

quadratus femoris m   sciatic n

gluteus maximus m   obturator externus m

origin of hamstring mm   ischium

ischiocavernosus m   ischiopubic ramus

ischiorectal fossa   puborectalis m

internal and external anal sphincters   anus

vagina   interpubic disk

femoral v   femoral a

l femur   vastus lateralis m
                                                           and r femur

sciatic n   origin of hamstring mm

gluteus maximus m   anus

Section 3 from above.

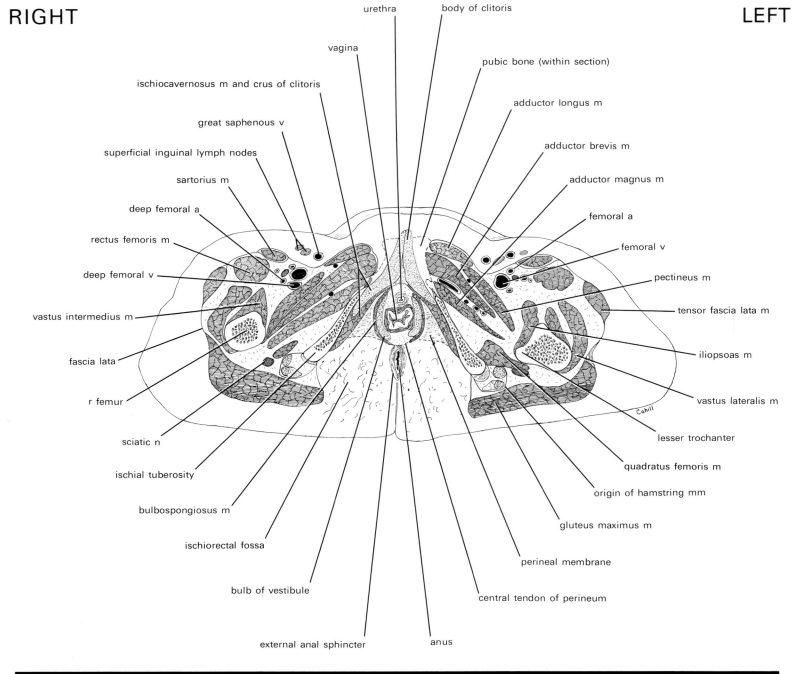

urethra

body of clitoris

vagina

pubic bone (within section)

ischiocavernosus m and crus of clitoris

adductor longus m

great saphenous v

adductor brevis m

superficial inguinal lymph nodes

adductor magnus m

sartorius m

femoral a

deep femoral a

femoral v

rectus femoris m

pectineus m

deep femoral v

tensor fascia lata m

vastus intermedius m

iliopsoas m

fascia lata

vastus lateralis m

r femur

lesser trochanter

sciatic n

quadratus femoris m

ischial tuberosity

origin of hamstring mm

bulbospongiosus m

gluteus maximus m

ischiorectal fossa

perineal membrane

bulb of vestibule

central tendon of perineum

external anal sphincter

anus

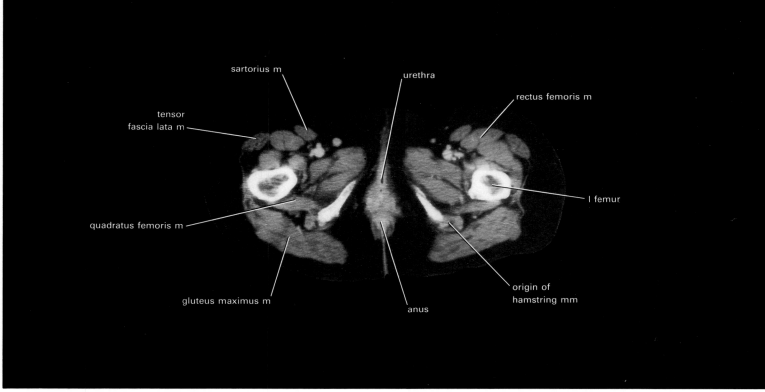

sartorius m

urethra

rectus femoris m

tensor fascia lata m

quadratus femoris m

l femur

gluteus maximus m

anus

origin of hamstring mm

Section 3 from below.

LEFT   RIGHT

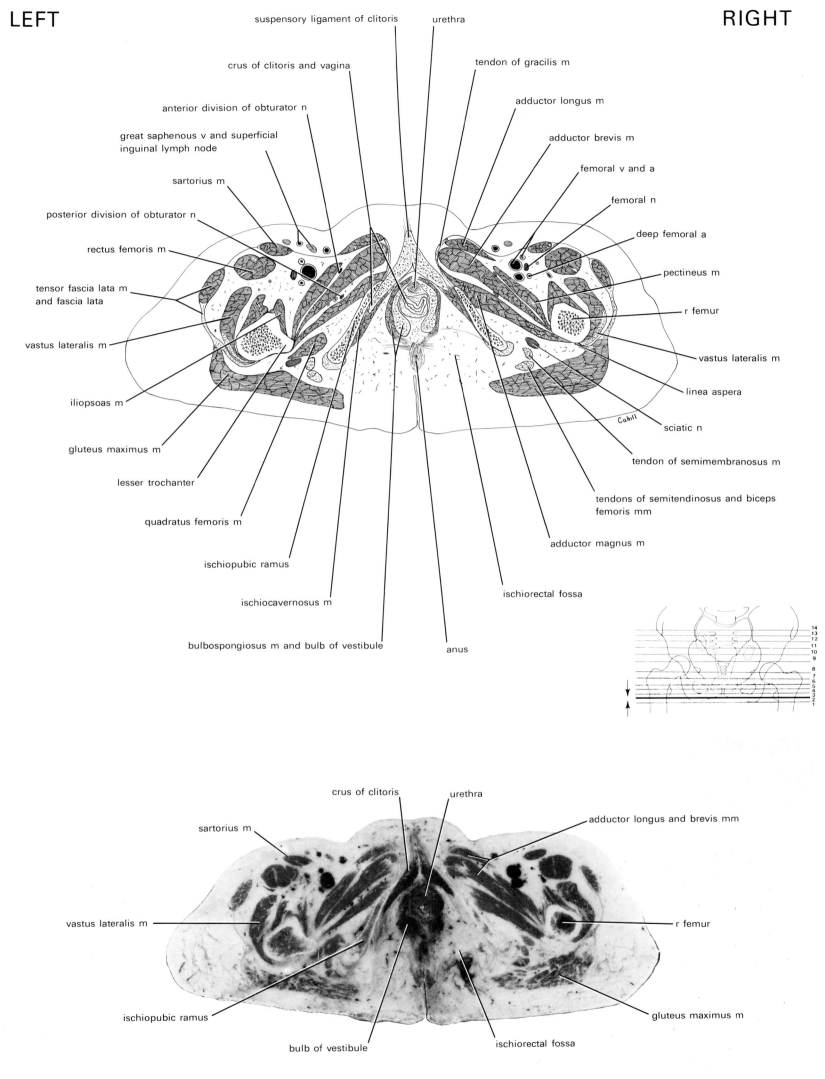

suspensory ligament of clitoris — urethra

crus of clitoris and vagina

tendon of gracilis m

anterior division of obturator n

adductor longus m

great saphenous v and superficial
inguinal lymph node

adductor brevis m

femoral v and a

sartorius m

femoral n

posterior division of obturator n

deep femoral a

rectus femoris m

pectineus m

tensor fascia lata m
and fascia lata

r femur

vastus lateralis m

vastus lateralis m

linea aspera

iliopsoas m

sciatic n

gluteus maximus m

tendon of semimembranosus m

lesser trochanter

tendons of semitendinosus and biceps
femoris mm

quadratus femoris m

adductor magnus m

ischiopubic ramus

ischiorectal fossa

ischiocavernosus m

bulbospongiosus m and bulb of vestibule

anus

crus of clitoris — urethra

sartorius m

adductor longus and brevis mm

vastus lateralis m

r femur

ischiopubic ramus

gluteus maximus m

bulb of vestibule

ischiorectal fossa

Section 2 from above.

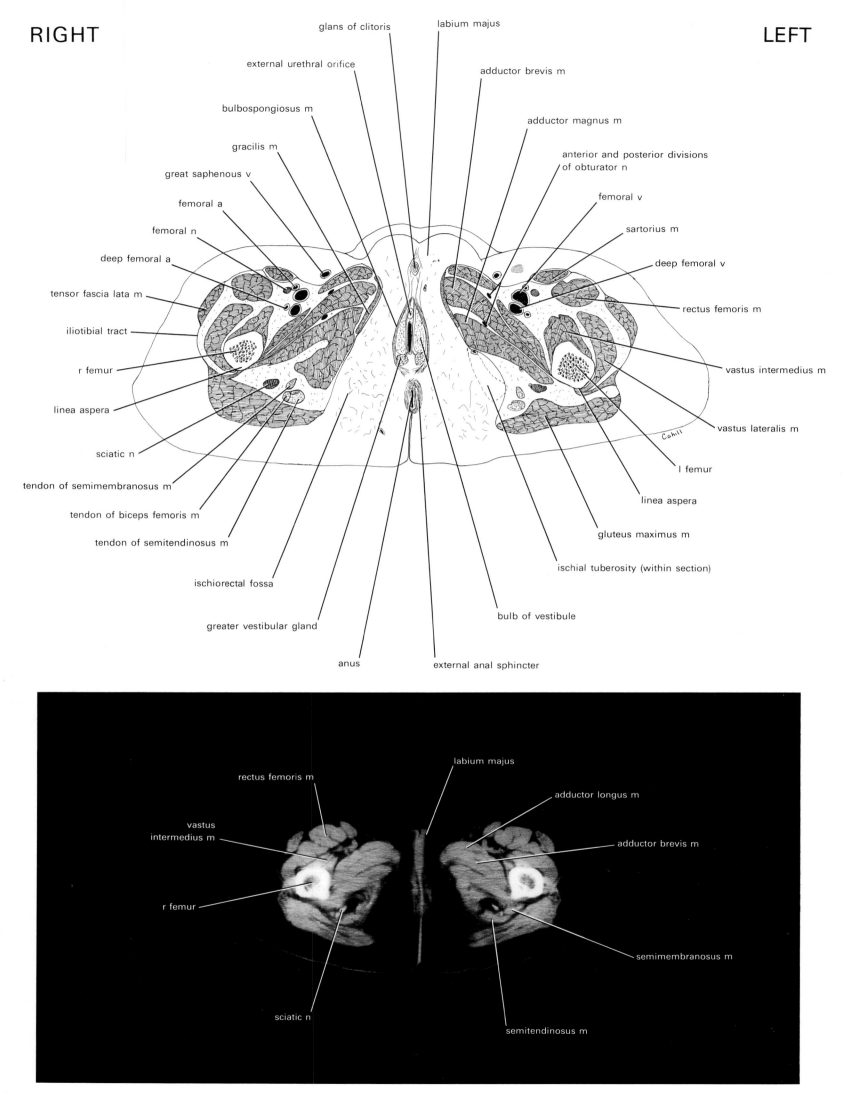

glans of clitoris

labium majus

external urethral orifice

adductor brevis m

bulbospongiosus m

adductor magnus m

gracilis m

anterior and posterior divisions
of obturator n

great saphenous v

femoral v

femoral a

sartorius m

femoral n

deep femoral a

deep femoral v

tensor fascia lata m

rectus femoris m

iliotibial tract

r femur

vastus intermedius m

linea aspera

sciatic n

vastus lateralis m

tendon of semimembranosus m

l femur

tendon of biceps femoris m

linea aspera

tendon of semitendinosus m

gluteus maximus m

ischial tuberosity (within section)

ischiorectal fossa

greater vestibular gland

bulb of vestibule

anus

external anal sphincter

rectus femoris m

labium majus

adductor longus m

vastus
intermedius m

adductor brevis m

r femur

semimembranosus m

sciatic n

semitendinosus m

Section 2 from below.

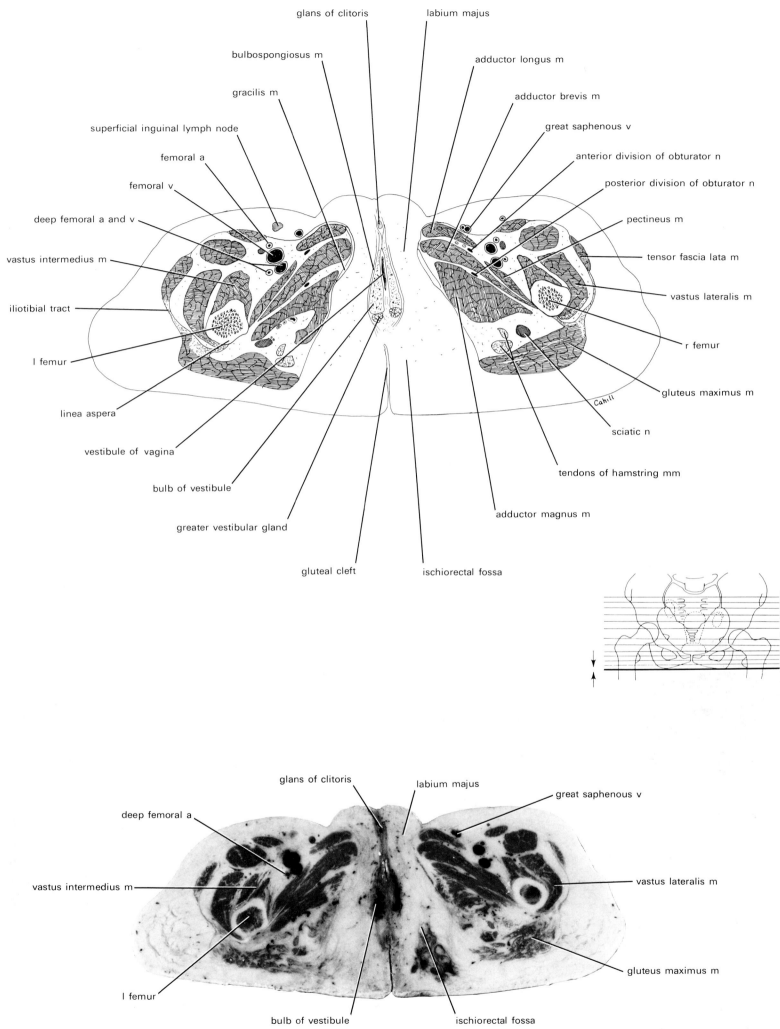

Section 1 from above.

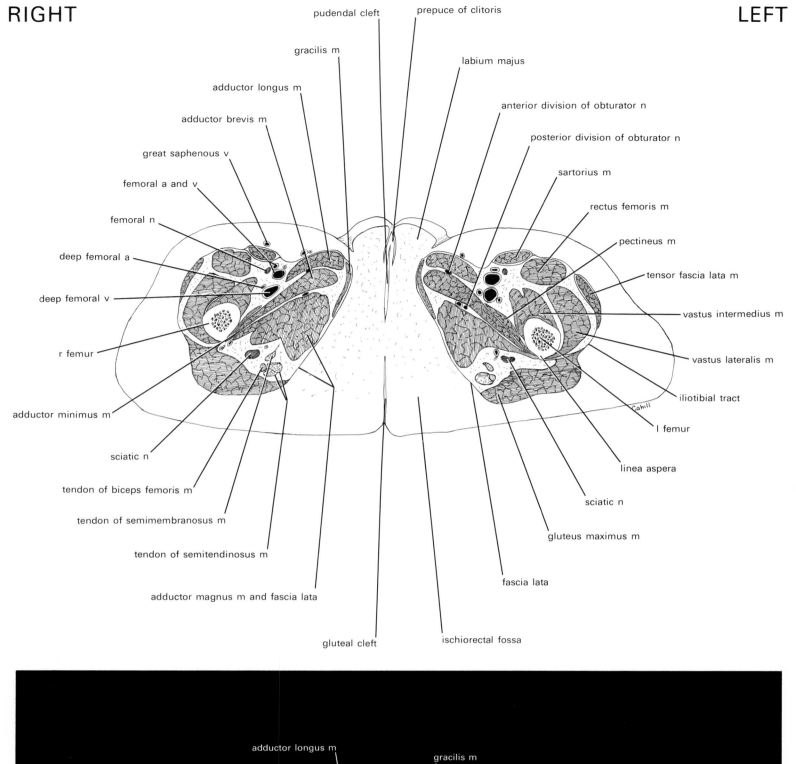

pudendal cleft

prepuce of clitoris

gracilis m

labium majus

adductor longus m

anterior division of obturator n

adductor brevis m

posterior division of obturator n

great saphenous v

sartorius m

femoral a and v

rectus femoris m

femoral n

pectineus m

deep femoral a

tensor fascia lata m

deep femoral v

vastus intermedius m

r femur

vastus lateralis m

adductor minimus m

iliotibial tract

sciatic n

l femur

tendon of biceps femoris m

linea aspera

tendon of semimembranosus m

sciatic n

tendon of semitendinosus m

gluteus maximus m

adductor magnus m and fascia lata

fascia lata

gluteal cleft

ischiorectal fossa

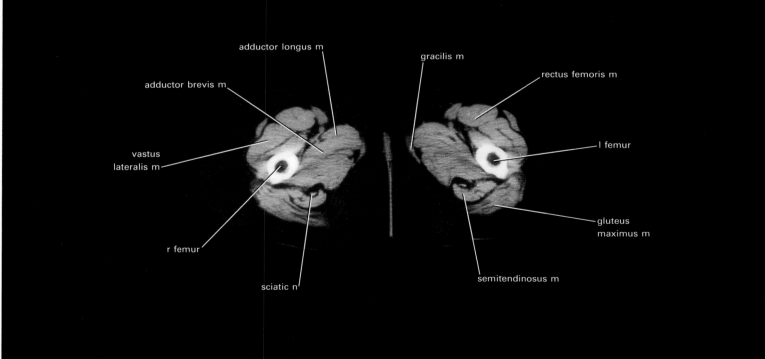

adductor longus m

gracilis m

adductor brevis m

rectus femoris m

vastus
lateralis m

l femur

r femur

gluteus
maximus m

sciatic n

semitendinosus m

Section 1 from below.

The Right Lower Limb

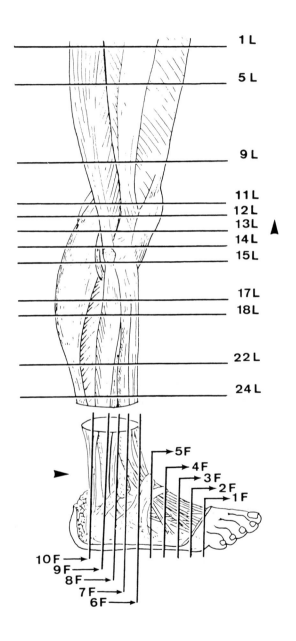

1 L

5 L

9 L

11 L
12 L
13 L
14 L
15 L

17 L
18 L

22 L

24 L

5 F

4 F

3 F

2 F

1 F

10 F

9 F

8 F

7 F

6 F

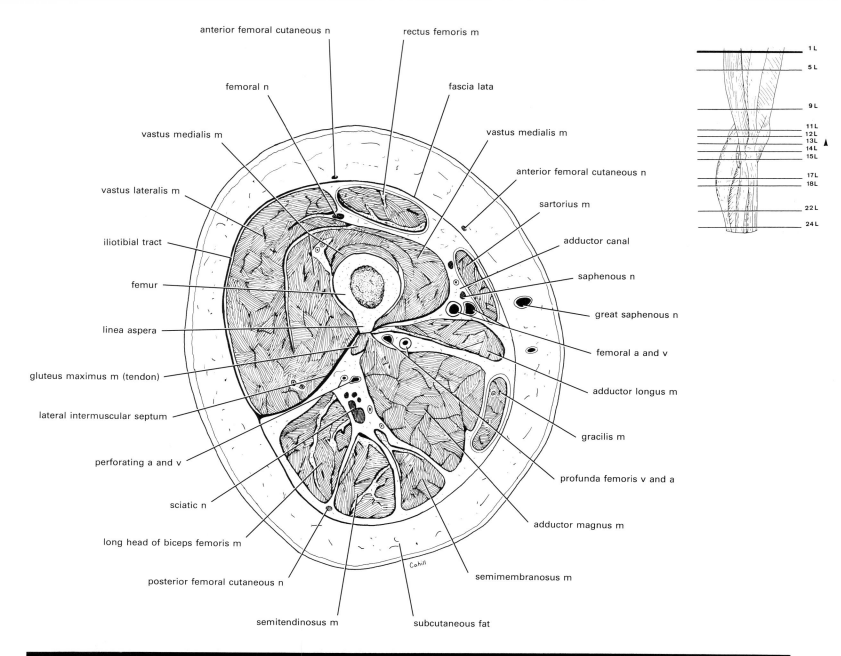

anterior femoral cutaneous n

rectus femoris m

femoral n

fascia lata

vastus medialis m

vastus medialis m

anterior femoral cutaneous n

vastus lateralis m

sartorius m

iliotibial tract

adductor canal

saphenous n

femur

great saphenous n

linea aspera

femoral a and v

gluteus maximus m (tendon)

adductor longus m

lateral intermuscular septum

gracilis m

perforating a and v

profunda femoris v and a

sciatic n

adductor magnus m

long head of biceps femoris m

semimembranosus m

posterior femoral cutaneous n

semitendinosus m

subcutaneous fat

Cahill

1 L
5 L
9 L
11 L
12 L
13 L
14 L
15 L
17 L
18 L
22 L
24 L

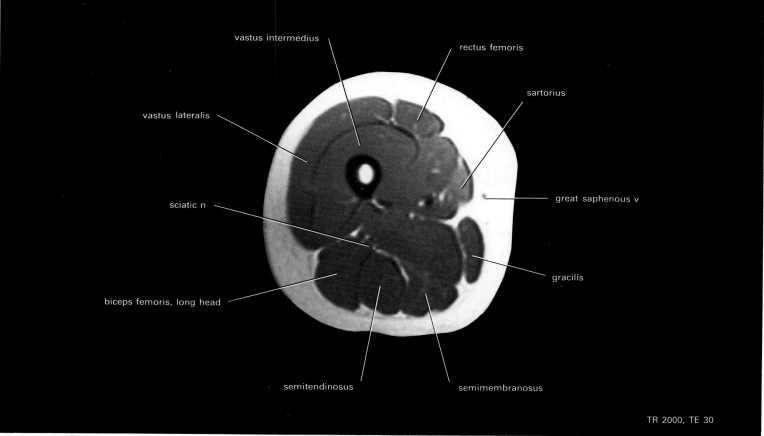

vastus intermedius

rectus femoris

vastus lateralis

sartorius

sciatic n

great saphenous v

biceps femoris, long head

gracilis

semitendinosus

semimembranosus

TR 2000, TE 30

Section 1L from below.

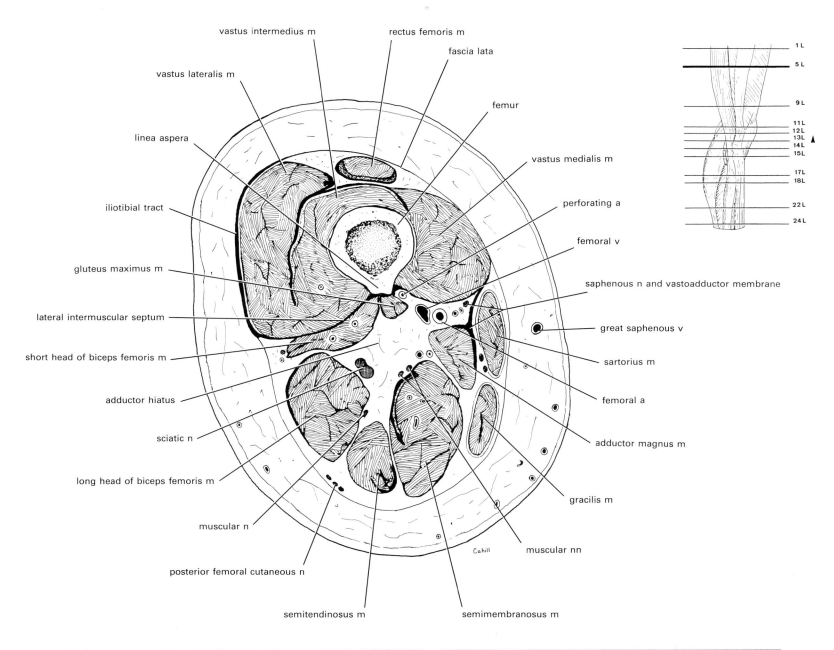

vastus intermedius m
rectus femoris m
fascia lata
vastus lateralis m
femur
vastus medialis m
linea aspera
perforating a
iliotibial tract
femoral v
gluteus maximus m
saphenous n and vastoadductor membrane
lateral intermuscular septum
great saphenous v
short head of biceps femoris m
sartorius m
adductor hiatus
femoral a
sciatic n
adductor magnus m
long head of biceps femoris m
gracilis m
muscular n
muscular nn
posterior femoral cutaneous n
semitendinosus m
semimembranosus m

1 L
5 L
9 L
11 L
12 L
13L
14L
15L
17L
18L
22 L
24 L

Cahill

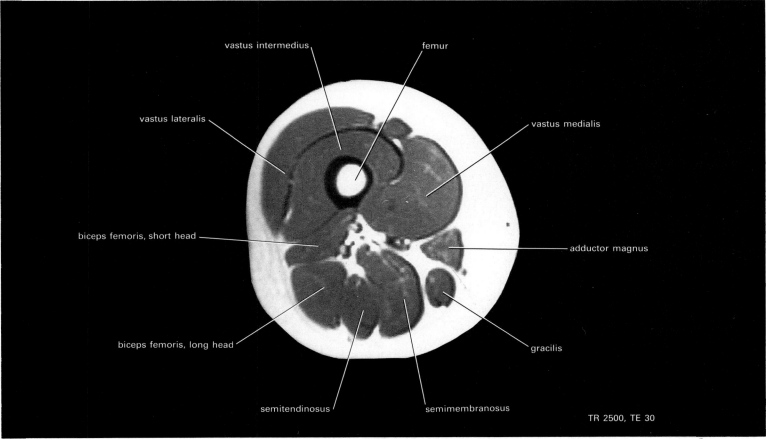

vastus intermedius
femur
vastus lateralis
vastus medialis
biceps femoris, short head
adductor magnus
biceps femoris, long head
gracilis
semitendinosus
semimembranosus
TR 2500, TE 30

Section 5L from below.

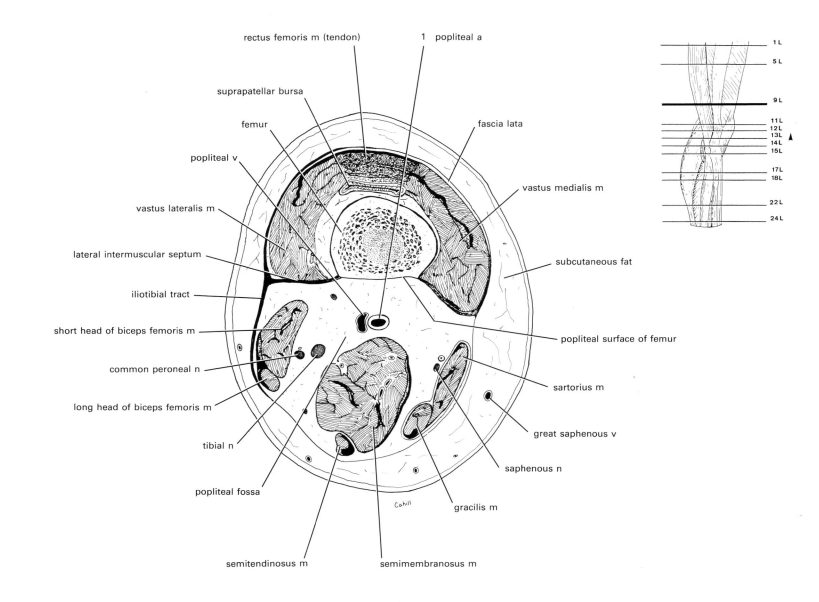

rectus femoris m (tendon)

1 popliteal a

suprapatellar bursa

fascia lata

femur

popliteal v

vastus medialis m

vastus lateralis m

lateral intermuscular septum

iliotibial tract

subcutaneous fat

short head of biceps femoris m

popliteal surface of femur

common peroneal n

long head of biceps femoris m

sartorius m

tibial n

great saphenous v

popliteal fossa

saphenous n

gracilis m

Cahill

semitendinosus m

semimembranosus m

1 L
5 L
9 L
11 L
12 L
13 L
14 L
15 L
17 L
18 L
22 L
24 L

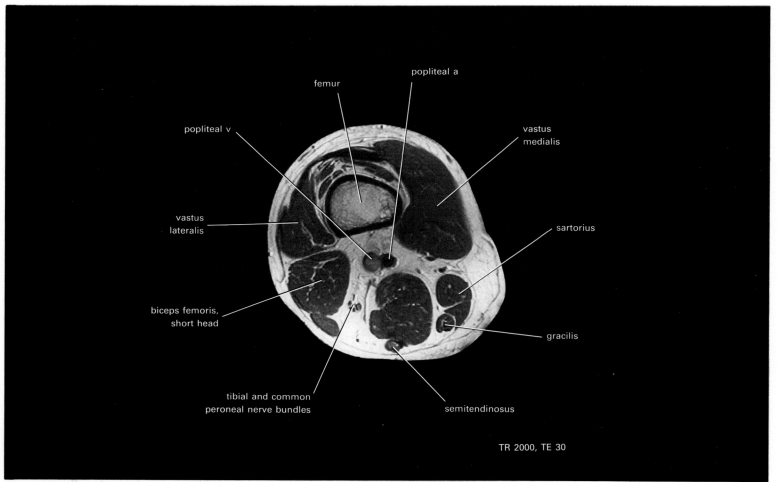

femur

popliteal a

popliteal v

vastus medialis

vastus lateralis

sartorius

biceps femoris, short head

gracilis

tibial and common peroneal nerve bundles

semitendinosus

TR 2000, TE 30

Section 9L from below.

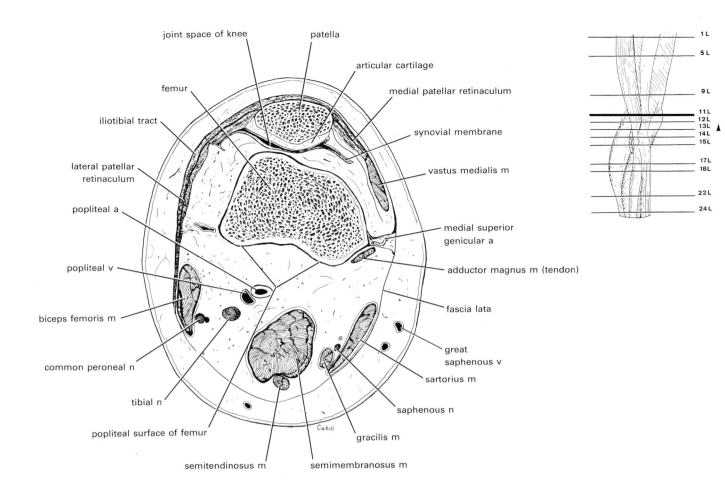

joint space of knee · patella · articular cartilage · medial patellar retinaculum · femur · synovial membrane · iliotibial tract · vastus medialis m · lateral patellar retinaculum · medial superior genicular a · popliteal a · adductor magnus m (tendon) · popliteal v · fascia lata · biceps femoris m · great saphenous v · common peroneal n · sartorius m · tibial n · saphenous n · popliteal surface of femur · gracilis m · semitendinosus m · semimembranosus m

Cahill

1 L
5 L
9 L
11 L
12 L
13 L
14 L
15 L
17 L
18 L
22 L
24 L

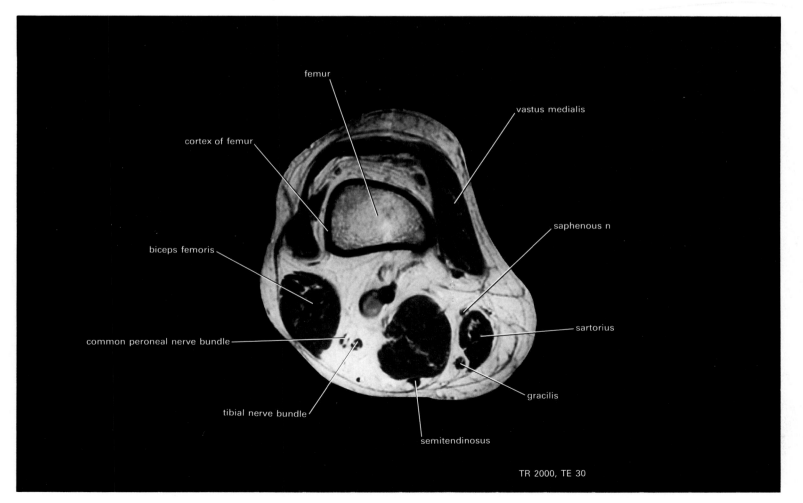

femur · vastus medialis · cortex of femur · biceps femoris · saphenous n · common peroneal nerve bundle · sartorius · tibial nerve bundle · gracilis · semitendinosus

TR 2000, TE 30

Section 11L from below.

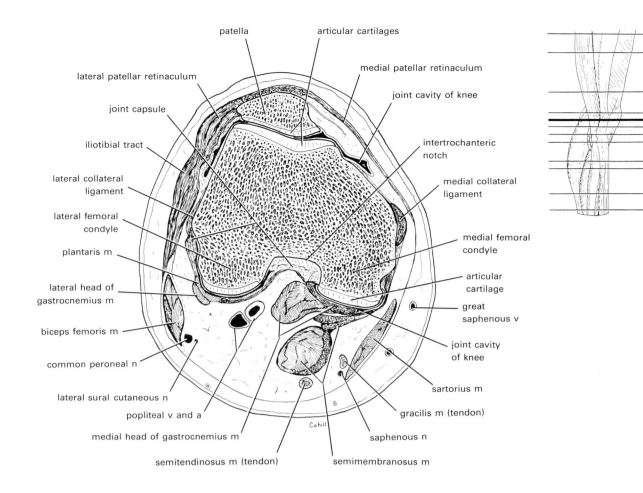

patella

articular cartilages

medial patellar retinaculum

joint cavity of knee

intertrochanteric notch

medial collateral ligament

medial femoral condyle

articular cartilage

great saphenous v

joint cavity of knee

sartorius m

gracilis m (tendon)

saphenous n

semimembranosus m

lateral patellar retinaculum

joint capsule

iliotibial tract

lateral collateral ligament

lateral femoral condyle

plantaris m

lateral head of gastrocnemius m

biceps femoris m

common peroneal n

lateral sural cutaneous n

popliteal v and a

medial head of gastrocnemius m

semitendinosus m (tendon)

Cahill

1 L
5 L
9 L
11 L
12 L
13 L
14 L
15 L
17 L
18 L
22 L
24 L

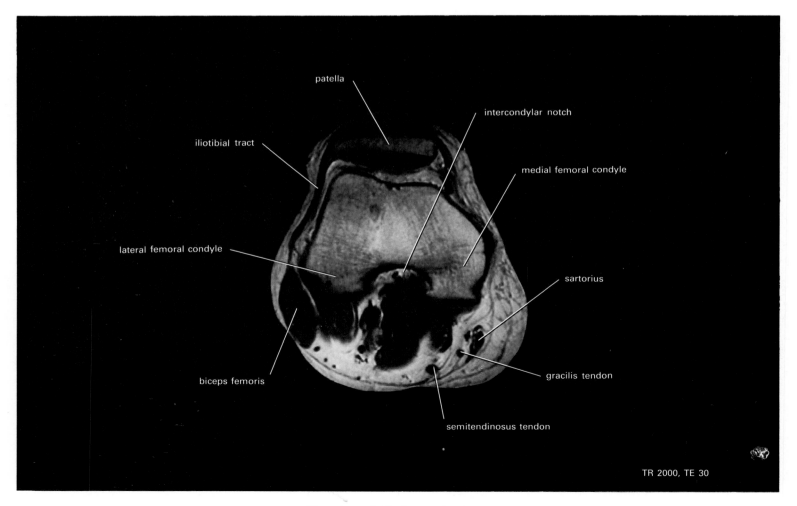

patella

intercondylar notch

iliotibial tract

medial femoral condyle

lateral femoral condyle

sartorius

biceps femoris

gracilis tendon

semitendinosus tendon

TR 2000, TE 30

Section 12L from below.

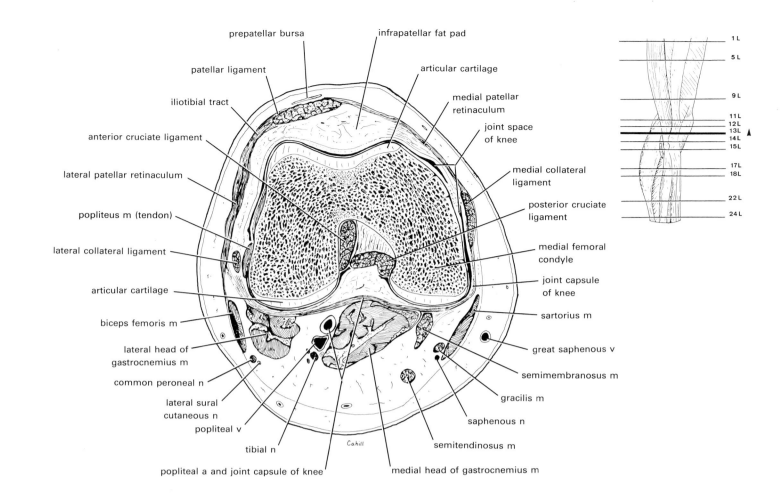

prepatellar bursa

infrapatellar fat pad

patellar ligament

articular cartilage

iliotibial tract

medial patellar
retinaculum

anterior cruciate ligament

joint space
of knee

lateral patellar retinaculum

medial collateral
ligament

popliteus m (tendon)

posterior cruciate
ligament

lateral collateral ligament

medial femoral
condyle

articular cartilage

joint capsule
of knee

biceps femoris m

sartorius m

lateral head of
gastrocnemius m

great saphenous v

common peroneal n

semimembranosus m

lateral sural
cutaneous n

gracilis m

popliteal v

saphenous n

tibial n

semitendinosus m

popliteal a and joint capsule of knee

medial head of gastrocnemius m

Cahill

1 L
5 L
9 L
11 L
12 L
13 L
14 L
15 L
17 L
18 L
22 L
24 L

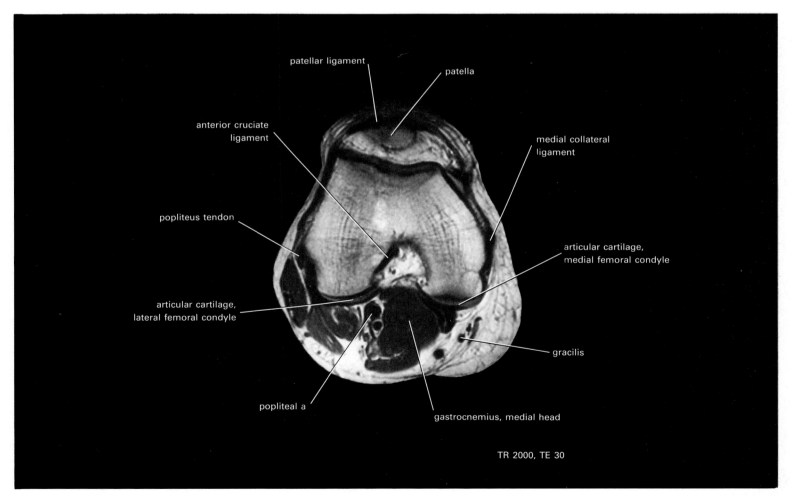

patellar ligament

patella

anterior cruciate
ligament

medial collateral
ligament

popliteus tendon

articular cartilage,
medial femoral condyle

articular cartilage,
lateral femoral condyle

gracilis

popliteal a

gastrocnemius, medial head

TR 2000, TE 30

Section 13L from below.

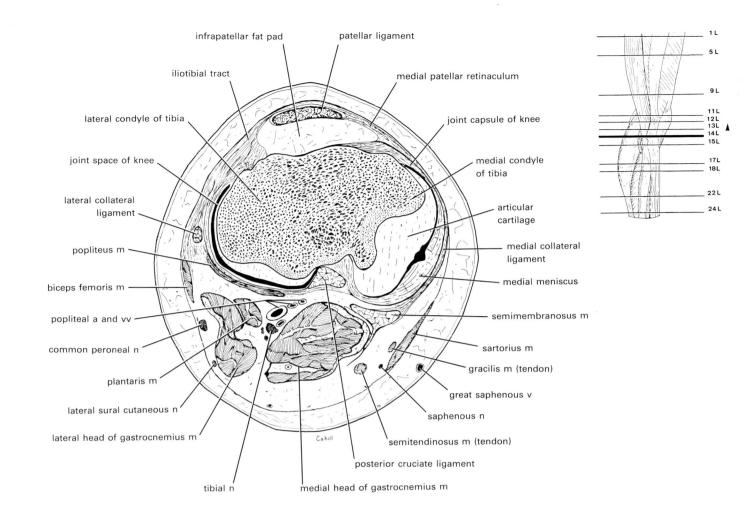

infrapatellar fat pad

patellar ligament

iliotibial tract

medial patellar retinaculum

lateral condyle of tibia

joint capsule of knee

joint space of knee

medial condyle of tibia

lateral collateral ligament

articular cartilage

popliteus m

medial collateral ligament

biceps femoris m

medial meniscus

popliteal a and vv

semimembranosus m

common peroneal n

sartorius m

plantaris m

gracilis m (tendon)

lateral sural cutaneous n

great saphenous v

lateral head of gastrocnemius m

saphenous n

semitendinosus m (tendon)

posterior cruciate ligament

tibial n

medial head of gastrocnemius m

Cahill

1 L
5 L
9 L
11 L
12 L
13 L
14 L
15 L
17 L
18 L
22 L
24 L

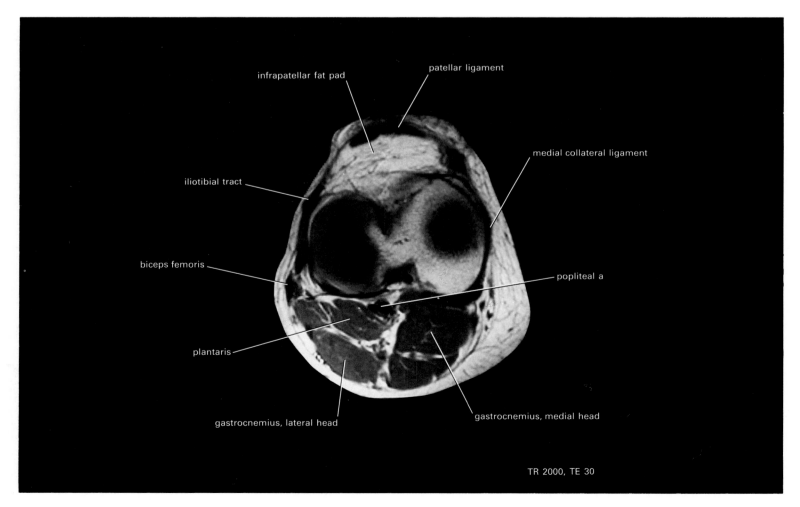

infrapatellar fat pad

patellar ligament

medial collateral ligament

iliotibial tract

biceps femoris

popliteal a

plantaris

gastrocnemius, medial head

gastrocnemius, lateral head

TR 2000, TE 30

Section 14L from below.

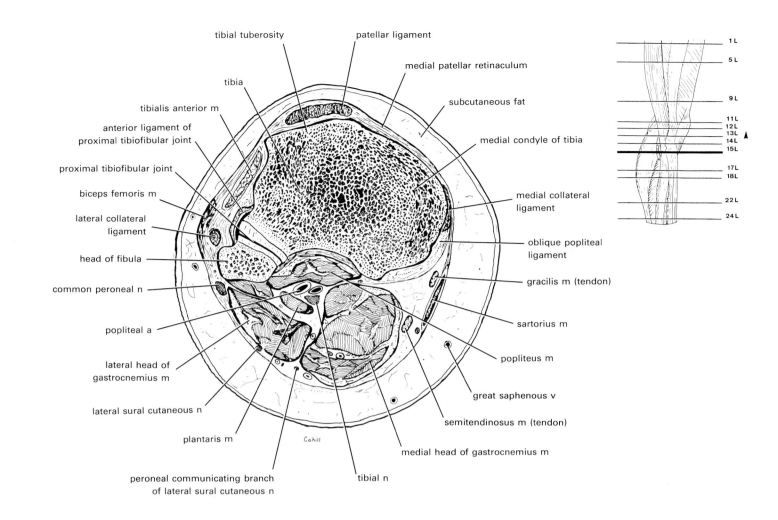

tibial tuberosity

patellar ligament

medial patellar retinaculum

tibia

subcutaneous fat

tibialis anterior m

anterior ligament of
proximal tibiofibular joint

medial condyle of tibia

proximal tibiofibular joint

medial collateral
ligament

biceps femoris m

lateral collateral
ligament

oblique popliteal
ligament

head of fibula

gracilis m (tendon)

common peroneal n

sartorius m

popliteal a

popliteus m

lateral head of
gastrocnemius m

great saphenous v

lateral sural cutaneous n

semitendinosus m (tendon)

plantaris m

medial head of gastrocnemius m

peroneal communicating branch
of lateral sural cutaneous n

tibial n

1 L
5 L
9 L
11 L
12 L
13L
14 L
15L
17 L
18 L
22 L
24 L

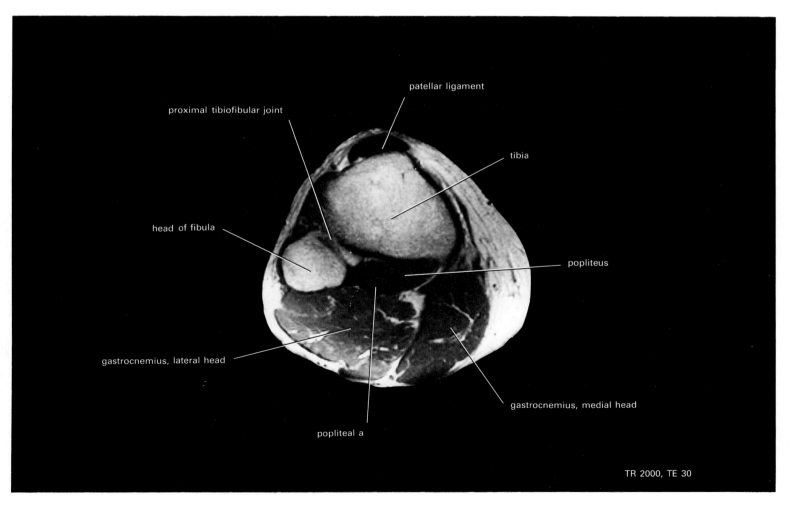

patellar ligament

proximal tibiofibular joint

tibia

head of fibula

popliteus

gastrocnemius, lateral head

gastrocnemius, medial head

popliteal a

TR 2000, TE 30

Section 15L from below.

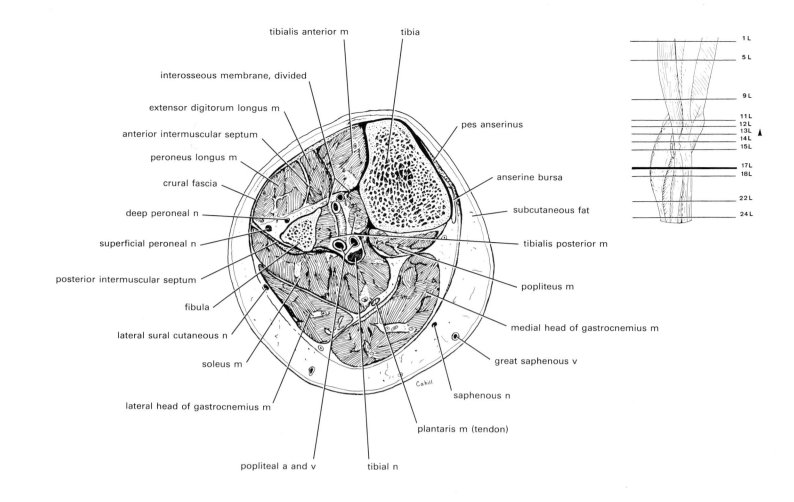

tibialis anterior m | tibia
interosseous membrane, divided
extensor digitorum longus m
anterior intermuscular septum
peroneus longus m
crural fascia
deep peroneal n
superficial peroneal n
posterior intermuscular septum
fibula
lateral sural cutaneous n
soleus m
lateral head of gastrocnemius m

pes anserinus
anserine bursa
subcutaneous fat
tibialis posterior m
popliteus m
medial head of gastrocnemius m
great saphenous v
saphenous n
plantaris m (tendon)

popliteal a and v | tibial n

1 L
5 L
9 L
11 L
12 L
13 L
14 L
15 L
17 L
18 L
22 L
24 L

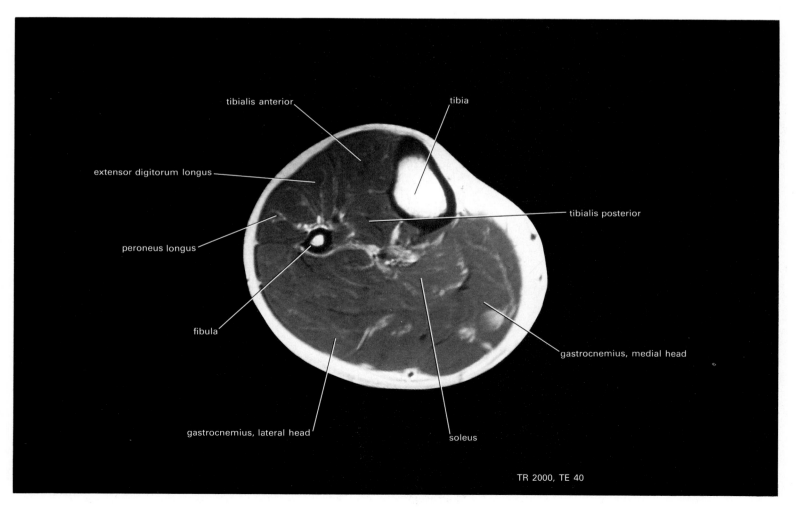

tibialis anterior | tibia
extensor digitorum longus
peroneus longus
tibialis posterior
fibula
gastrocnemius, medial head
gastrocnemius, lateral head
soleus

TR 2000, TE 40

Section 17L from below.

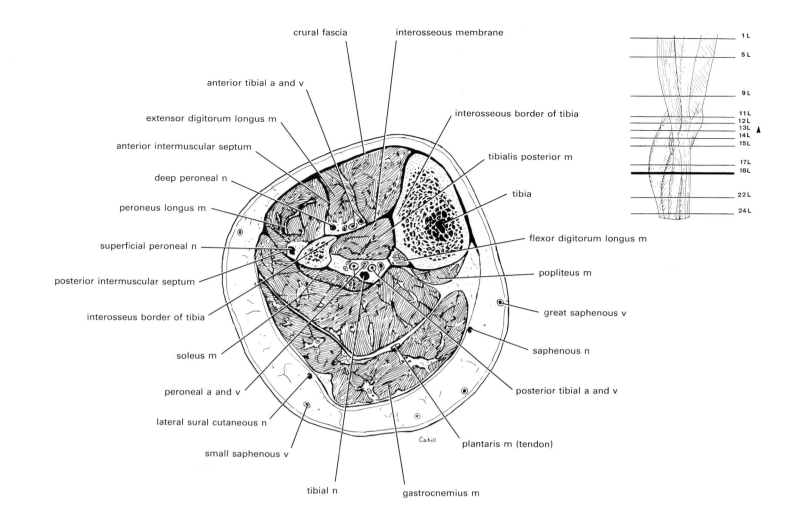

crural fascia
interosseous membrane

anterior tibial a and v

extensor digitorum longus m
interosseous border of tibia

anterior intermuscular septum
tibialis posterior m

deep peroneal n
tibia

peroneus longus m

superficial peroneal n
flexor digitorum longus m

posterior intermuscular septum
popliteus m

interosseus border of tibia
great saphenous v

soleus m
saphenous n

peroneal a and v
posterior tibial a and v

lateral sural cutaneous n
plantaris m (tendon)

small saphenous v

tibial n
gastrocnemius m

Cahill

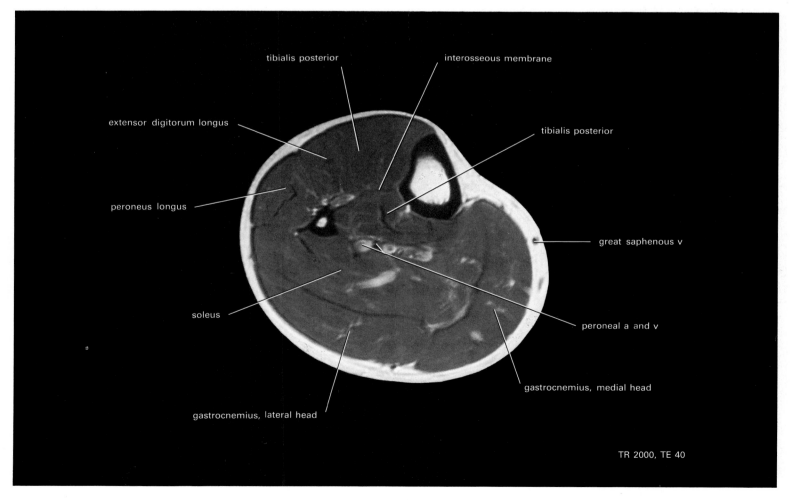

tibialis posterior
interosseous membrane

extensor digitorum longus
tibialis posterior

peroneus longus

great saphenous v

soleus
peroneal a and v

gastrocnemius, medial head

gastrocnemius, lateral head

TR 2000, TE 40

Section 18L from below.

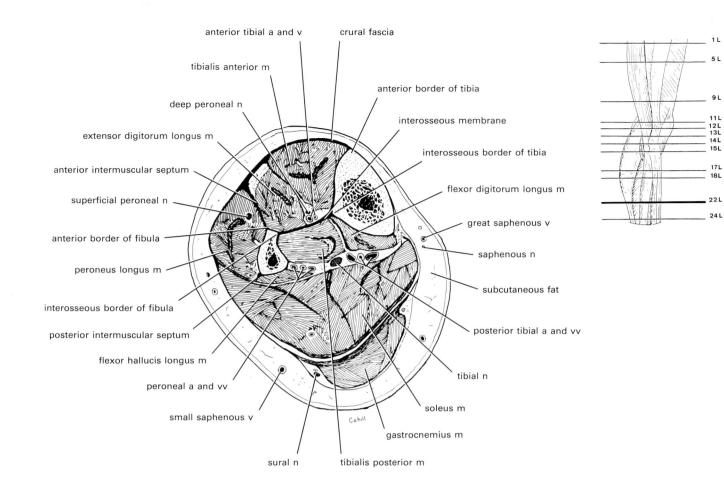

anterior tibial a and v

crural fascia

tibialis anterior m

anterior border of tibia

deep peroneal n

interosseous membrane

extensor digitorum longus m

interosseous border of tibia

anterior intermuscular septum

flexor digitorum longus m

superficial peroneal n

great saphenous v

anterior border of fibula

saphenous n

peroneus longus m

subcutaneous fat

interosseous border of fibula

posterior tibial a and vv

posterior intermuscular septum

tibial n

flexor hallucis longus m

soleus m

peroneal a and vv

gastrocnemius m

small saphenous v

Cahill

sural n   tibialis posterior m

1 L
5 L
9 L
11 L
12 L
13L
14 L
15 L
17 L
18 L
22L
24 L

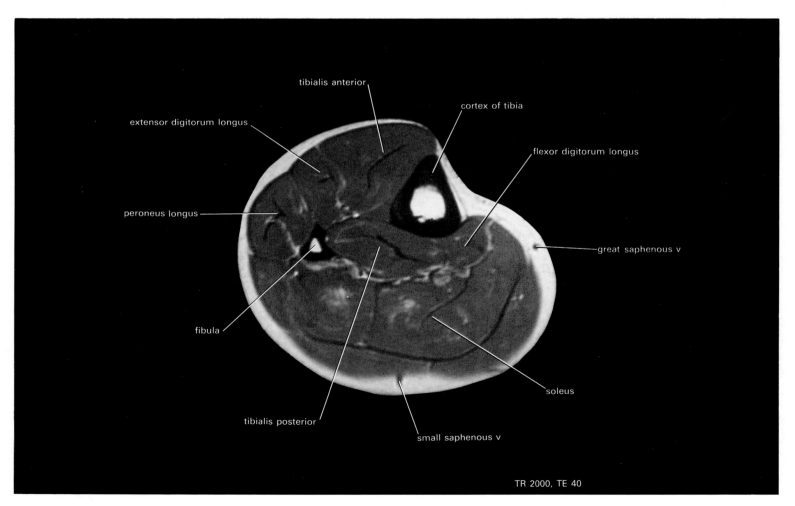

tibialis anterior

cortex of tibia

extensor digitorum longus

flexor digitorum longus

peroneus longus

great saphenous v

fibula

soleus

tibialis posterior

small saphenous v

TR 2000, TE 40

Section 22L from below.

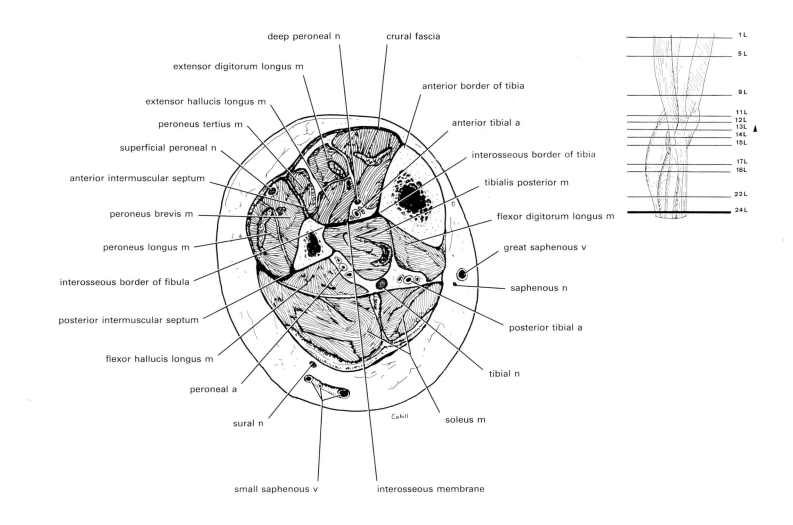

deep peroneal n
crural fascia
extensor digitorum longus m
anterior border of tibia
extensor hallucis longus m
anterior tibial a
peroneus tertius m
anterior tibial a
superficial peroneal n
interosseous border of tibia
anterior intermuscular septum
tibialis posterior m
peroneus brevis m
flexor digitorum longus m
peroneus longus m
great saphenous v
interosseous border of fibula
saphenous n
posterior intermuscular septum
posterior tibial a
flexor hallucis longus m
tibial n
peroneal a
sural n
soleus m
Cahill

small saphenous v
interosseous membrane

1 L
5 L
9 L
11 L
12 L
13 L
14 L
15 L
17 L
18 L
22 L
24 L

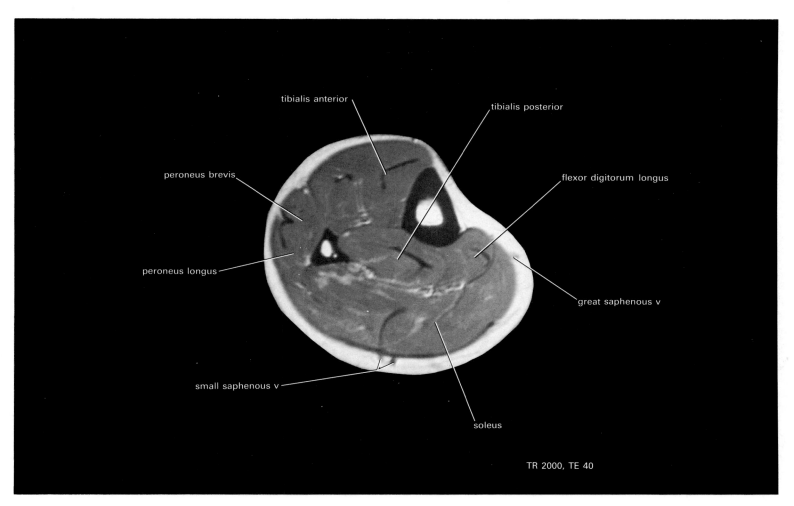

tibialis anterior
tibialis posterior
peroneus brevis
flexor digitorum longus
peroneus longus
great saphenous v
small saphenous v
soleus

TR 2000, TE 40

Section 24L from below.

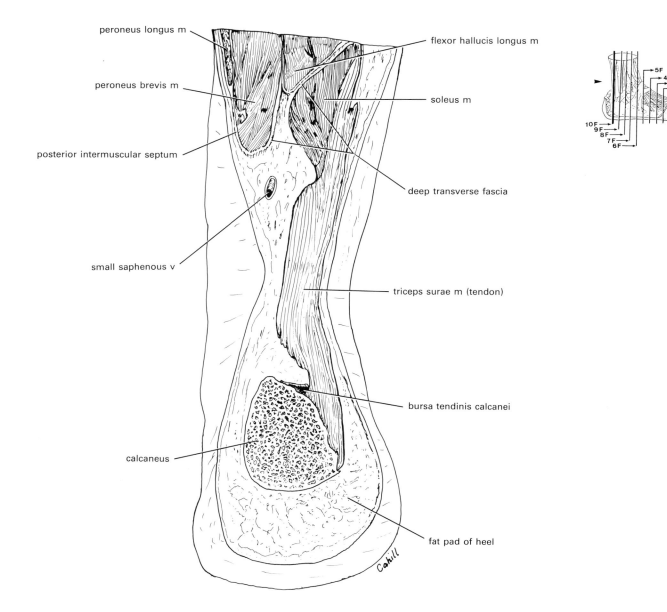

peroneus longus m

flexor hallucis longus m

peroneus brevis m

soleus m

posterior intermuscular septum

deep transverse fascia

small saphenous v

triceps surae m (tendon)

bursa tendinis calcanei

calcaneus

fat pad of heel

Cahill

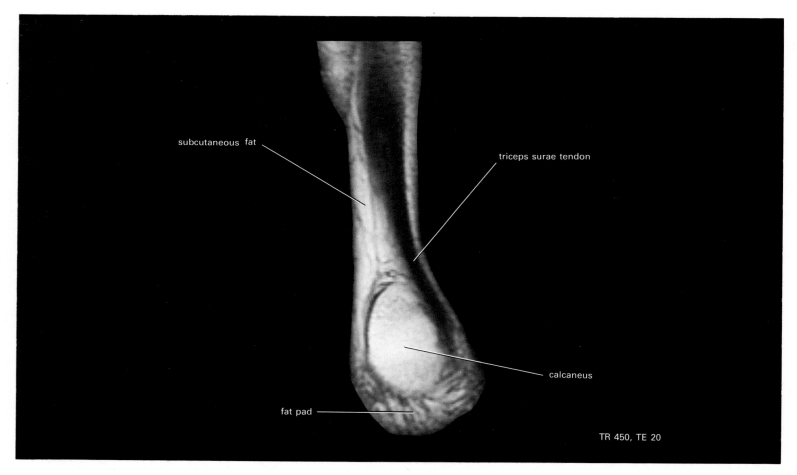

subcutaneos fat

triceps surae tendon

calcaneus

fat pad

TR 450, TE 20

Section 10F from behind.

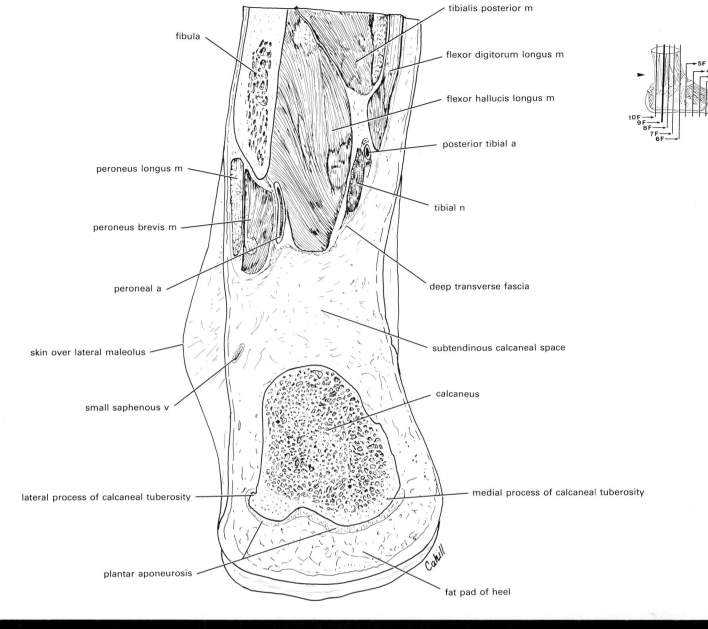

fibula

tibialis posterior m

flexor digitorum longus m

flexor hallucis longus m

posterior tibial a

peroneus longus m

tibial n

peroneus brevis m

peroneal a

deep transverse fascia

subtendinous calcaneal space

skin over lateral maleolus

small saphenous v

calcaneus

lateral process of calcaneal tuberosity

medial process of calcaneal tuberosity

plantar aponeurosis

fat pad of heel

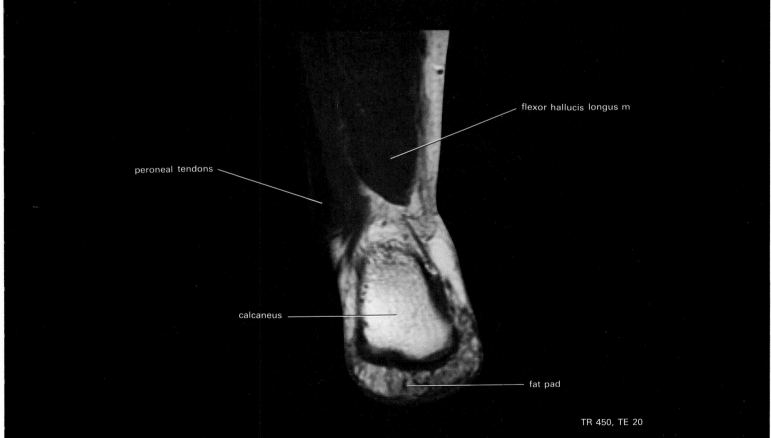

flexor hallucis longus m

peroneal tendons

calcaneus

fat pad

TR 450, TE 20

Section 9F from behind.

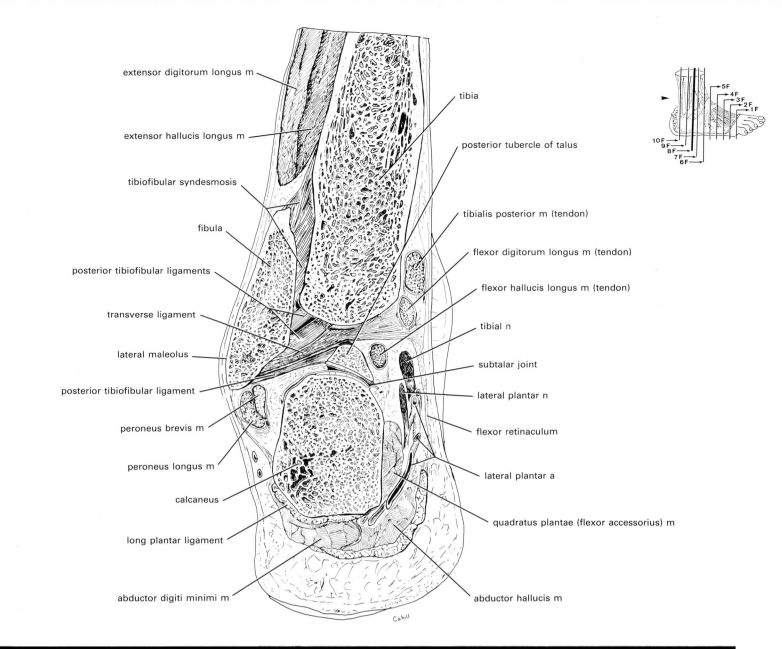

extensor digitorum longus m

extensor hallucis longus m

tibiofibular syndesmosis

fibula

posterior tibiofibular ligaments

transverse ligament

lateral maleolus

posterior tibiofibular ligament

peroneus brevis m

peroneus longus m

calcaneus

long plantar ligament

abductor digiti minimi m

tibia

posterior tubercle of talus

tibialis posterior m (tendon)

flexor digitorum longus m (tendon)

flexor hallucis longus m (tendon)

tibial n

subtalar joint

lateral plantar n

flexor retinaculum

lateral plantar a

quadratus plantae (flexor accessorius) m

abductor hallucis m

Cahill

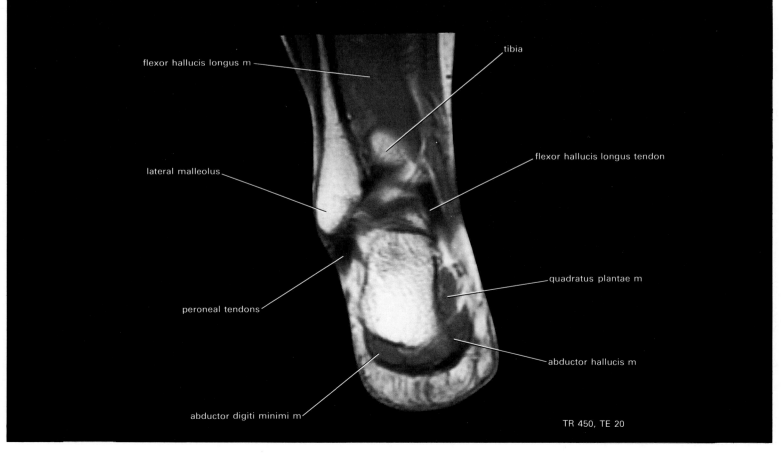

flexor hallucis longus m

lateral malleolus

peroneal tendons

abductor digiti minimi m

tibia

flexor hallucis longus tendon

quadratus plantae m

abductor hallucis m

TR 450, TE 20

Section 8F from behind.

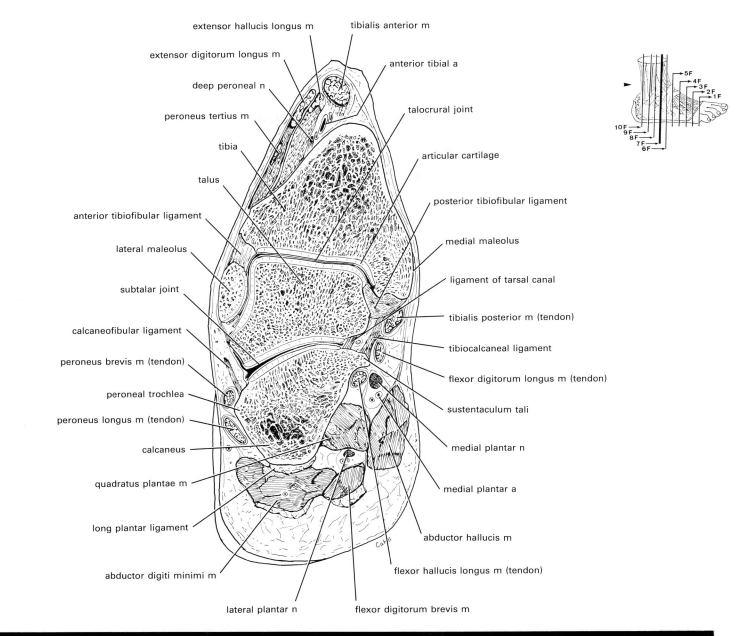

extensor hallucis longus m

tibialis anterior m

extensor digitorum longus m

anterior tibial a

deep peroneal n

peroneus tertius m

talocrural joint

tibia

articular cartilage

talus

posterior tibiofibular ligament

anterior tibiofibular ligament

medial maleolus

lateral maleolus

ligament of tarsal canal

subtalar joint

tibialis posterior m (tendon)

calcaneofibular ligament

tibiocalcaneal ligament

peroneus brevis m (tendon)

flexor digitorum longus m (tendon)

peroneal trochlea

sustentaculum tali

peroneus longus m (tendon)

calcaneus

medial plantar n

quadratus plantae m

medial plantar a

long plantar ligament

abductor hallucis m

abductor digiti minimi m

flexor hallucis longus m (tendon)

lateral plantar n

flexor digitorum brevis m

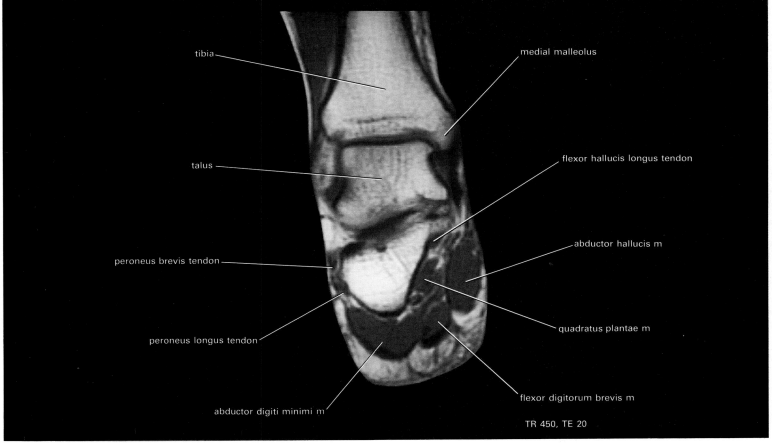

tibia

medial malleolus

talus

flexor hallucis longus tendon

peroneus brevis tendon

abductor hallucis m

peroneus longus tendon

quadratus plantae m

abductor digiti minimi m

flexor digitorum brevis m

TR 450, TE 20

Section 7F from behind.

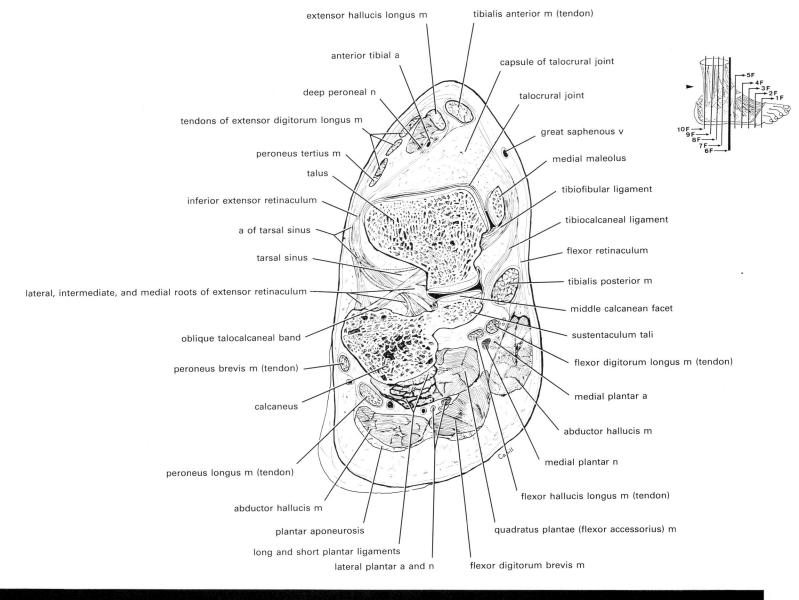

extensor hallucis longus m

tibialis anterior m (tendon)

anterior tibial a

capsule of talocrural joint

deep peroneal n

talocrural joint

tendons of extensor digitorum longus m

great saphenous v

peroneus tertius m

medial maleolus

talus

tibiofibular ligament

inferior extensor retinaculum

tibiocalcaneal ligament

a of tarsal sinus

flexor retinaculum

tarsal sinus

tibialis posterior m

lateral, intermediate, and medial roots of extensor retinaculum

middle calcanean facet

oblique talocalcaneal band

sustentaculum tali

peroneus brevis m (tendon)

flexor digitorum longus m (tendon)

calcaneus

medial plantar a

abductor hallucis m

peroneus longus m (tendon)

medial plantar n

abductor hallucis m

flexor hallucis longus m (tendon)

plantar aponeurosis

quadratus plantae (flexor accessorius) m

long and short plantar ligaments

lateral plantar a and n

flexor digitorum brevis m

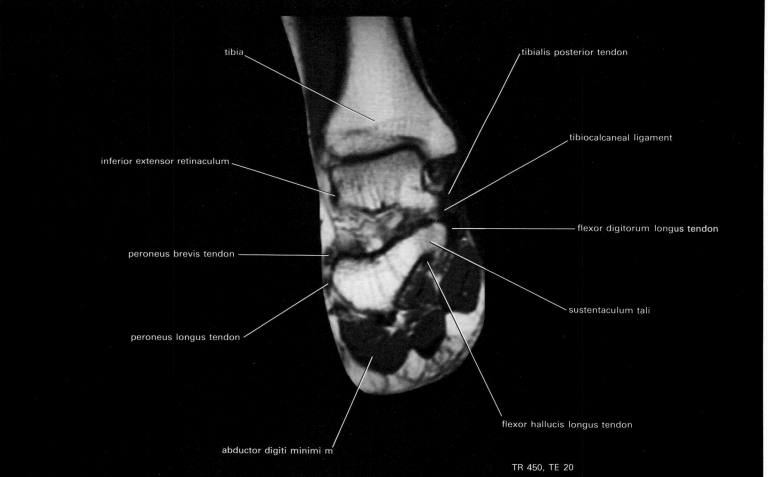

tibia

tibialis posterior tendon

tibiocalcaneal ligament

inferior extensor retinaculum

flexor digitorum longus tendon

peroneus brevis tendon

sustentaculum tali

peroneus longus tendon

flexor hallucis longus tendon

abductor digiti minimi m

TR 450, TE 20

Section 6F from behind.

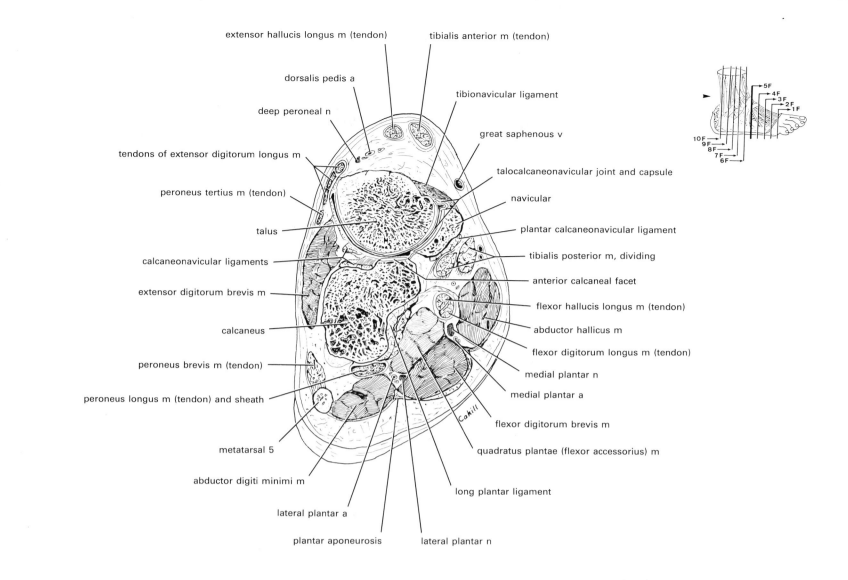

extensor hallucis longus m (tendon)

tibialis anterior m (tendon)

dorsalis pedis a

tibionavicular ligament

deep peroneal n

great saphenous v

tendons of extensor digitorum longus m

talocalcaneonavicular joint and capsule

peroneus tertius m (tendon)

navicular

talus

plantar calcaneonavicular ligament

calcaneonavicular ligaments

tibialis posterior m, dividing

extensor digitorum brevis m

anterior calcaneal facet

flexor hallucis longus m (tendon)

abductor hallicus m

calcaneus

flexor digitorum longus m (tendon)

peroneus brevis m (tendon)

medial plantar n

peroneus longus m (tendon) and sheath

medial plantar a

flexor digitorum brevis m

metatarsal 5

quadratus plantae (flexor accessorius) m

abductor digiti minimi m

long plantar ligament

lateral plantar a

plantar aponeurosis

lateral plantar n

Cahill

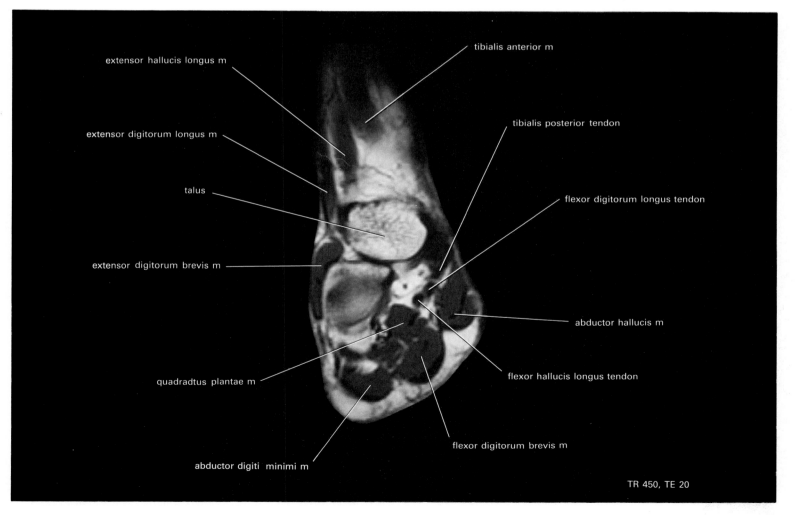

extensor hallucis longus m

tibialis anterior m

extensor digitorum longus m

tibialis posterior tendon

talus

flexor digitorum longus tendon

extensor digitorum brevis m

abductor hallucis m

quadradtus plantae m

flexor hallucis longus tendon

abductor digiti minimi m

flexor digitorum brevis m

TR 450, TE 20

Section 5F from behind.

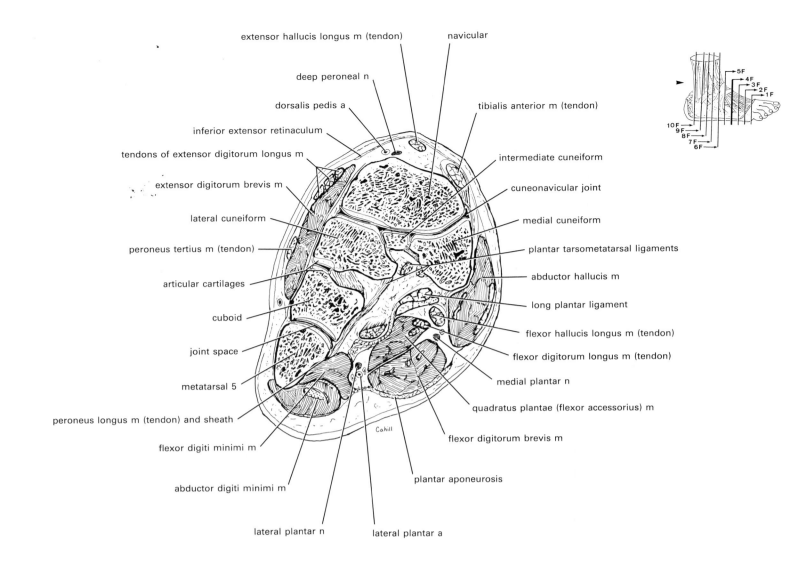

extensor hallucis longus m (tendon)
navicular
deep peroneal n
dorsalis pedis a
tibialis anterior m (tendon)
inferior extensor retinaculum
tendons of extensor digitorum longus m
intermediate cuneiform
extensor digitorum brevis m
cuneonavicular joint
lateral cuneiform
medial cuneiform
peroneus tertius m (tendon)
plantar tarsometatarsal ligaments
abductor hallucis m
articular cartilages
long plantar ligament
cuboid
flexor hallucis longus m (tendon)
joint space
flexor digitorum longus m (tendon)
metatarsal 5
medial plantar n
peroneus longus m (tendon) and sheath
quadratus plantae (flexor accessorius) m
flexor digiti minimi m
flexor digitorum brevis m
abductor digiti minimi m
plantar aponeurosis
lateral plantar n
lateral plantar a

Cahill

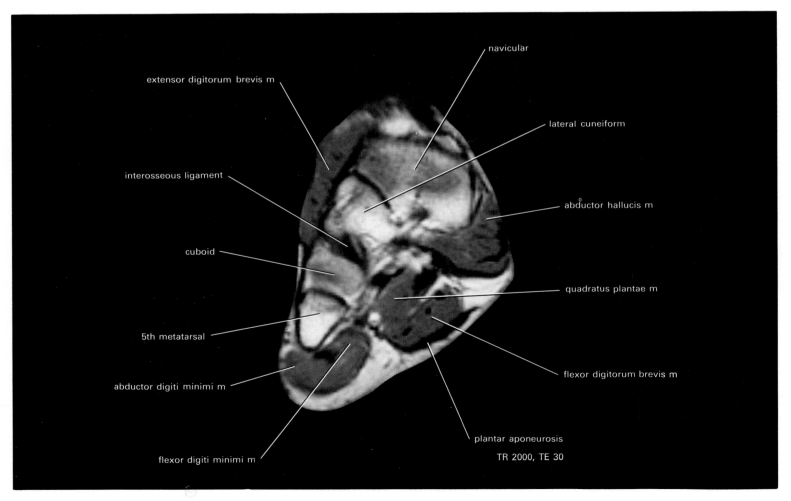

navicular
extensor digitorum brevis m
lateral cuneiform
interosseous ligament
abductor hallucis m
cuboid
quadratus plantae m
5th metatarsal
flexor digitorum brevis m
abductor digiti minimi m
plantar aponeurosis
flexor digiti minimi m
TR 2000, TE 30

Section 4F from behind.

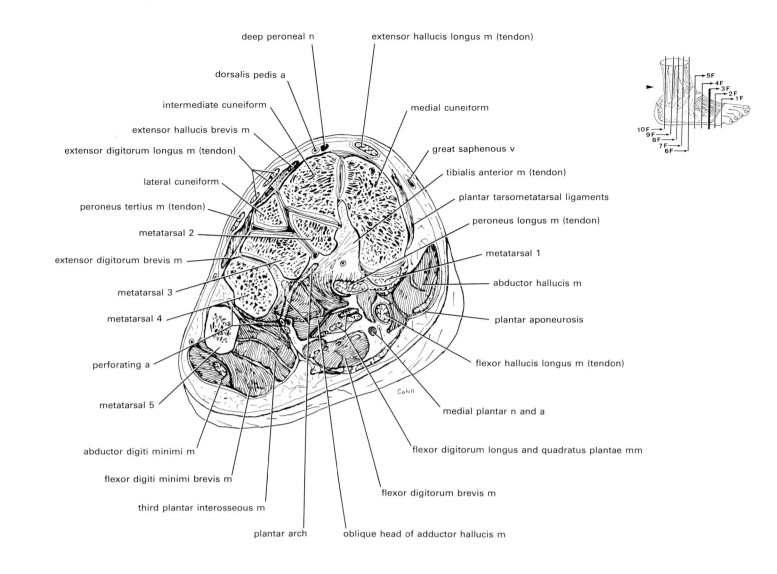

deep peroneal n

extensor hallucis longus m (tendon)

dorsalis pedis a

intermediate cuneiform

medial cuneiform

extensor hallucis brevis m

extensor digitorum longus m (tendon)

great saphenous v

tibialis anterior m (tendon)

lateral cuneiform

plantar tarsometatarsal ligaments

peroneus tertius m (tendon)

peroneus longus m (tendon)

metatarsal 2

metatarsal 1

extensor digitorum brevis m

abductor hallucis m

metatarsal 3

metatarsal 4

plantar aponeurosis

perforating a

flexor hallucis longus m (tendon)

metatarsal 5

medial plantar n and a

abductor digiti minimi m

flexor digitorum longus and quadratus plantae mm

flexor digiti minimi brevis m

flexor digitorum brevis m

third plantar interosseous m

plantar arch

oblique head of adductor hallucis m

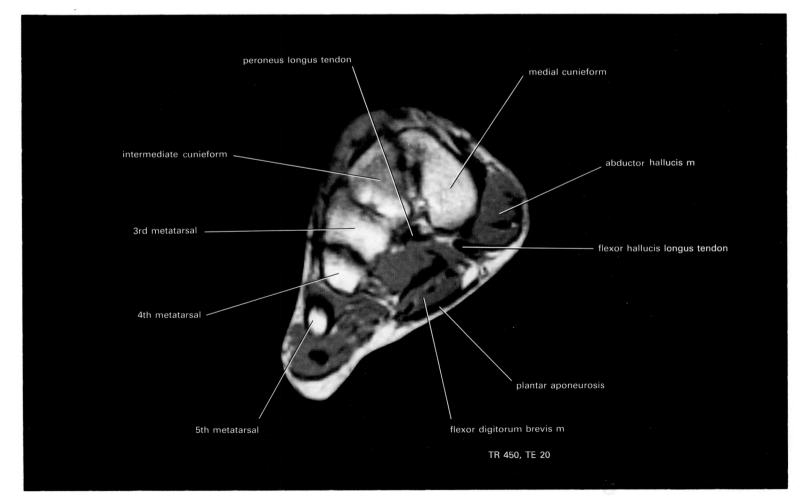

peroneus longus tendon

medial cunieform

intermediate cunieform

abductor hallucis m

3rd metatarsal

flexor hallucis longus tendon

4th metatarsal

plantar aponeurosis

5th metatarsal

flexor digitorum brevis m

TR 450, TE 20

Section 3F from behind.

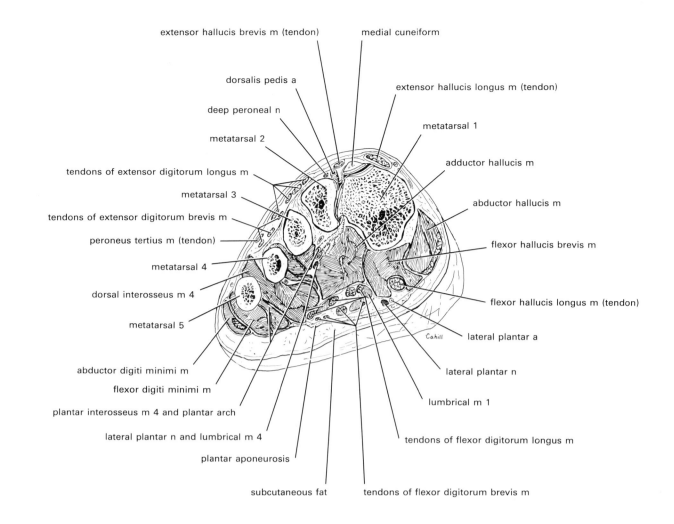

extensor hallucis brevis m (tendon)

medial cuneiform

dorsalis pedis a

extensor hallucis longus m (tendon)

deep peroneal n

metatarsal 1

metatarsal 2

adductor hallucis m

tendons of extensor digitorum longus m

metatarsal 3

abductor hallucis m

tendons of extensor digitorum brevis m

peroneus tertius m (tendon)

flexor hallucis brevis m

metatarsal 4

dorsal interosseus m 4

flexor hallucis longus m (tendon)

metatarsal 5

lateral plantar a

abductor digiti minimi m

lateral plantar n

flexor digiti minimi m

lumbrical m 1

plantar interosseus m 4 and plantar arch

lateral plantar n and lumbrical m 4

tendons of flexor digitorum longus m

plantar aponeurosis

subcutaneous fat

tendons of flexor digitorum brevis m

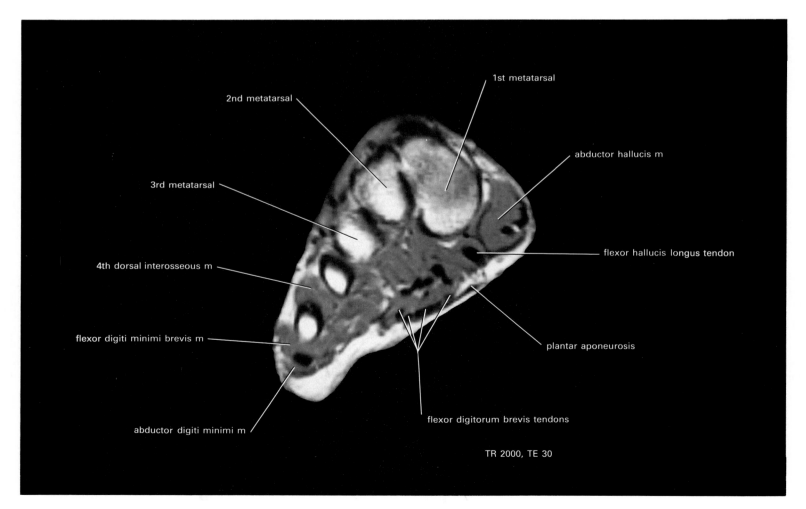

1st metatarsal

2nd metatarsal

3rd metatarsal

abductor hallucis m

4th dorsal interosseous m

flexor hallucis longus tendon

flexor digiti minimi brevis m

plantar aponeurosis

abductor digiti minimi m

flexor digitorum brevis tendons

TR 2000, TE 30

Section 2F from behind.

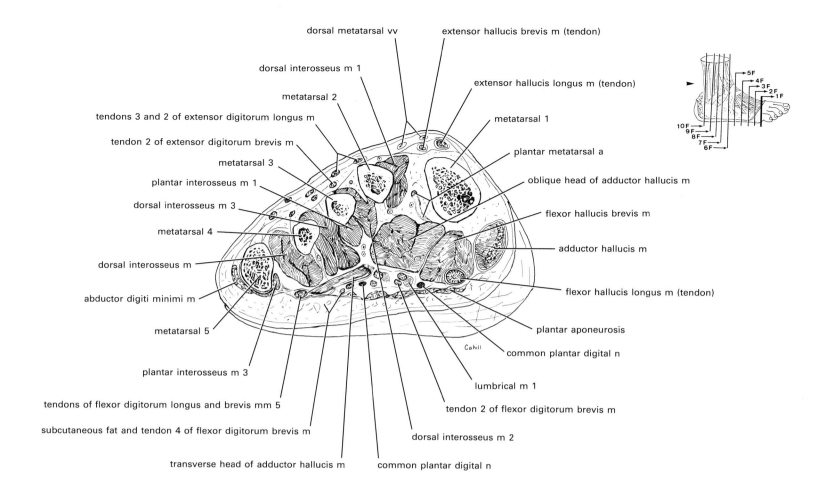

dorsal metatarsal vv

extensor hallucis brevis m (tendon)

dorsal interosseus m 1

extensor hallucis longus m (tendon)

metatarsal 2

tendons 3 and 2 of extensor digitorum longus m

metatarsal 1

tendon 2 of extensor digitorum brevis m

plantar metatarsal a

metatarsal 3

oblique head of adductor hallucis m

plantar interosseus m 1

dorsal interosseus m 3

flexor hallucis brevis m

metatarsal 4

adductor hallucis m

dorsal interosseus m

flexor hallucis longus m (tendon)

abductor digiti minimi m

metatarsal 5

plantar aponeurosis

common plantar digital n

plantar interosseus m 3

lumbrical m 1

tendons of flexor digitorum longus and brevis mm 5

tendon 2 of flexor digitorum brevis m

subcutaneous fat and tendon 4 of flexor digitorum brevis m

dorsal interosseus m 2

transverse head of adductor hallucis m

common plantar digital n

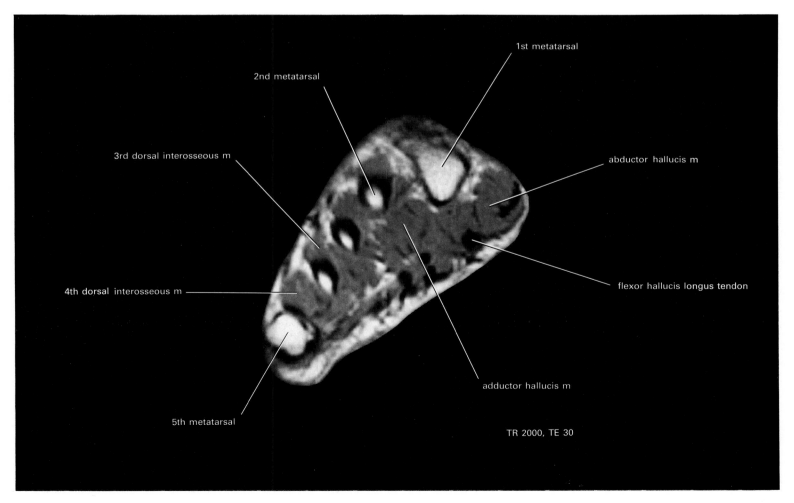

1st metatarsal

2nd metatarsal

3rd dorsal interosseous m

abductor hallucis m

4th dorsal interosseous m

flexor hallucis longus tendon

5th metatarsal

adductor hallucis m

TR 2000, TE 30

Section 1F from behind.

The Right Upper Limb With Hand in Pronation

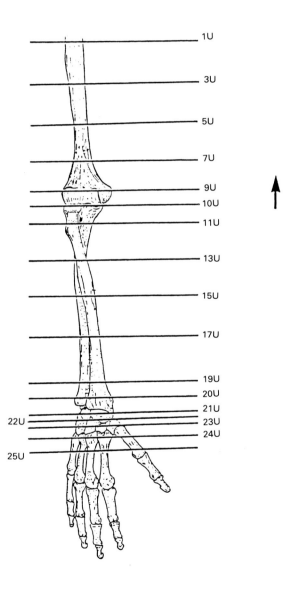

1U
3U
5U
7U
9U
10U
11U
13U
15U
17U
19U
20U
21U
22U
23U
24U
25U

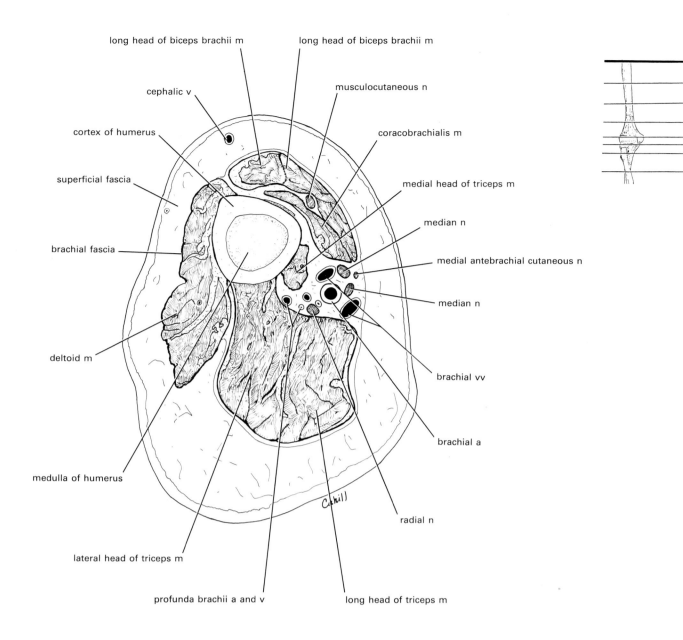

long head of biceps brachii m

long head of biceps brachii m

cephalic v

musculocutaneous n

cortex of humerus

coracobrachialis m

superficial fascia

medial head of triceps m

median n

brachial fascia

medial antebrachial cutaneous n

median n

deltoid m

brachial vv

brachial a

medulla of humerus

radial n

lateral head of triceps m

profunda brachii a and v

long head of triceps m

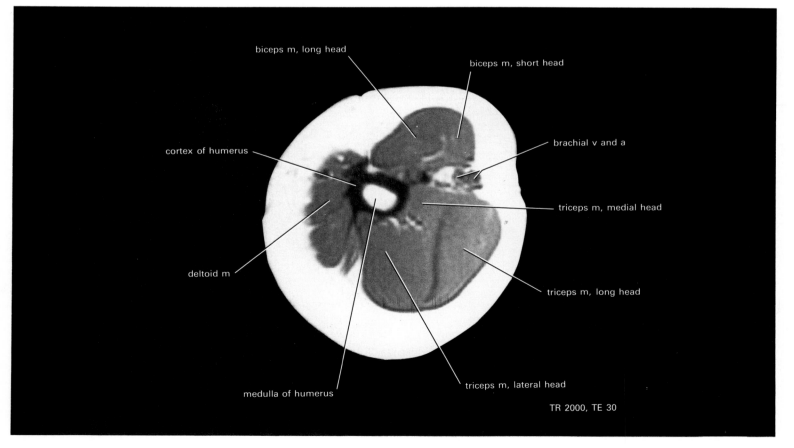

biceps m, long head

biceps m, short head

cortex of humerus

brachial v and a

deltoid m

triceps m, medial head

triceps m, long head

medulla of humerus

triceps m, lateral head

TR 2000, TE 30

Section 1U from below.

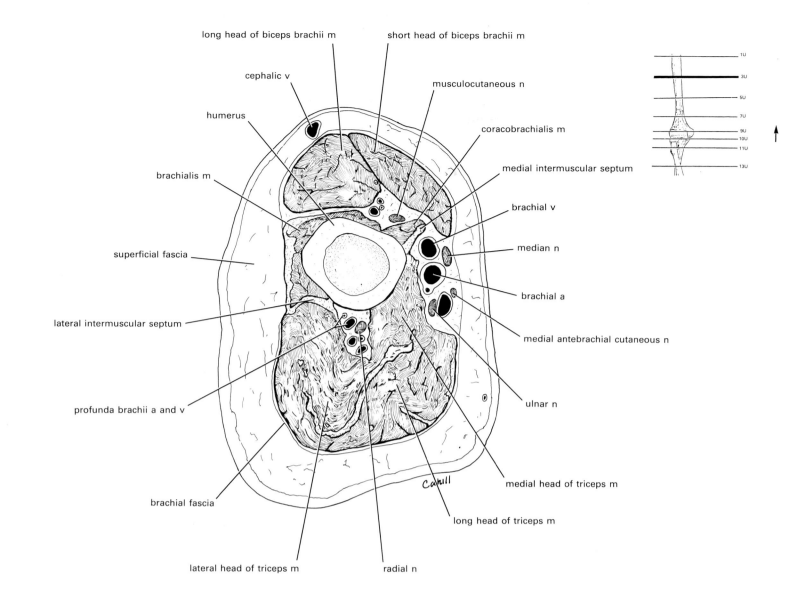

long head of biceps brachii m

short head of biceps brachii m

cephalic v

musculocutaneous n

humerus

coracobrachialis m

brachialis m

medial intermuscular septum

brachial v

median n

superficial fascia

brachial a

lateral intermuscular septum

medial antebrachial cutaneous n

profunda brachii a and v

ulnar n

medial head of triceps m

brachial fascia

long head of triceps m

lateral head of triceps m

radial n

Cahill

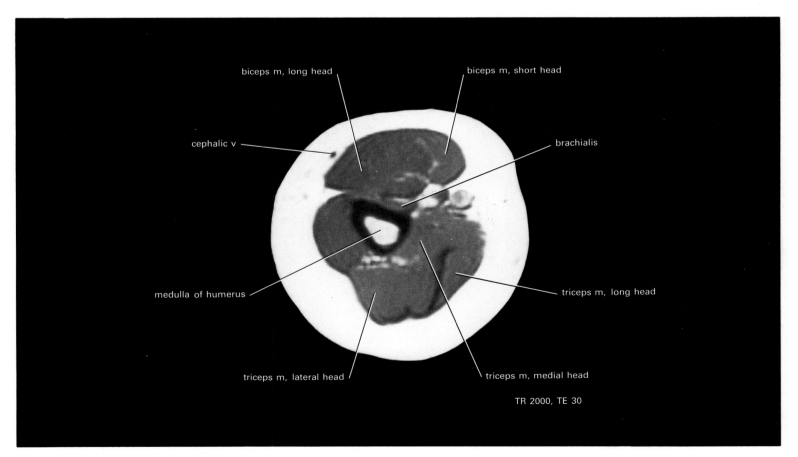

biceps m, long head

biceps m, short head

cephalic v

brachialis

medulla of humerus

triceps m, long head

triceps m, lateral head

triceps m, medial head

TR 2000, TE 30

Section 3U from below.

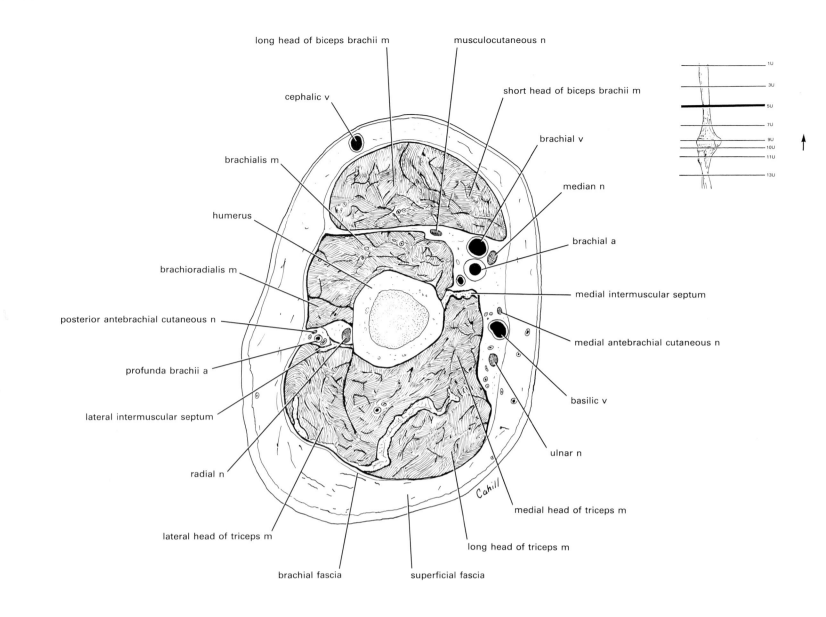

long head of biceps brachii m

musculocutaneous n

cephalic v

short head of biceps brachii m

brachial v

brachialis m

median n

humerus

brachial a

brachioradialis m

medial intermuscular septum

posterior antebrachial cutaneous n

medial antebrachial cutaneous n

profunda brachii a

basilic v

lateral intermuscular septum

ulnar n

radial n

medial head of triceps m

lateral head of triceps m

long head of triceps m

brachial fascia

superficial fascia

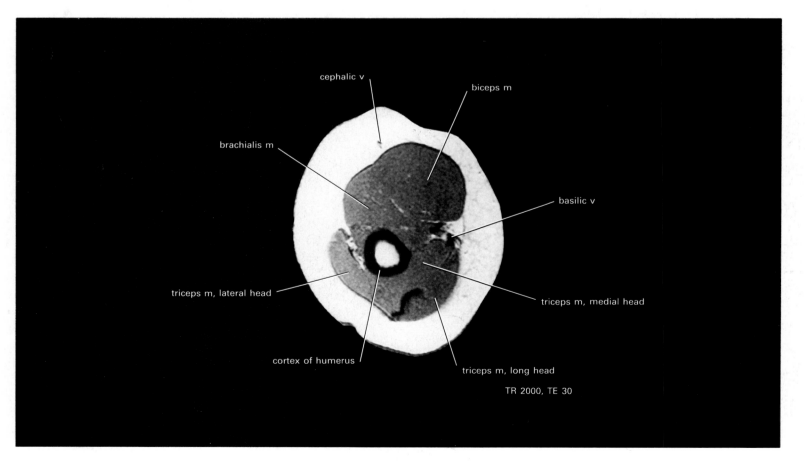

cephalic v

biceps m

brachialis m

basilic v

triceps m, lateral head

triceps m, medial head

cortex of humerus

triceps m, long head

TR 2000, TE 30

Section 5U from below.

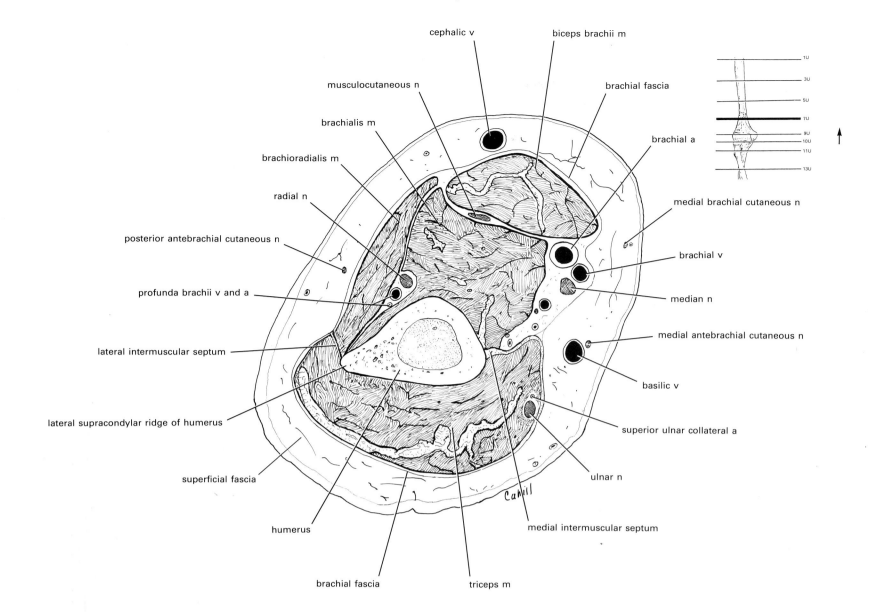

cephalic v

biceps brachii m

musculocutaneous n

brachial fascia

brachialis m

brachial a

brachioradialis m

medial brachial cutaneous n

radial n

posterior antebrachial cutaneous n

brachial v

profunda brachii v and a

median n

lateral intermuscular septum

medial antebrachial cutaneous n

basilic v

lateral supracondylar ridge of humerus

superior ulnar collateral a

superficial fascia

ulnar n

humerus

medial intermuscular septum

brachial fascia

triceps m

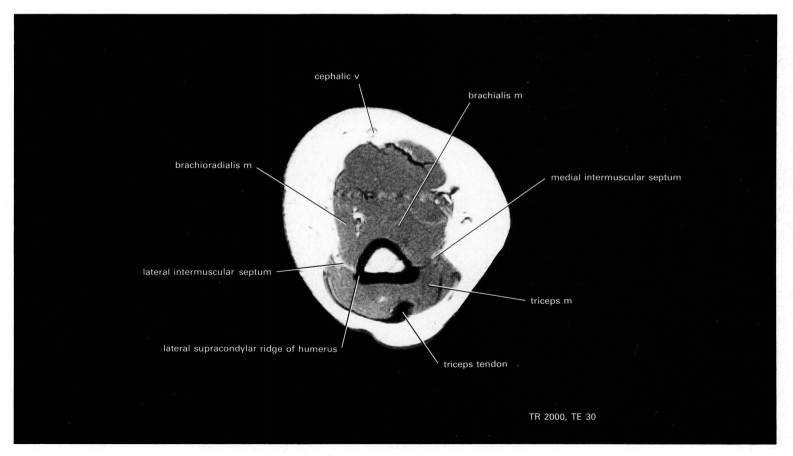

cephalic v

brachialis m

brachioradialis m

medial intermuscular septum

lateral intermuscular septum

triceps m

lateral supracondylar ridge of humerus

triceps tendon

TR 2000, TE 30

Section 7U from below.

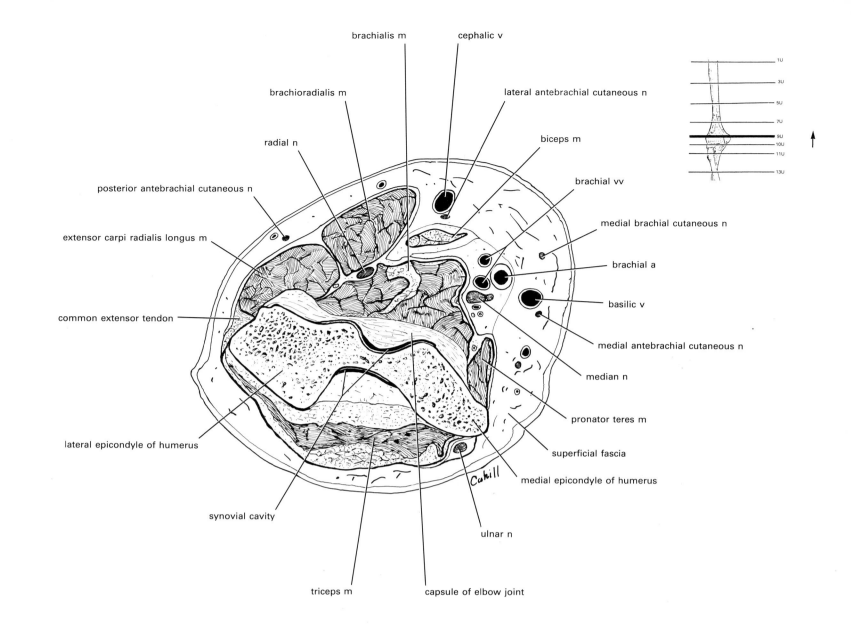

brachialis m
cephalic v
brachioradialis m
lateral antebrachial cutaneous n
radial n
biceps m
posterior antebrachial cutaneous n
brachial vv
extensor carpi radialis longus m
medial brachial cutaneous n
brachial a
common extensor tendon
basilic v
medial antebrachial cutaneous n
median n
pronator teres m
lateral epicondyle of humerus
superficial fascia
medial epicondyle of humerus
synovial cavity
ulnar n
triceps m
capsule of elbow joint

Cahill

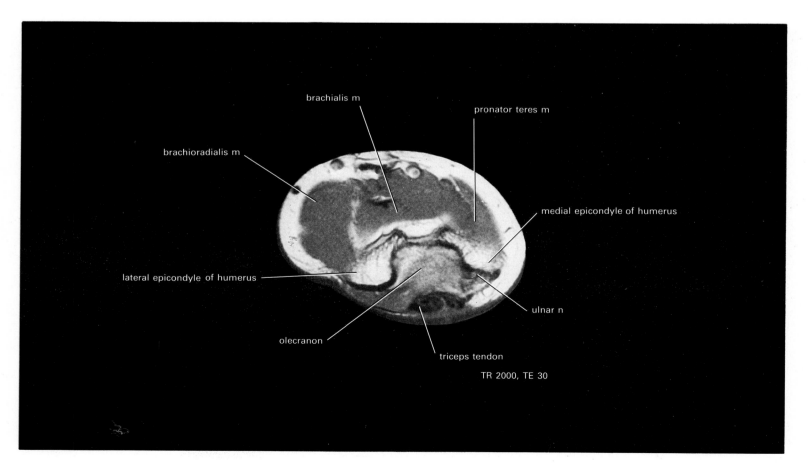

brachialis m
pronator teres m
brachioradialis m
medial epicondyle of humerus
lateral epicondyle of humerus
ulnar n
olecranon
triceps tendon
TR 2000, TE 30

Section 9U from below.

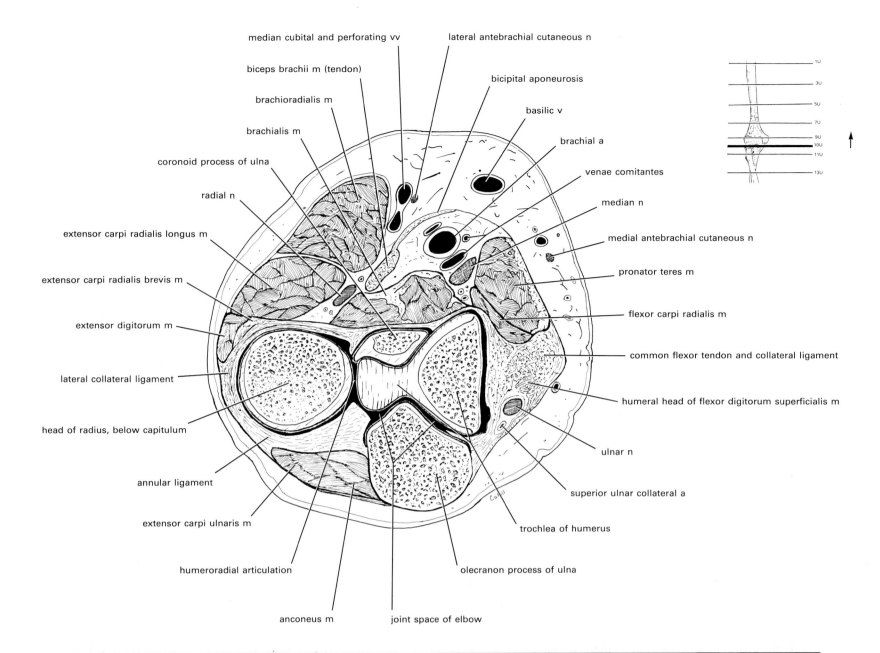

median cubital and perforating vv

biceps brachii m (tendon)

brachioradialis m

brachialis m

coronoid process of ulna

radial n

extensor carpi radialis longus m

extensor carpi radialis brevis m

extensor digitorum m

lateral collateral ligament

head of radius, below capitulum

annular ligament

extensor carpi ulnaris m

humeroradial articulation

anconeus m

lateral antebrachial cutaneous n

bicipital aponeurosis

basilic v

brachial a

venae comitantes

median n

medial antebrachial cutaneous n

pronator teres m

flexor carpi radialis m

common flexor tendon and collateral ligament

humeral head of flexor digitorum superficialis m

ulnar n

superior ulnar collateral a

trochlea of humerus

olecranon process of ulna

joint space of elbow

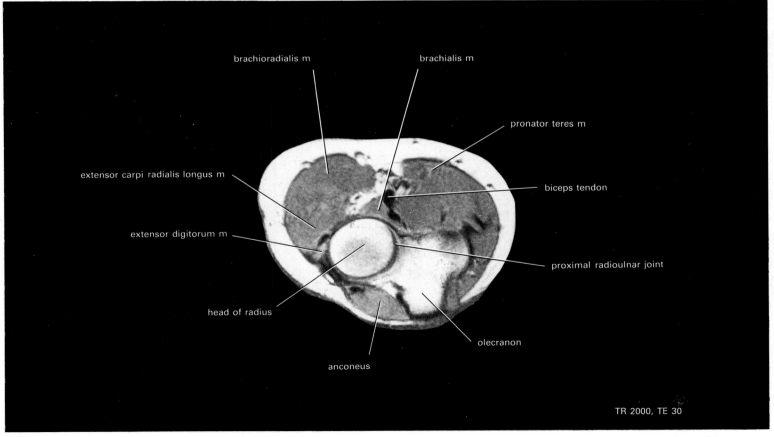

brachioradialis m

brachialis m

pronator teres m

extensor carpi radialis longus m

biceps tendon

extensor digitorum m

proximal radioulnar joint

head of radius

olecranon

anconeus

TR 2000, TE 30

Section 10U from below.

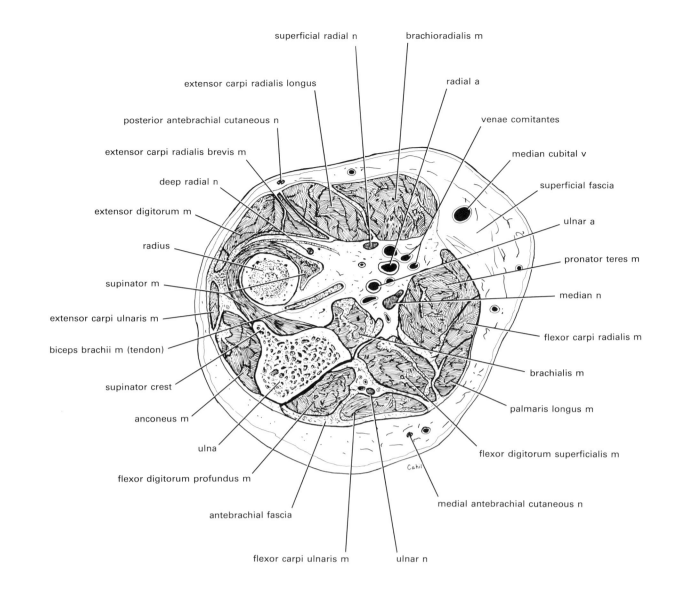

superficial radial n
brachioradialis m
extensor carpi radialis longus
radial a
posterior antebrachial cutaneous n
venae comitantes
extensor carpi radialis brevis m
median cubital v
deep radial n
superficial fascia
extensor digitorum m
ulnar a
radius
pronator teres m
supinator m
median n
extensor carpi ulnaris m
flexor carpi radialis m
biceps brachii m (tendon)
brachialis m
supinator crest
palmaris longus m
anconeus m
ulna
flexor digitorum superficialis m
flexor digitorum profundus m
medial antebrachial cutaneous n
antebrachial fascia
flexor carpi ulnaris m
ulnar n

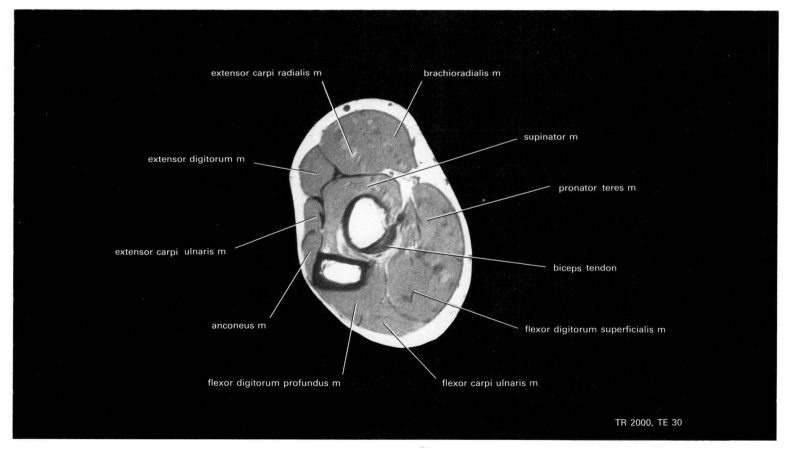

extensor carpi radialis m
brachioradialis m
extensor digitorum m
supinator m
pronator teres m
extensor carpi ulnaris m
biceps tendon
anconeus m
flexor digitorum superficialis m
flexor digitorum profundus m
flexor carpi ulnaris m

TR 2000, TE 30

Section 11U from below.

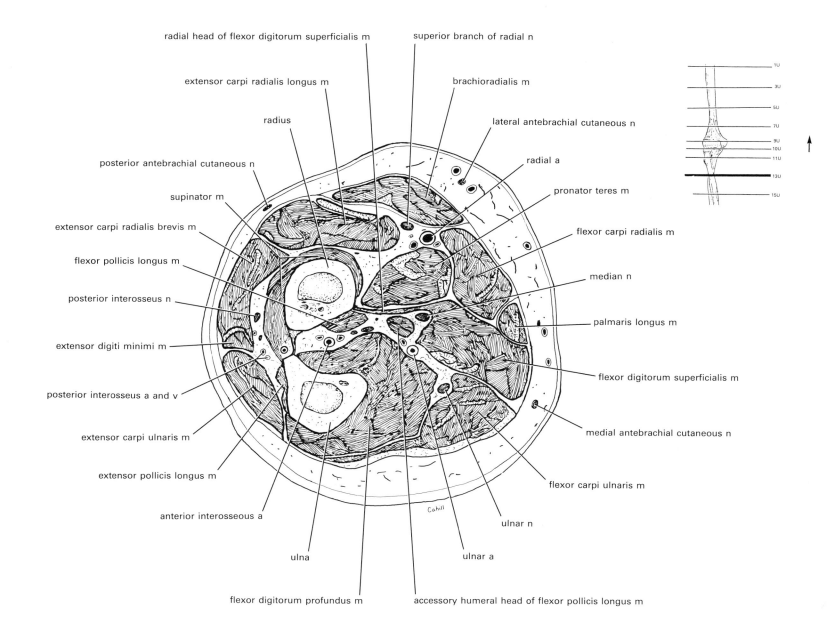

radial head of flexor digitorum superficialis m

superior branch of radial n

extensor carpi radialis longus m

brachioradialis m

radius

lateral antebrachial cutaneous n

posterior antebrachial cutaneous n

radial a

supinator m

pronator teres m

extensor carpi radialis brevis m

flexor carpi radialis m

flexor pollicis longus m

median n

posterior interosseus n

palmaris longus m

extensor digiti minimi m

posterior interosseus a and v

flexor digitorum superficialis m

extensor carpi ulnaris m

medial antebrachial cutaneous n

extensor pollicis longus m

flexor carpi ulnaris m

anterior interosseous a

ulnar n

ulna

ulnar a

flexor digitorum profundus m

accessory humeral head of flexor pollicis longus m

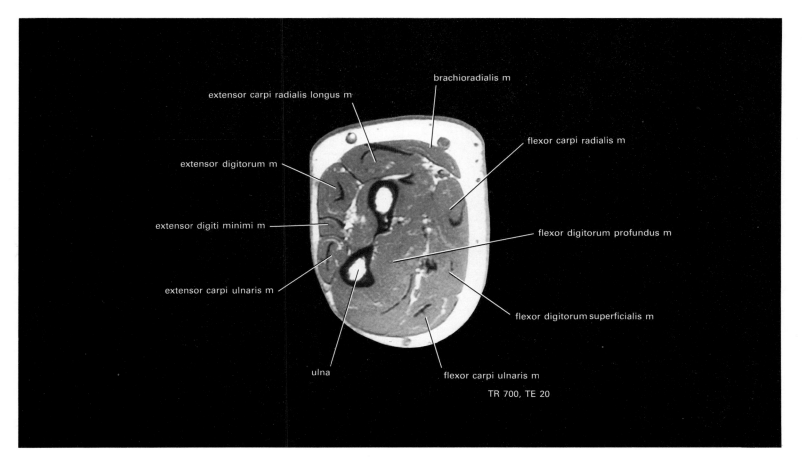

brachioradialis m

extensor carpi radialis longus m

flexor carpi radialis m

extensor digitorum m

extensor digiti minimi m

flexor digitorum profundus m

extensor carpi ulnaris m

flexor digitorum superficialis m

ulna

flexor carpi ulnaris m

TR 700, TE 20

Section 13U from below.

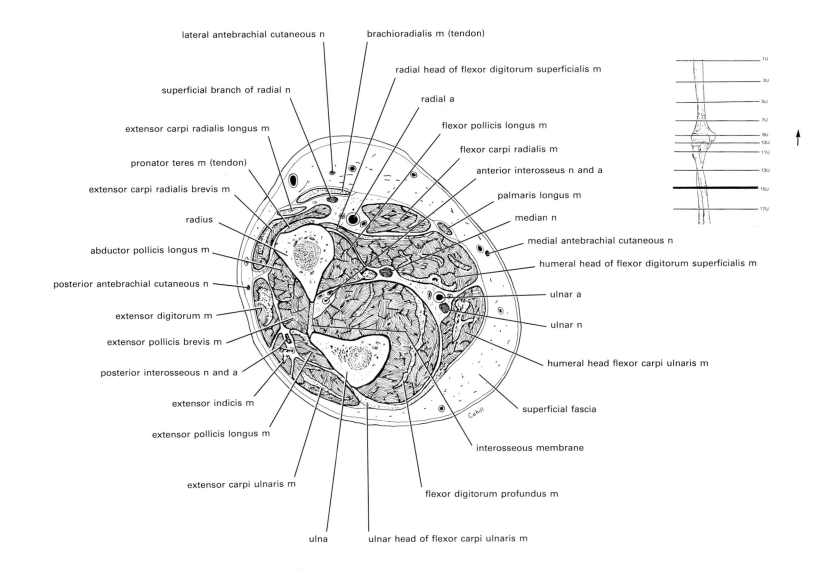

lateral antebrachial cutaneous n

brachioradialis m (tendon)

radial head of flexor digitorum superficialis m

superficial branch of radial n

radial a

extensor carpi radialis longus m

flexor pollicis longus m

pronator teres m (tendon)

flexor carpi radialis m

extensor carpi radialis brevis m

anterior interosseus n and a

radius

palmaris longus m

abductor pollicis longus m

median n

posterior antebrachial cutaneous n

medial antebrachial cutaneous n

humeral head of flexor digitorum superficialis m

extensor digitorum m

ulnar a

extensor pollicis brevis m

ulnar n

posterior interosseous n and a

humeral head flexor carpi ulnaris m

extensor indicis m

superficial fascia

extensor pollicis longus m

interosseous membrane

extensor carpi ulnaris m

flexor digitorum profundus m

ulna

ulnar head of flexor carpi ulnaris m

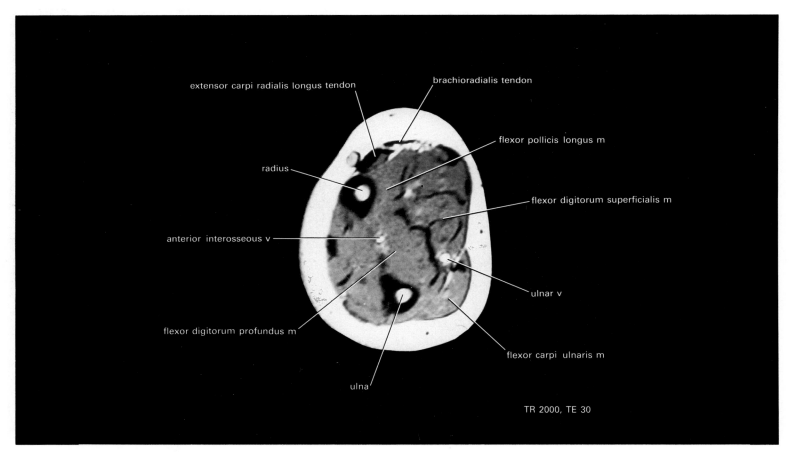

extensor carpi radialis longus tendon

brachioradialis tendon

flexor pollicis longus m

radius

flexor digitorum superficialis m

anterior interosseous v

ulnar v

flexor digitorum profundus m

flexor carpi ulnaris m

ulna

TR 2000, TE 30

Section 15U from below.

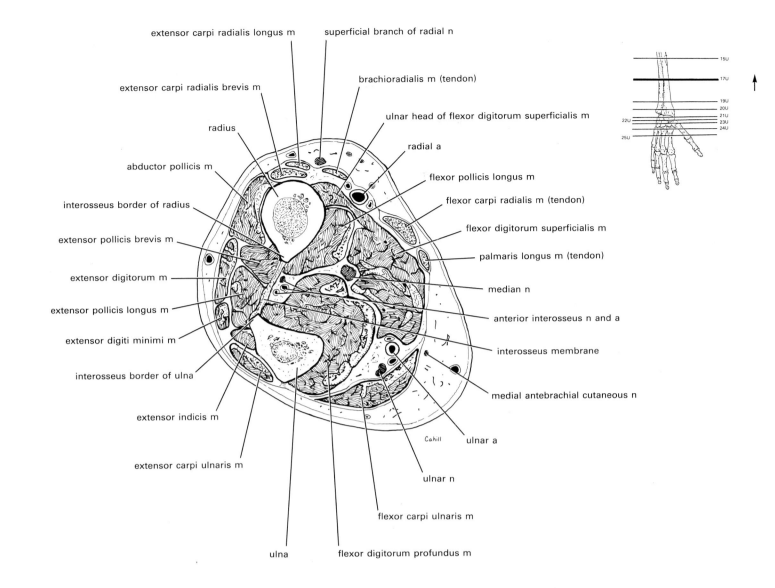

extensor carpi radialis longus m — superficial branch of radial n

extensor carpi radialis brevis m — brachioradialis m (tendon)

radius — ulnar head of flexor digitorum superficialis m

abductor pollicis m — radial a

interosseus border of radius — flexor pollicis longus m

extensor pollicis brevis m — flexor carpi radialis m (tendon)

extensor digitorum m — flexor digitorum superficialis m

extensor pollicis longus m — palmaris longus m (tendon)

extensor digiti minimi m — median n

interosseus border of ulna — anterior interosseus n and a

extensor indicis m — interosseus membrane

extensor carpi ulnaris m — medial antebrachial cutaneous n

Cahill — ulnar a

ulnar n

flexor carpi ulnaris m

ulna — flexor digitorum profundus m

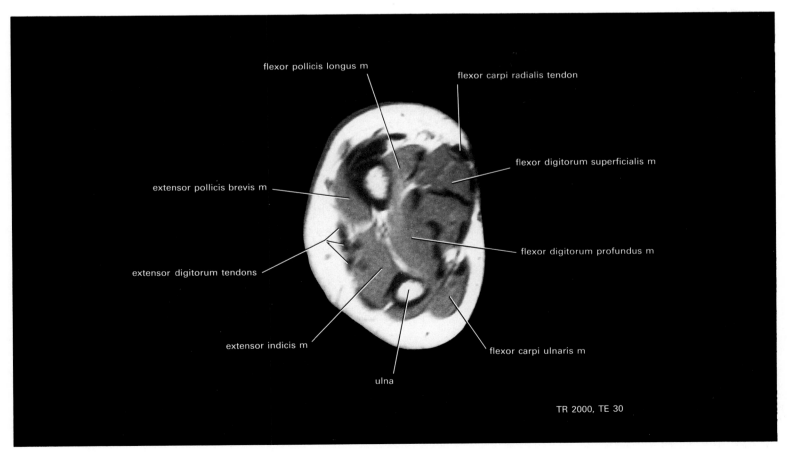

flexor pollicis longus m — flexor carpi radialis tendon

extensor pollicis brevis m — flexor digitorum superficialis m

flexor digitorum profundus m

extensor digitorum tendons

extensor indicis m — flexor carpi ulnaris m

ulna

TR 2000, TE 30

Section 17U from below.

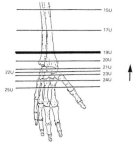

superficial branch of radial n

brachialis m (tendon)

abductor pollicis longus m (tendon)

radial a

extensor carpi radialis brevis m (tendon)

extensor pollicis brevis m (tendon)

flexor pollicis longus m

extensor carpi radialis brevis m (tendon)

flexor carpi radialis m (tendon)

median n

radius

palmaris longus m (tendon)

extensor pollicis longus m

pronator quadratus m

tendons of extensor digitorum m

flexor digitorum superficialis m

flexor digitorum profundus m

extensor digiti minimi m (tendon)

ulnar a

extensor indicis m

ulnar n

ulna

flexor carpi ulnaris m

interosseus membrane

extensor carpi ulnaris m

anterior interosseus n

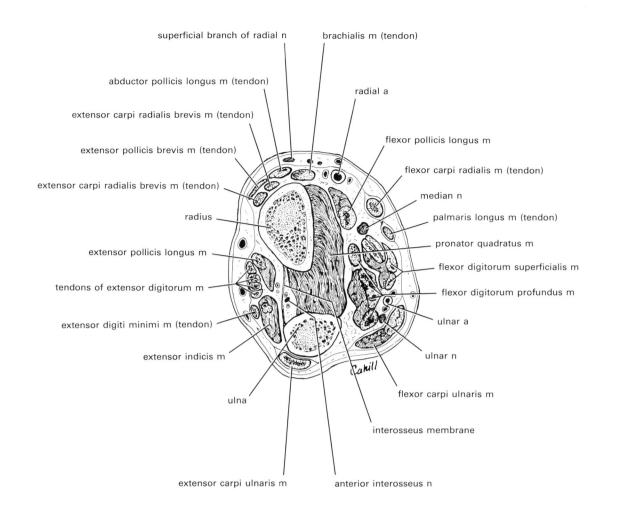

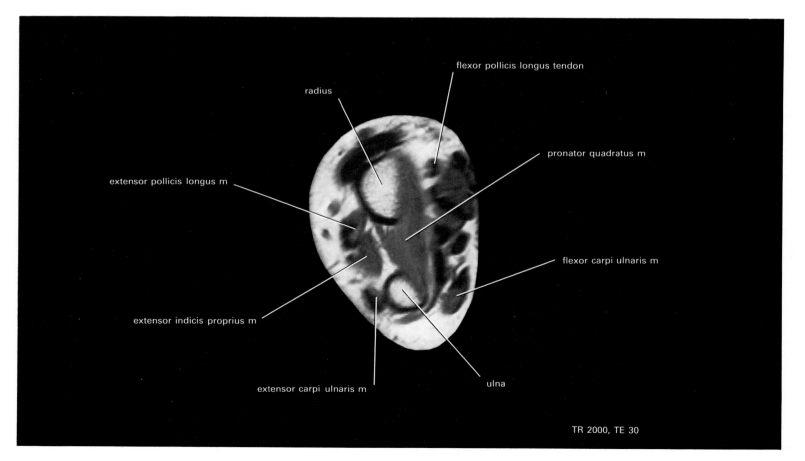

flexor pollicis longus tendon

radius

pronator quadratus m

extensor pollicis longus m

extensor indicis proprius m

flexor carpi ulnaris m

extensor carpi ulnaris m

ulna

TR 2000, TE 30

Section 19U from below.

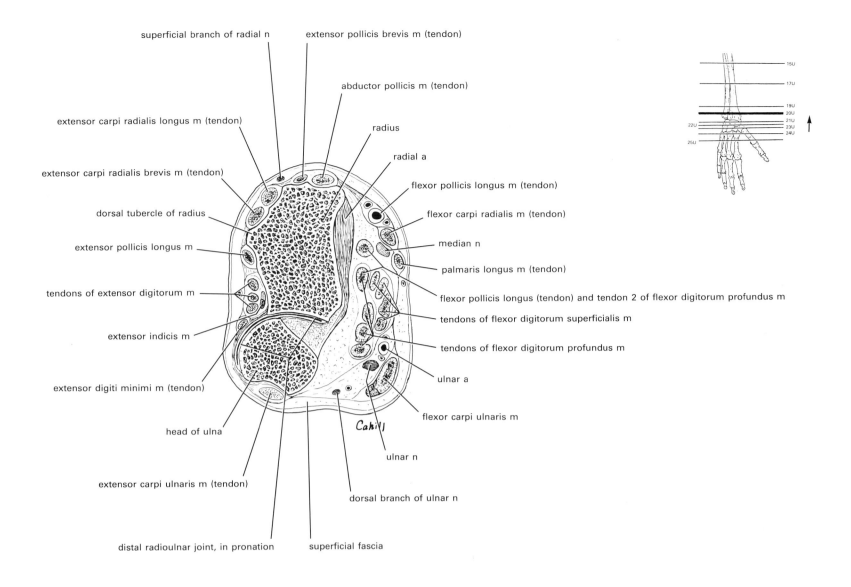

superficial branch of radial n

extensor pollicis brevis m (tendon)

abductor pollicis m (tendon)

extensor carpi radialis longus m (tendon)

radius

extensor carpi radialis brevis m (tendon)

radial a

dorsal tubercle of radius

flexor pollicis longus m (tendon)

extensor pollicis longus m

flexor carpi radialis m (tendon)

tendons of extensor digitorum m

median n

extensor indicis m

palmaris longus m (tendon)

flexor pollicis longus (tendon) and tendon 2 of flexor digitorum profundus m

tendons of flexor digitorum superficialis m

tendons of flexor digitorum profundus m

extensor digiti minimi m (tendon)

ulnar a

head of ulna

flexor carpi ulnaris m

extensor carpi ulnaris m (tendon)

ulnar n

dorsal branch of ulnar n

distal radioulnar joint, in pronation

superficial fascia

15U

17U

19U
20U
21U
22U
23U
24U

25U

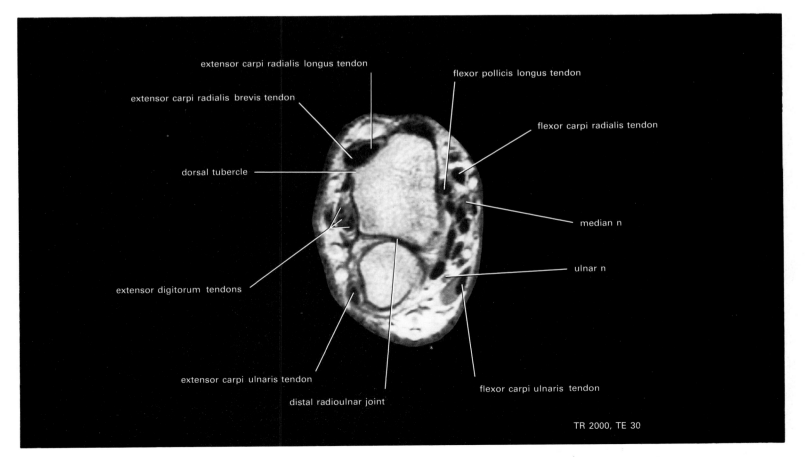

extensor carpi radialis longus tendon

flexor pollicis longus tendon

extensor carpi radialis brevis tendon

flexor carpi radialis tendon

dorsal tubercle

median n

extensor digitorum tendons

ulnar n

extensor carpi ulnaris tendon

flexor carpi ulnaris tendon

distal radioulnar joint

TR 2000, TE 30

Section 20U from below.

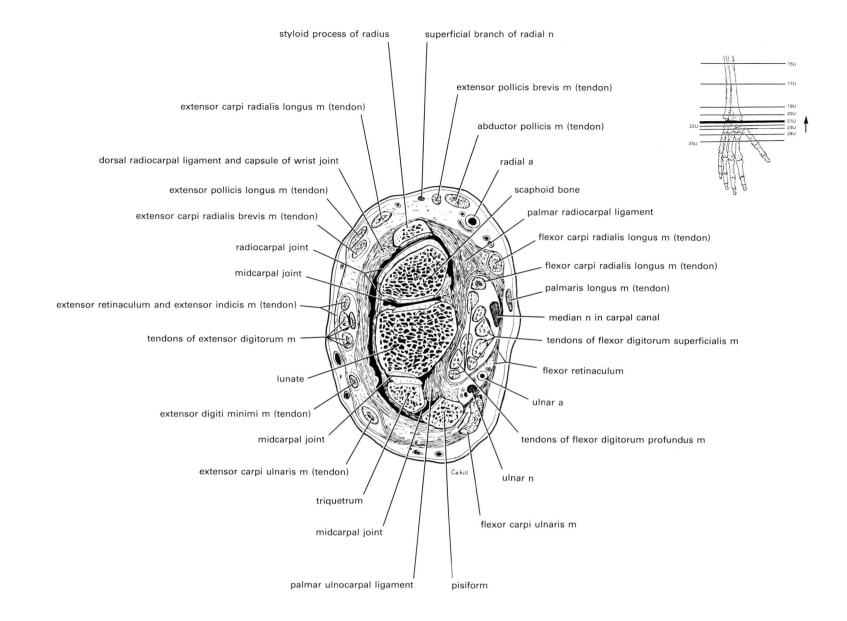

styloid process of radius

superficial branch of radial n

extensor pollicis brevis m (tendon)

extensor carpi radialis longus m (tendon)

abductor pollicis m (tendon)

dorsal radiocarpal ligament and capsule of wrist joint

radial a

scaphoid bone

extensor pollicis longus m (tendon)

palmar radiocarpal ligament

extensor carpi radialis brevis m (tendon)

flexor carpi radialis longus m (tendon)

radiocarpal joint

flexor carpi radialis longus m (tendon)

midcarpal joint

palmaris longus m (tendon)

extensor retinaculum and extensor indicis m (tendon)

median n in carpal canal

tendons of extensor digitorum m

tendons of flexor digitorum superficialis m

flexor retinaculum

lunate

ulnar a

extensor digiti minimi m (tendon)

tendons of flexor digitorum profundus m

midcarpal joint

extensor carpi ulnaris m (tendon)

ulnar n

triquetrum

flexor carpi ulnaris m

midcarpal joint

palmar ulnocarpal ligament

pisiform

Cahill

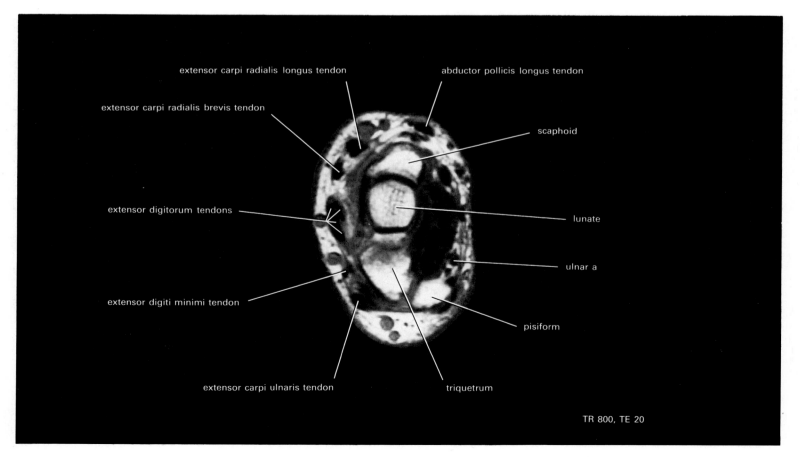

extensor carpi radialis longus tendon

abductor pollicis longus tendon

extensor carpi radialis brevis tendon

scaphoid

extensor digitorum tendons

lunate

ulnar a

extensor digiti minimi tendon

pisiform

extensor carpi ulnaris tendon

triquetrum

TR 800, TE 20

Section 21U from below.

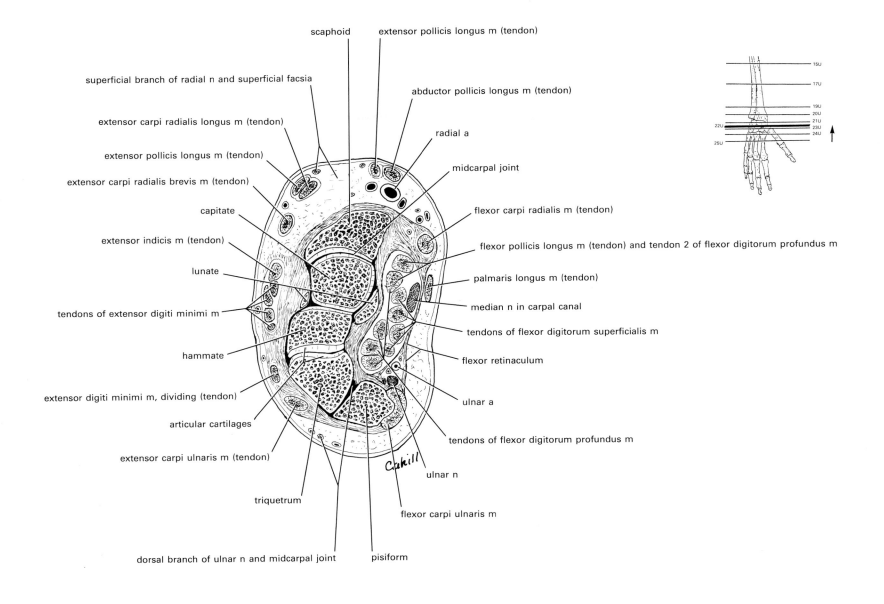

scaphoid

extensor pollicis longus m (tendon)

superficial branch of radial n and superficial facsia

abductor pollicis longus m (tendon)

extensor carpi radialis longus m (tendon)

radial a

extensor pollicis longus m (tendon)

midcarpal joint

extensor carpi radialis brevis m (tendon)

capitate

flexor carpi radialis m (tendon)

extensor indicis m (tendon)

flexor pollicis longus m (tendon) and tendon 2 of flexor digitorum profundus m

lunate

palmaris longus m (tendon)

tendons of extensor digiti minimi m

median n in carpal canal

tendons of flexor digitorum superficialis m

hammate

flexor retinaculum

extensor digiti minimi m, dividing (tendon)

articular cartilages

tendons of flexor digitorum profundus m

extensor carpi ulnaris m (tendon)

ulnar a

triquetrum

ulnar n

flexor carpi ulnaris m

dorsal branch of ulnar n and midcarpal joint

pisiform

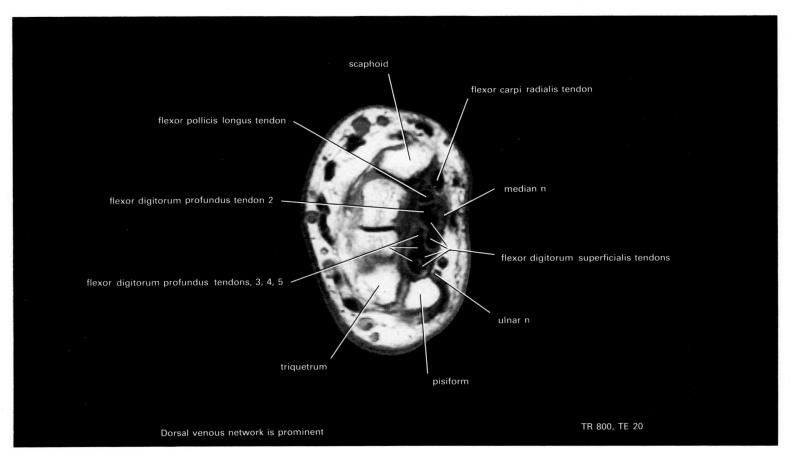

scaphoid

flexor carpi radialis tendon

flexor pollicis longus tendon

median n

flexor digitorum profundus tendon 2

flexor digitorum superficialis tendons

flexor digitorum profundus tendons, 3, 4, 5

ulnar n

triquetrum

pisiform

Dorsal venous network is prominent

TR 800, TE 20

Section 22U from below.

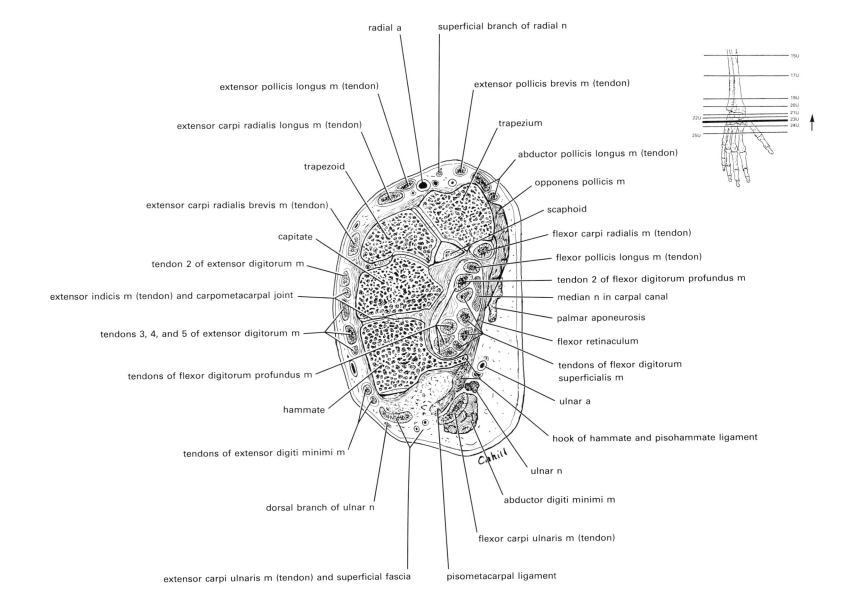

radial a
superficial branch of radial n
extensor pollicis longus m (tendon)
extensor pollicis brevis m (tendon)
extensor carpi radialis longus m (tendon)
trapezium
trapezoid
abductor pollicis longus m (tendon)
opponens pollicis m
extensor carpi radialis brevis m (tendon)
scaphoid
capitate
flexor carpi radialis m (tendon)
tendon 2 of extensor digitorum m
flexor pollicis longus m (tendon)
extensor indicis m (tendon) and carpometacarpal joint
tendon 2 of flexor digitorum profundus m
median n in carpal canal
tendons 3, 4, and 5 of extensor digitorum m
palmar aponeurosis
flexor retinaculum
tendons of flexor digitorum profundus m
tendons of flexor digitorum superficialis m
hammate
ulnar a
tendons of extensor digiti minimi m
hook of hammate and pisohammate ligament
ulnar n
dorsal branch of ulnar n
abductor digiti minimi m
flexor carpi ulnaris m (tendon)
extensor carpi ulnaris m (tendon) and superficial fascia
pisometacarpal ligament

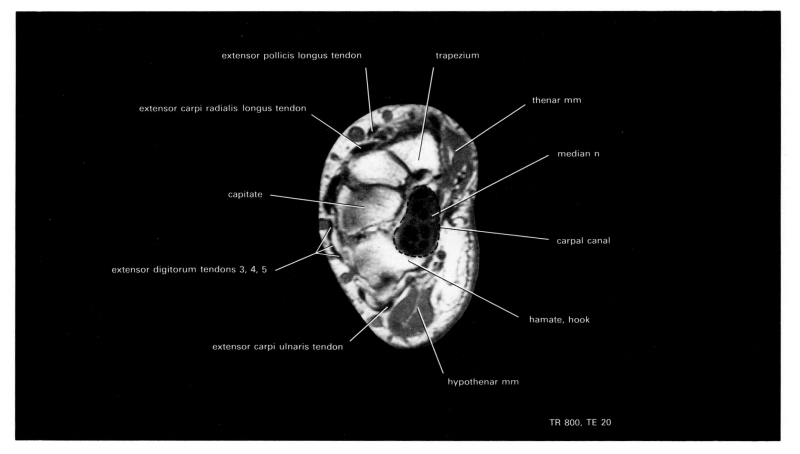

extensor pollicis longus tendon
trapezium
thenar mm
extensor carpi radialis longus tendon
capitate
median n
carpal canal
extensor digitorum tendons 3, 4, 5
hamate, hook
extensor carpi ulnaris tendon
hypothenar mm

TR 800, TE 20

Section 23U from below.

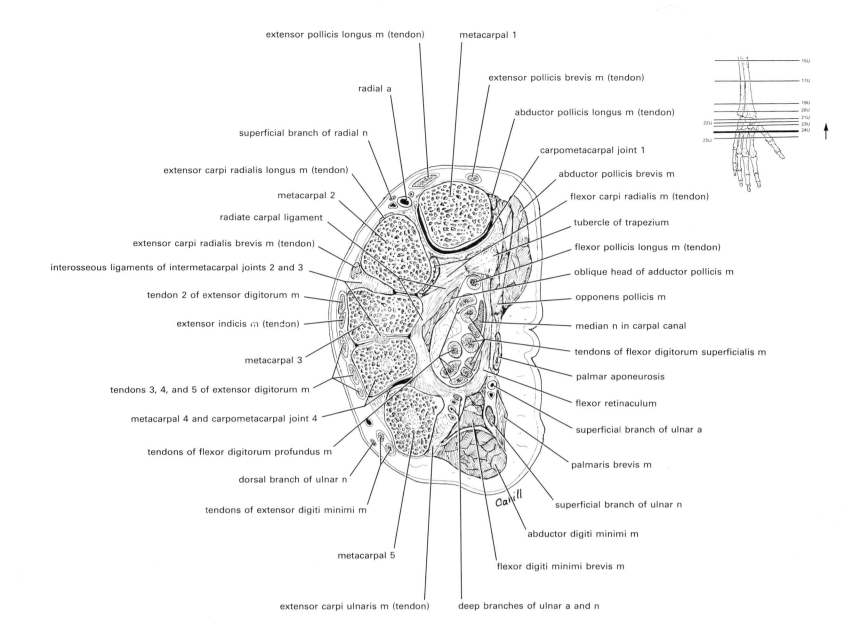

extensor pollicis longus m (tendon)
metacarpal 1
radial a
extensor pollicis brevis m (tendon)
abductor pollicis longus m (tendon)
superficial branch of radial n
carpometacarpal joint 1
extensor carpi radialis longus m (tendon)
abductor pollicis brevis m
metacarpal 2
flexor carpi radialis m (tendon)
radiate carpal ligament
tubercle of trapezium
extensor carpi radialis brevis m (tendon)
flexor pollicis longus m (tendon)
interosseous ligaments of intermetacarpal joints 2 and 3
oblique head of adductor pollicis m
tendon 2 of extensor digitorum m
opponens pollicis m
extensor indicis m (tendon)
median n in carpal canal
metacarpal 3
tendons of flexor digitorum superficialis m
palmar aponeurosis
tendons 3, 4, and 5 of extensor digitorum m
flexor retinaculum
metacarpal 4 and carpometacarpal joint 4
superficial branch of ulnar a
tendons of flexor digitorum profundus m
palmaris brevis m
dorsal branch of ulnar n
superficial branch of ulnar n
tendons of extensor digiti minimi m
abductor digiti minimi m
metacarpal 5
flexor digiti minimi brevis m
extensor carpi ulnaris m (tendon)
deep branches of ulnar a and n

Cahill

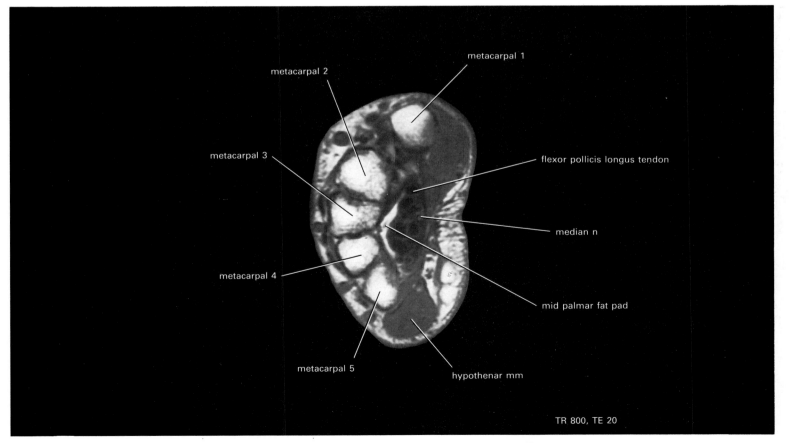

metacarpal 1
metacarpal 2
metacarpal 3
flexor pollicis longus tendon
metacarpal 4
median n
metacarpal 5
mid palmar fat pad
hypothenar mm

TR 800, TE 20

Section 24U from below.

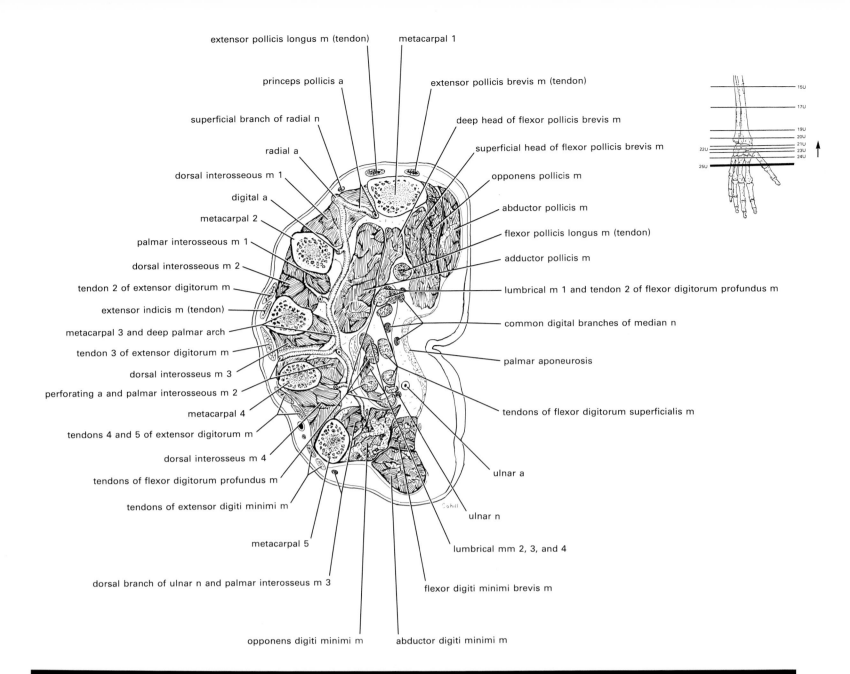

extensor pollicis longus m (tendon)

metacarpal 1

princeps pollicis a

extensor pollicis brevis m (tendon)

superficial branch of radial n

deep head of flexor pollicis brevis m

radial a

superficial head of flexor pollicis brevis m

dorsal interosseous m 1

opponens pollicis m

digital a

abductor pollicis m

metacarpal 2

flexor pollicis longus m (tendon)

palmar interosseous m 1

adductor pollicis m

dorsal interosseous m 2

lumbrical m 1 and tendon 2 of flexor digitorum profundus m

tendon 2 of extensor digitorum m

extensor indicis m (tendon)

common digital branches of median n

metacarpal 3 and deep palmar arch

tendon 3 of extensor digitorum m

palmar aponeurosis

dorsal interosseus m 3

perforating a and palmar interosseous m 2

metacarpal 4

tendons of flexor digitorum superficialis m

tendons 4 and 5 of extensor digitorum m

dorsal interosseus m 4

tendons of flexor digitorum profundus m

ulnar a

tendons of extensor digiti minimi m

ulnar n

metacarpal 5

lumbrical mm 2, 3, and 4

dorsal branch of ulnar n and palmar interosseus m 3

flexor digiti minimi brevis m

opponens digiti minimi m

abductor digiti minimi m

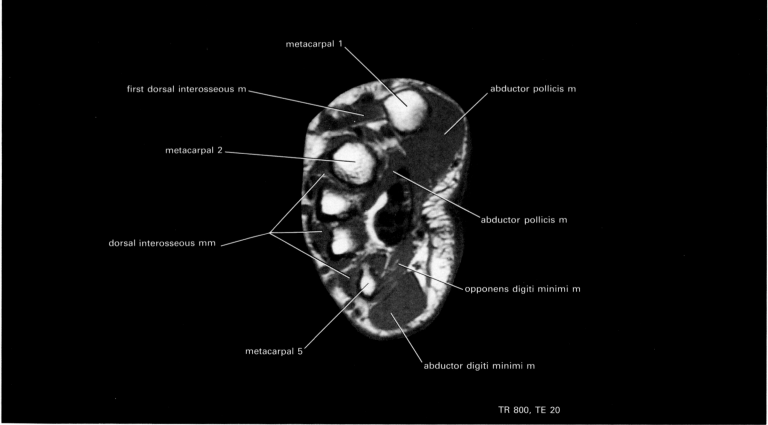

metacarpal 1

first dorsal interosseous m

abductor pollicis m

metacarpal 2

abductor pollicis m

dorsal interosseous mm

opponens digiti minimi m

metacarpal 5

abductor digiti minimi m

TR 800, TE 20

Section 25U from below.

The Head 20° From Orbitomeatal Plane

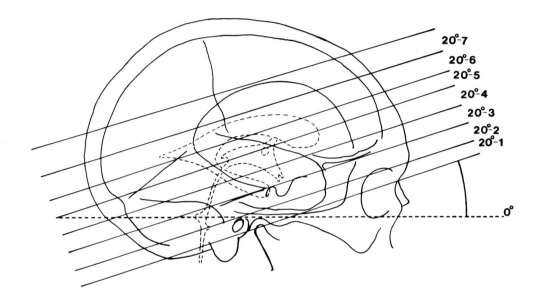

20°-7
20°-6
20°-5
20°-4
20°-3
20°-2
20°-1

0°

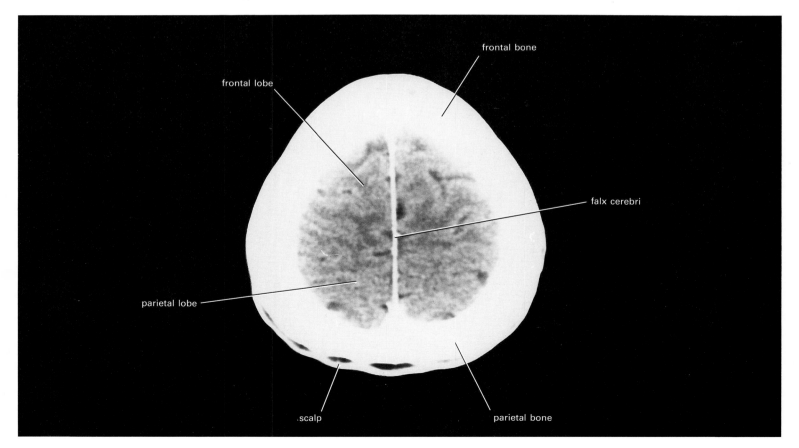

LEFT   RIGHT

superior sagittal sinus

superior frontal gyrus

frontal lobe

frontal radiation of corpus callosum

falx cerebri

middle frontal gyrus

precentral gyrus

precentral gyrus

central sulcus

postcentral gyrus

postcentral sulcus

parietal radiation of corpus callosum

inferior parietal lobule

subarachnoid space

paracentral lobule

parietal bone

superior parietal lobule

precuneus

longitudinal fissure

superior sagittal sinus

frontal bone

frontal lobe

falx cerebri

parietal lobe

.scalp

parietal bone

Section 7 from above.

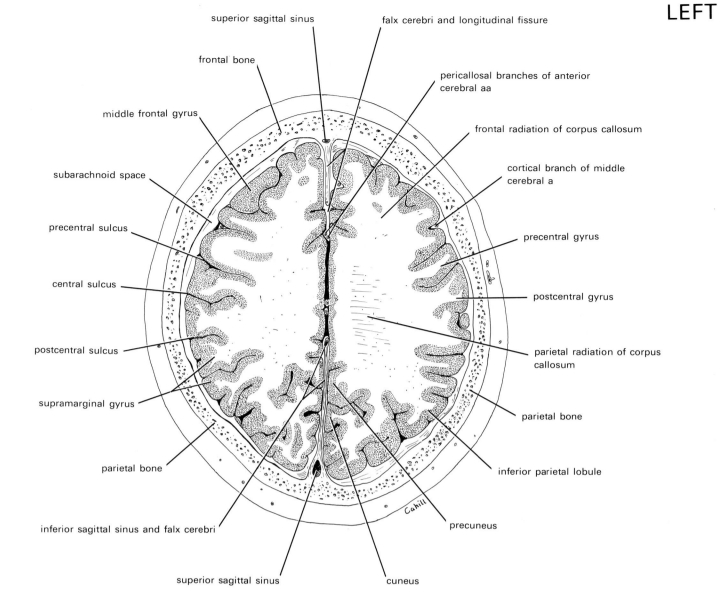

superior sagittal sinus

falx cerebri and longitudinal fissure

frontal bone

pericallosal branches of anterior cerebral aa

middle frontal gyrus

frontal radiation of corpus callosum

subarachnoid space

cortical branch of middle cerebral a

precentral sulcus

precentral gyrus

central sulcus

postcentral gyrus

postcentral sulcus

parietal radiation of corpus callosum

supramarginal gyrus

parietal bone

parietal bone

inferior parietal lobule

inferior sagittal sinus and falx cerebri

precuneus

superior sagittal sinus

cuneus

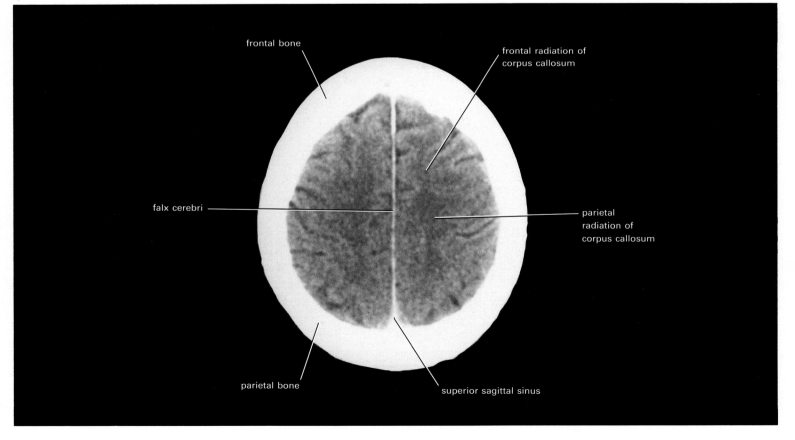

frontal bone

frontal radiation of corpus callosum

falx cerebri

parietal radiation of corpus callosum

parietal bone

superior sagittal sinus

Section 7 from below.

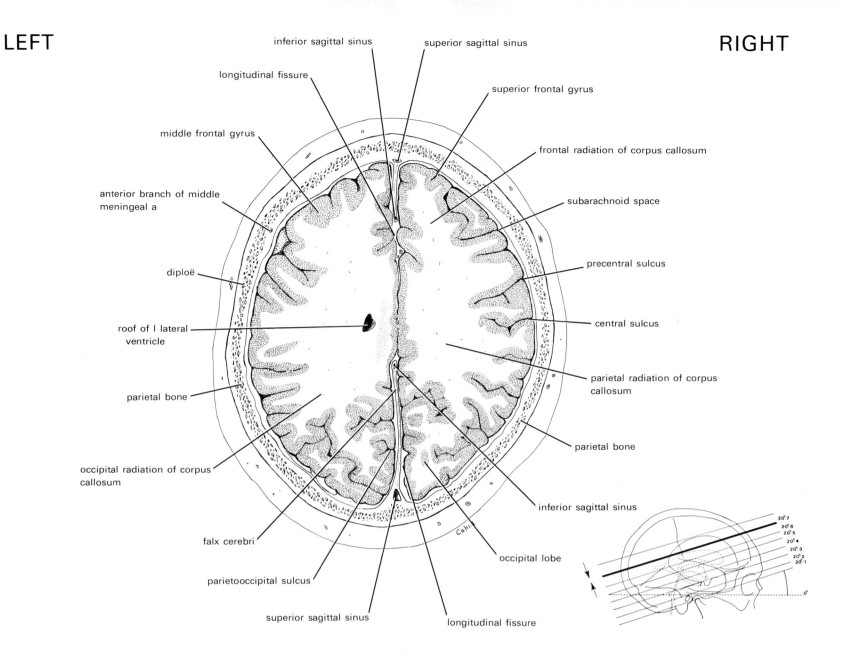

inferior sagittal sinus
superior sagittal sinus
longitudinal fissure
superior frontal gyrus
middle frontal gyrus
frontal radiation of corpus callosum
anterior branch of middle meningeal a
subarachnoid space
diploë
precentral sulcus
roof of I lateral ventricle
central sulcus
parietal bone
parietal radiation of corpus callosum
occipital radiation of corpus callosum
parietal bone
inferior sagittal sinus
falx cerebri
parietooccipital sulcus
occipital lobe
superior sagittal sinus
longitudinal fissure

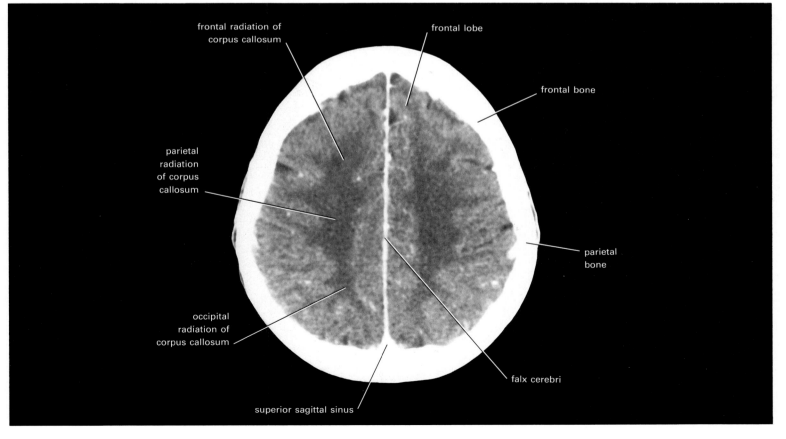

frontal radiation of corpus callosum
frontal lobe
frontal bone
parietal radiation of corpus callosum
occipital radiation of corpus callosum
parietal bone
falx cerebri
superior sagittal sinus

Section 6 from above.

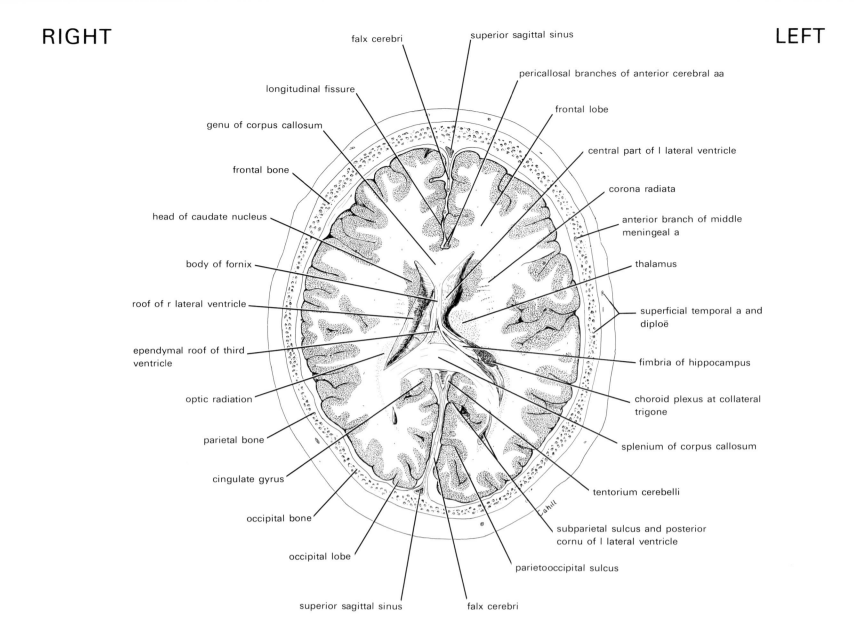

falx cerebri

superior sagittal sinus

pericallosal branches of anterior cerebral aa

longitudinal fissure

frontal lobe

genu of corpus callosum

central part of l lateral ventricle

frontal bone

corona radiata

head of caudate nucleus

anterior branch of middle meningeal a

body of fornix

thalamus

roof of r lateral ventricle

superficial temporal a and diploë

ependymal roof of third ventricle

fimbria of hippocampus

optic radiation

choroid plexus at collateral trigone

parietal bone

splenium of corpus callosum

cingulate gyrus

tentorium cerebelli

occipital bone

subparietal sulcus and posterior cornu of l lateral ventricle

occipital lobe

parietooccipital sulcus

superior sagittal sinus   falx cerebri

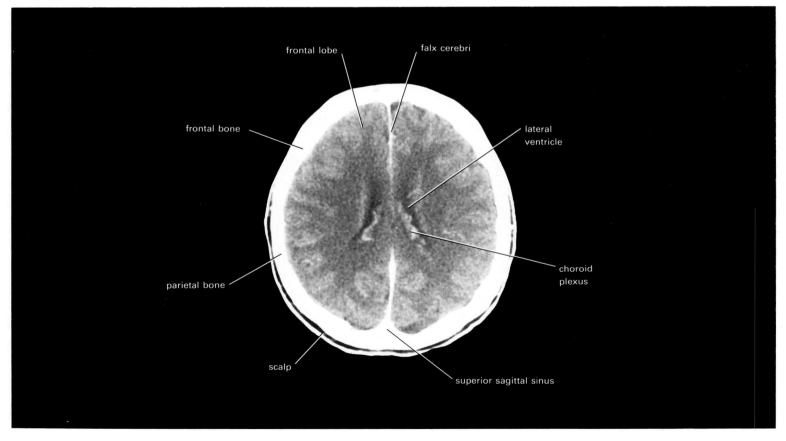

frontal lobe   falx cerebri

frontal bone

lateral ventricle

parietal bone

choroid plexus

scalp

superior sagittal sinus

Section 6 from below.

LEFT   RIGHT

cingulate gyrus   falx cerebri

genu of corpus callosum   pericallosal branches of anterior cerebral aa

anterior cornu of I lateral ventricle   frontal lobe

frontal bone

stria terminalis   head of caudate nucleus

body of fornix   choroid plexus

internal capsule   column of fornix

insula   thalamus

temporal operculum   parietal bone

tail of caudate nucleus   glomus choroideum

optic radiation   splenium of corpus callosum

crus of fornix   pineal body

posterior cornu of I
lateral ventricle   tentorium cerebelli

occipital bone   inferior sagittal sinus

occipital lobe

falx cerebri   superior sagittal sinus

anterior horn of
lateral ventricle   falx cerebri

head of
caudate
nucleus

internal capsule

choroid plexus
in posterior horn
of lateral
ventricle

paired internal
cerebral vv

optic radiation

great cerebral v   straight sinus

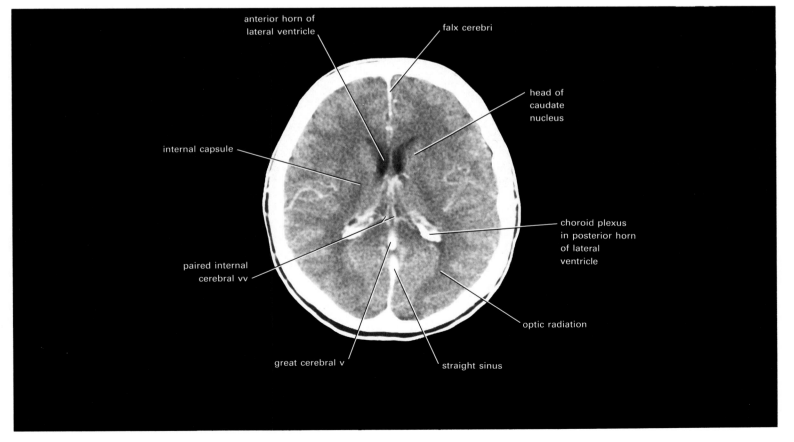

Section 5 from above.

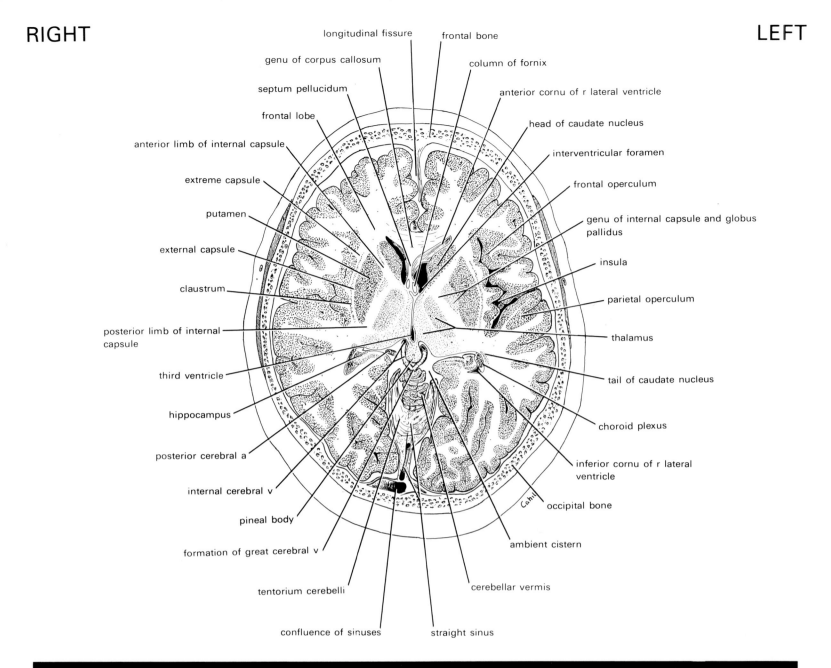

longitudinal fissure

frontal bone

genu of corpus callosum

column of fornix

septum pellucidum

anterior cornu of r lateral ventricle

frontal lobe

head of caudate nucleus

anterior limb of internal capsule

interventricular foramen

extreme capsule

frontal operculum

putamen

genu of internal capsule and globus pallidus

external capsule

insula

claustrum

parietal operculum

posterior limb of internal capsule

thalamus

third ventricle

tail of caudate nucleus

hippocampus

choroid plexus

posterior cerebral a

inferior cornu of r lateral ventricle

internal cerebral v

occipital bone

pineal body

formation of great cerebral v

ambient cistern

tentorium cerebelli

cerebellar vermis

confluence of sinuses

straight sinus

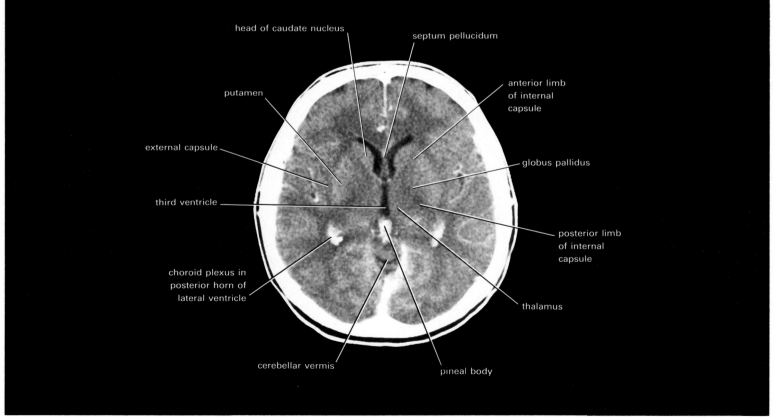

head of caudate nucleus

septum pellucidum

putamen

anterior limb of internal capsule

external capsule

globus pallidus

third ventricle

posterior limb of internal capsule

choroid plexus in posterior horn of lateral ventricle

thalamus

cerebellar vermis

pineal body

Section 5 from below.

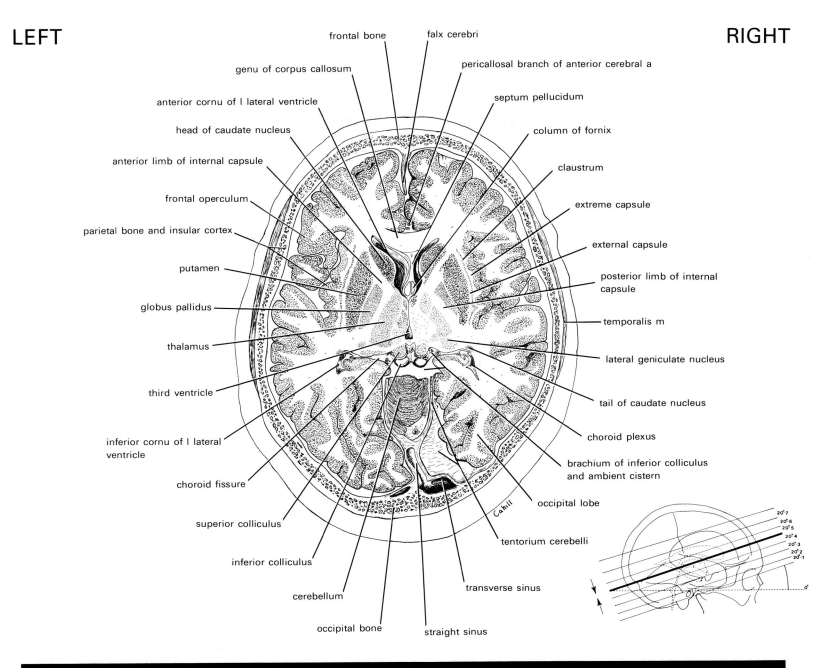

frontal bone
falx cerebri
genu of corpus callosum
pericallosal branch of anterior cerebral a
anterior cornu of I lateral ventricle
septum pellucidum
head of caudate nucleus
column of fornix
anterior limb of internal capsule
claustrum
frontal operculum
extreme capsule
parietal bone and insular cortex
external capsule
putamen
posterior limb of internal capsule
globus pallidus
temporalis m
thalamus
lateral geniculate nucleus
third ventricle
tail of caudate nucleus
inferior cornu of I lateral ventricle
choroid plexus
choroid fissure
brachium of inferior colliculus and ambient cistern
superior colliculus
occipital lobe
inferior colliculus
tentorium cerebelli
cerebellum
transverse sinus
occipital bone
straight sinus

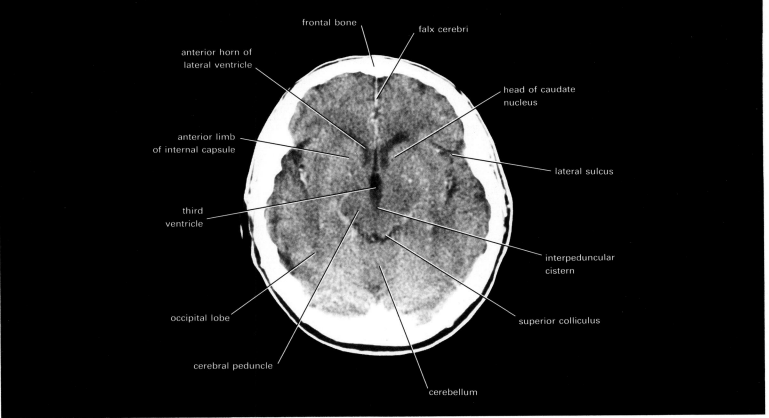

frontal bone
falx cerebri
anterior horn of lateral ventricle
head of caudate nucleus
anterior limb of internal capsule
lateral sulcus
third ventricle
interpeduncular cistern
occipital lobe
superior colliculus
cerebral peduncle
cerebellum

Section 4 from above.

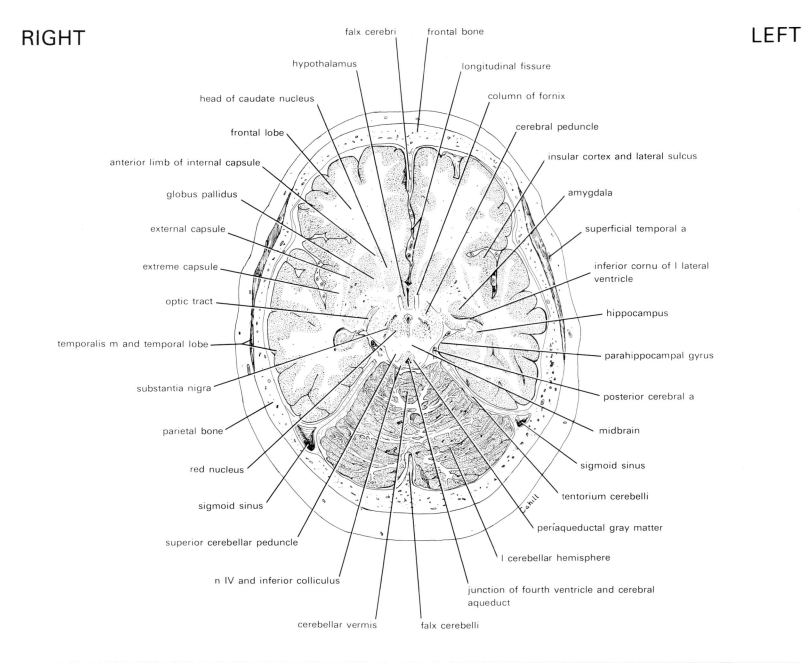

falx cerebri
frontal bone
hypothalamus
longitudinal fissure
head of caudate nucleus
column of fornix
frontal lobe
cerebral peduncle
anterior limb of internal capsule
insular cortex and lateral sulcus
globus pallidus
amygdala
external capsule
superficial temporal a
extreme capsule
inferior cornu of l lateral ventricle
optic tract
hippocampus
temporalis m and temporal lobe
parahippocampal gyrus
substantia nigra
posterior cerebral a
parietal bone
midbrain
red nucleus
sigmoid sinus
sigmoid sinus
tentorium cerebelli
superior cerebellar peduncle
periaqueductal gray matter
n IV and inferior colliculus
l cerebellar hemisphere
junction of fourth ventricle and cerebral aqueduct
cerebellar vermis
falx cerebelli

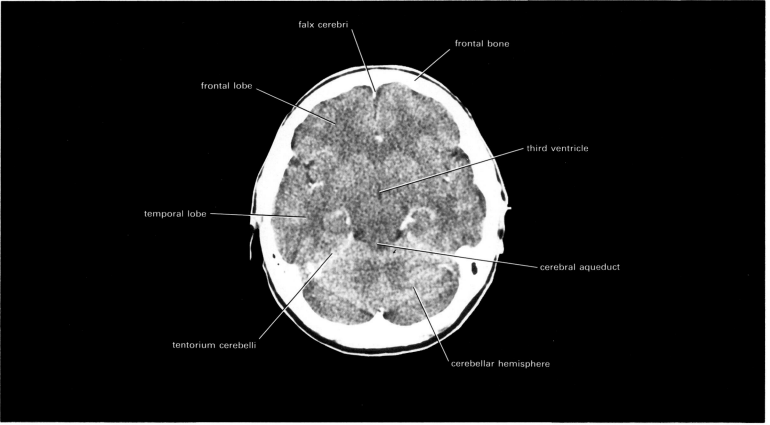

falx cerebri
frontal bone
frontal lobe
third ventricle
temporal lobe
cerebral aqueduct
tentorium cerebelli
cerebellar hemisphere

Section 4 from below.

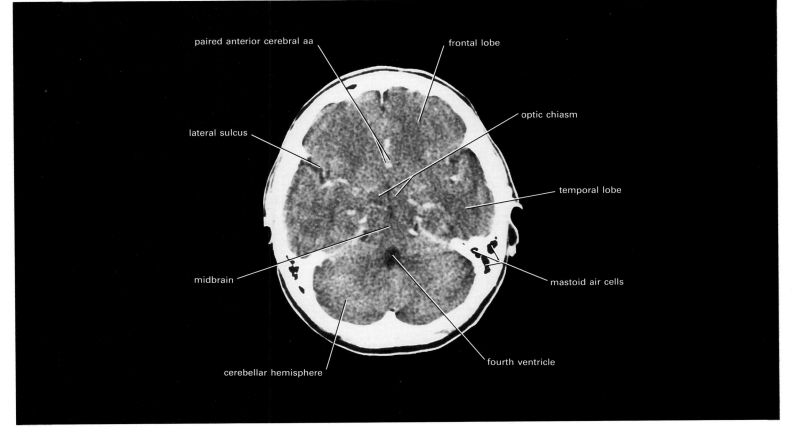

cingulate gyrus

falx cerebri

lamina terminalis

anterior cerebral aa

third ventricle

frontal lobe

mamillary body

head of caudate nucleus

putamen and lateral sulcus

temporalis m

claustrum

middle cerebral a and insula

optic tract

amygdala

cerebral peduncle

interpeduncular fossa

inferior cornu of r lateral ventricle

substantia nigra

l auricle

r auricle

midbrain

hippocampus

temporal lobe

parahippocampal gyrus

tentorium cerebelli

sigmoid sinus

superior cerebellar peduncle

posterior cerebral a

l cerebellar hemisphere

n IV

occipital bone

r cerebellar hemisphere

fourth ventricle

falx cerebelli

cerebellar vermis

Cahill

paired anterior cerebral aa

frontal lobe

optic chiasm

lateral sulcus

temporal lobe

midbrain

mastoid air cells

cerebellar hemisphere

fourth ventricle

Section 3 from above.

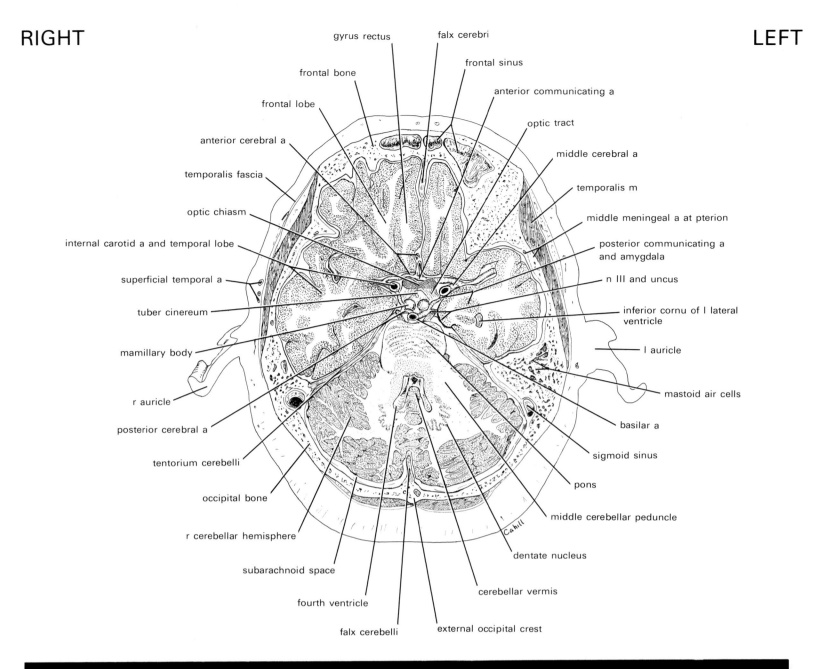

gyrus rectus
falx cerebri
frontal sinus
frontal bone
anterior communicating a
frontal lobe
optic tract
anterior cerebral a
middle cerebral a
temporalis fascia
temporalis m
optic chiasm
middle meningeal a at pterion
internal carotid a and temporal lobe
posterior communicating a and amygdala
superficial temporal a
n III and uncus
tuber cinereum
inferior cornu of l lateral ventricle
mamillary body
l auricle
r auricle
mastoid air cells
posterior cerebral a
basilar a
tentorium cerebelli
sigmoid sinus
occipital bone
pons
r cerebellar hemisphere
middle cerebellar peduncle
subarachnoid space
dentate nucleus
fourth ventricle
cerebellar vermis
falx cerebelli
external occipital crest

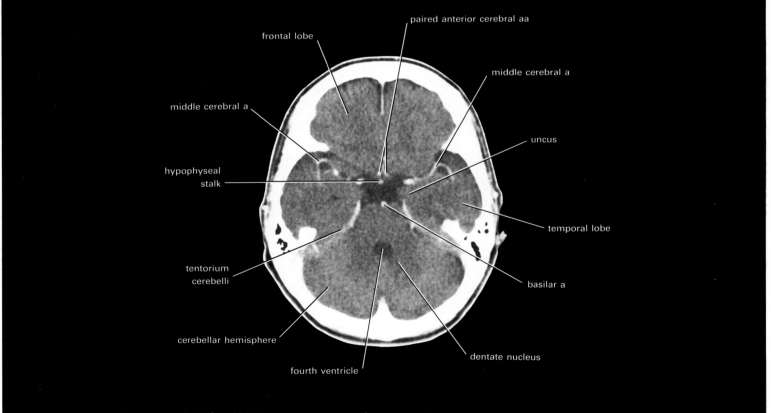

frontal lobe
paired anterior cerebral aa
middle cerebral a
middle cerebral a
uncus
hypophyseal stalk
temporal lobe
tentorium cerebelli
basilar a
cerebellar hemisphere
dentate nucleus
fourth ventricle

Section 3 from below.

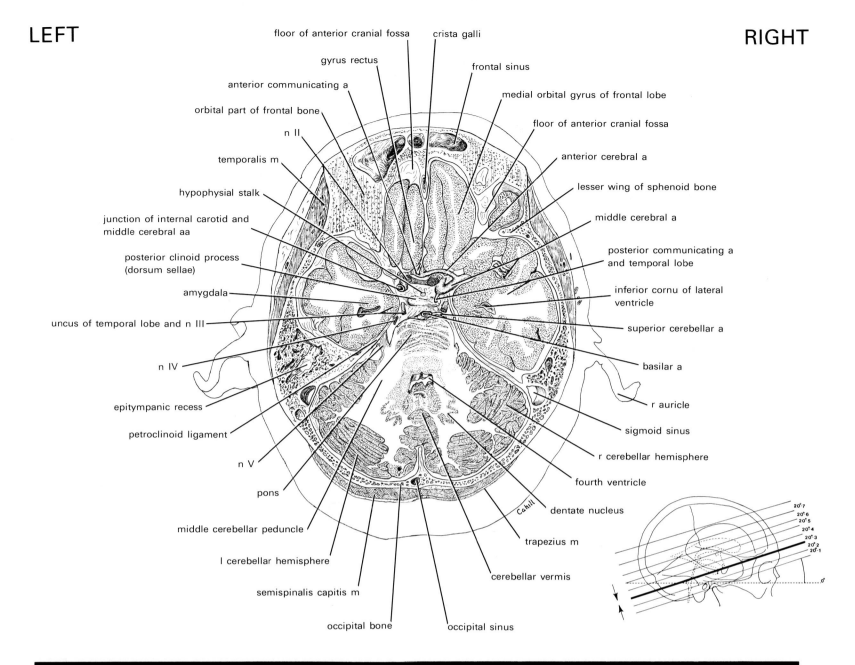

LEFT   RIGHT

floor of anterior cranial fossa
crista galli
gyrus rectus
frontal sinus
anterior communicating a
medial orbital gyrus of frontal lobe
orbital part of frontal bone
floor of anterior cranial fossa
n II
anterior cerebral a
temporalis m
lesser wing of sphenoid bone
hypophysial stalk
middle cerebral a
junction of internal carotid and
middle cerebral aa
posterior communicating a
and temporal lobe
posterior clinoid process
(dorsum sellae)
inferior cornu of lateral
ventricle
amygdala
uncus of temporal lobe and n III
superior cerebellar a
n IV
basilar a
epitympanic recess
r auricle
petroclinoid ligament
sigmoid sinus
n V
r cerebellar hemisphere
pons
fourth ventricle
middle cerebellar peduncle
dentate nucleus
l cerebellar hemisphere
trapezius m
semispinalis capitis m
cerebellar vermis
occipital bone
occipital sinus

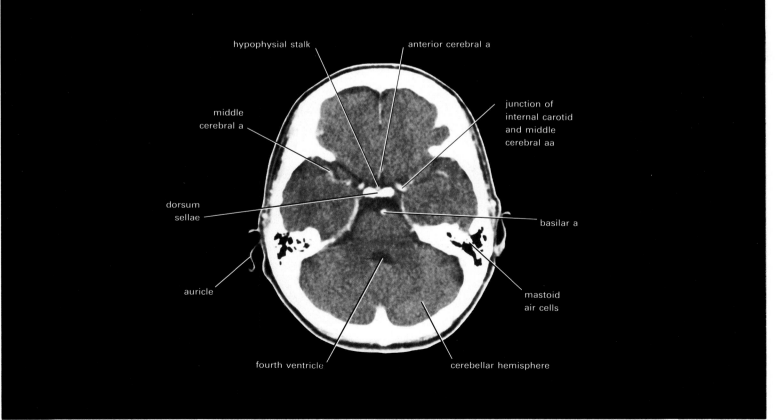

hypophysial stalk
anterior cerebral a
middle
cerebral a
junction of
internal carotid
and middle
cerebral aa
dorsum
sellae
basilar a
auricle
mastoid
air cells
fourth ventricle
cerebellar hemisphere

Section 2 from above.

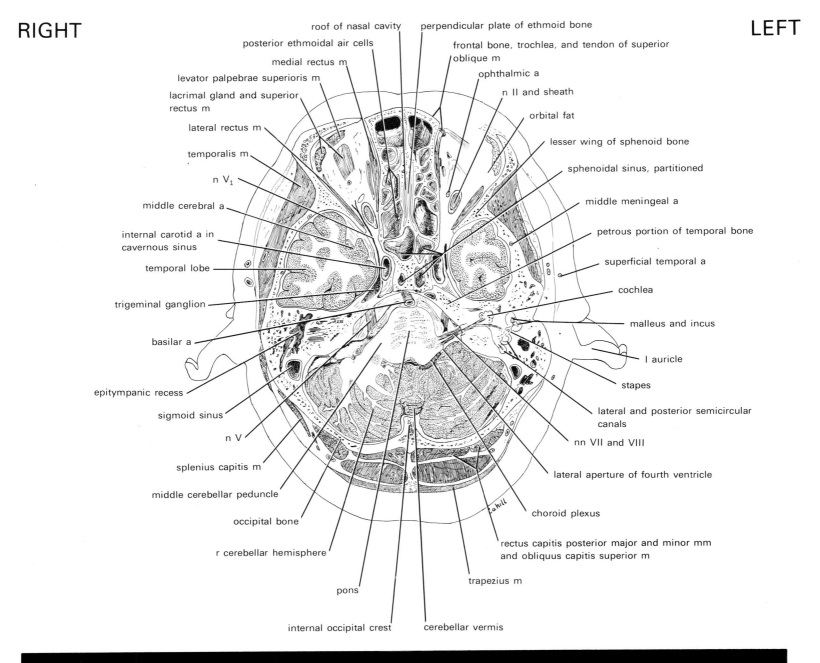

roof of nasal cavity
perpendicular plate of ethmoid bone
posterior ethmoidal air cells
frontal bone, trochlea, and tendon of superior oblique m
medial rectus m
ophthalmic a
levator palpebrae superioris m
n II and sheath
lacrimal gland and superior rectus m
orbital fat
lateral rectus m
lesser wing of sphenoid bone
temporalis m
sphenoidal sinus, partitioned
n V₁
middle meningeal a
middle cerebral a
petrous portion of temporal bone
internal carotid a in cavernous sinus
superficial temporal a
temporal lobe
cochlea
trigeminal ganglion
malleus and incus
basilar a
l auricle
epitympanic recess
stapes
sigmoid sinus
lateral and posterior semicircular canals
n V
nn VII and VIII
splenius capitis m
lateral aperture of fourth ventricle
middle cerebellar peduncle
occipital bone
choroid plexus
r cerebellar hemisphere
rectus capitis posterior major and minor mm and obliquus capitis superior m
trapezius m
pons
internal occipital crest
cerebellar vermis

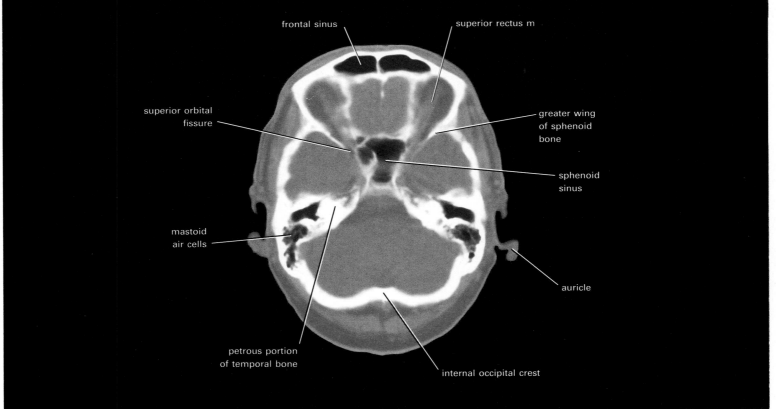

frontal sinus
superior rectus m
superior orbital fissure
greater wing of sphenoid bone
sphenoid sinus
mastoid air cells
auricle
petrous portion of temporal bone
internal occipital crest

Section 2 from below.

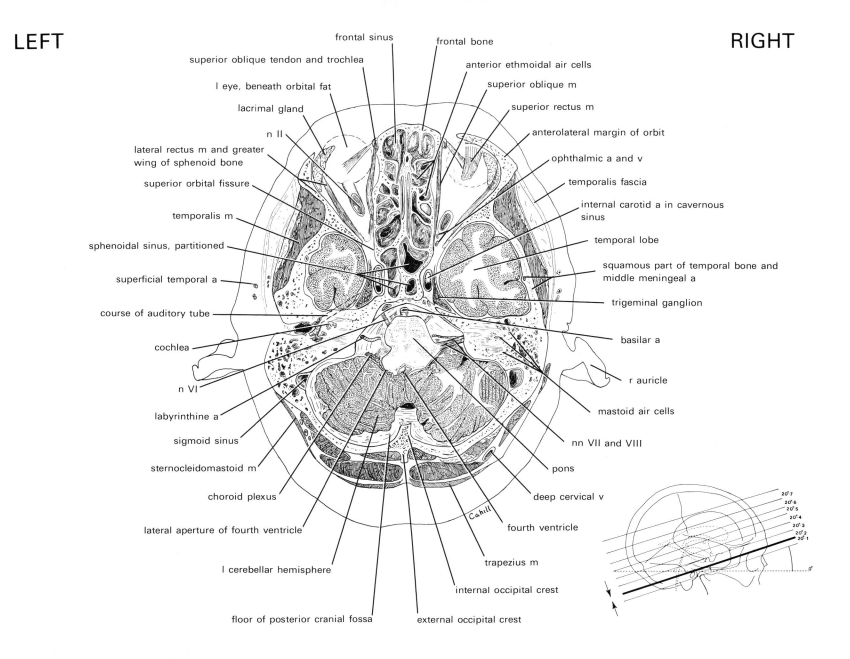

frontal sinus
frontal bone
superior oblique tendon and trochlea
anterior ethmoidal air cells
l eye, beneath orbital fat
superior oblique m
lacrimal gland
superior rectus m
n II
anterolateral margin of orbit
lateral rectus m and greater wing of sphenoid bone
ophthalmic a and v
superior orbital fissure
temporalis fascia
temporalis m
internal carotid a in cavernous sinus
sphenoidal sinus, partitioned
temporal lobe
superficial temporal a
squamous part of temporal bone and middle meningeal a
course of auditory tube
trigeminal ganglion
cochlea
basilar a
n VI
r auricle
labyrinthine a
mastoid air cells
sigmoid sinus
nn VII and VIII
sternocleidomastoid m
pons
choroid plexus
deep cervical v
lateral aperture of fourth ventricle
fourth ventricle
l cerebellar hemisphere
trapezius m
internal occipital crest
floor of posterior cranial fossa
external occipital crest

Cahill

20°·7
20°·6
20°·5
20°·4
20°·3
20°·2
20°·1
0°

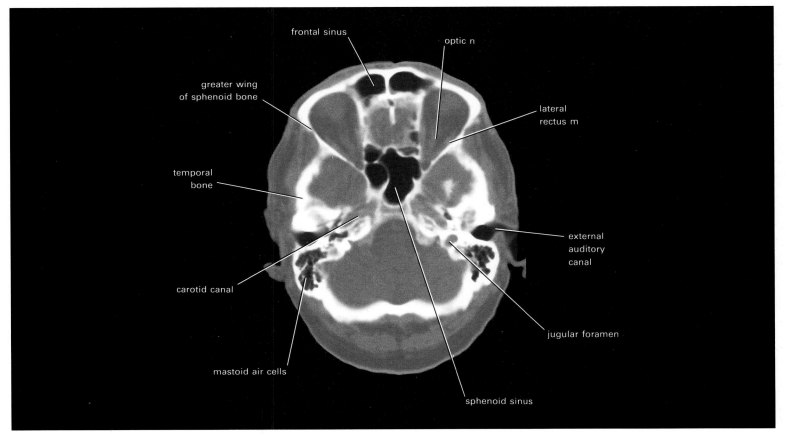

frontal sinus
optic n
greater wing of sphenoid bone
lateral rectus m
temporal bone
external auditory canal
carotid canal
jugular foramen
mastoid air cells
sphenoid sinus

Section 1 from above.

RIGHT   LEFT

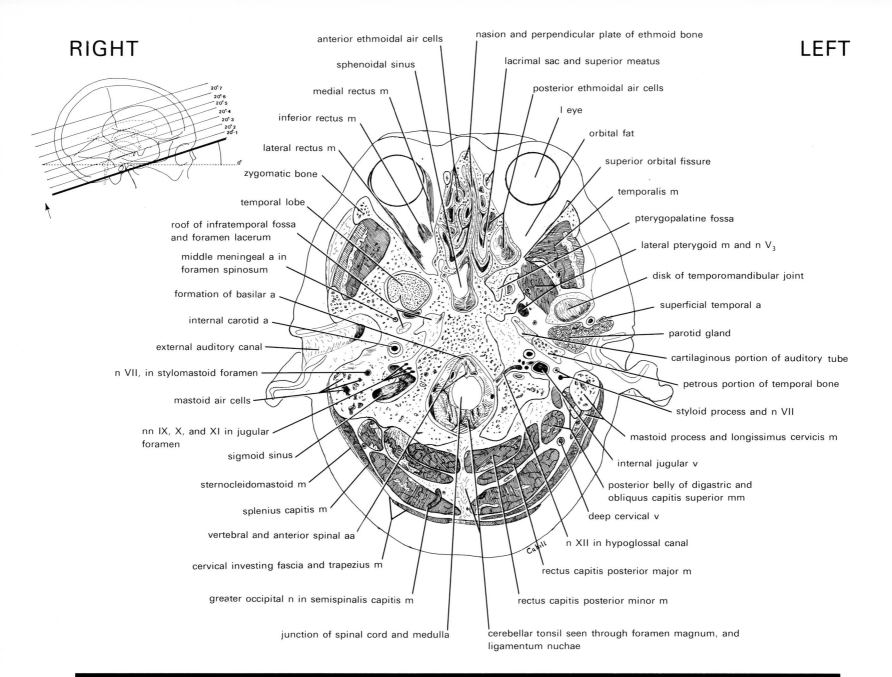

anterior ethmoidal air cells — nasion and perpendicular plate of ethmoid bone
sphenoidal sinus — lacrimal sac and superior meatus
medial rectus m — posterior ethmoidal air cells
inferior rectus m — l eye
lateral rectus m — orbital fat
zygomatic bone — superior orbital fissure
temporal lobe — temporalis m
roof of infratemporal fossa and foramen lacerum — pterygopalatine fossa
middle meningeal a in foramen spinosum — lateral pterygoid m and n V₃
formation of basilar a — disk of temporomandibular joint
internal carotid a — superficial temporal a
external auditory canal — parotid gland
n VII, in stylomastoid foramen — cartilaginous portion of auditory tube
mastoid air cells — petrous portion of temporal bone
nn IX, X, and XI in jugular foramen — styloid process and n VII
sigmoid sinus — mastoid process and longissimus cervicis m
sternocleidomastoid m — internal jugular v
splenius capitis m — posterior belly of digastric and obliquus capitis superior mm
vertebral and anterior spinal aa — deep cervical v
cervical investing fascia and trapezius m — n XII in hypoglossal canal
greater occipital n in semispinalis capitis m — rectus capitis posterior major m
— rectus capitis posterior minor m
junction of spinal cord and medulla — cerebellar tonsil seen through foramen magnum, and ligamentum nuchae

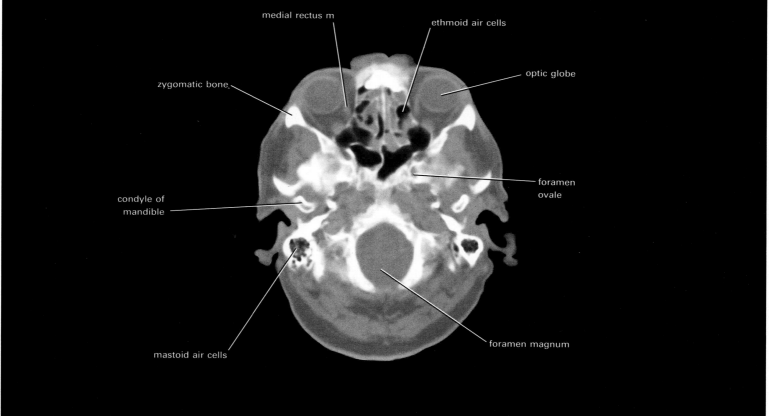

medial rectus m — ethmoid air cells
zygomatic bone — optic globe
condyle of mandible — foramen ovale
mastoid air cells — foramen magnum

Section 1 from below.

The Head and Neck 0° From Orbitomeatal Plane

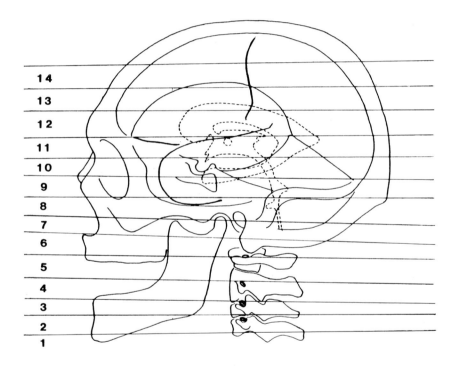

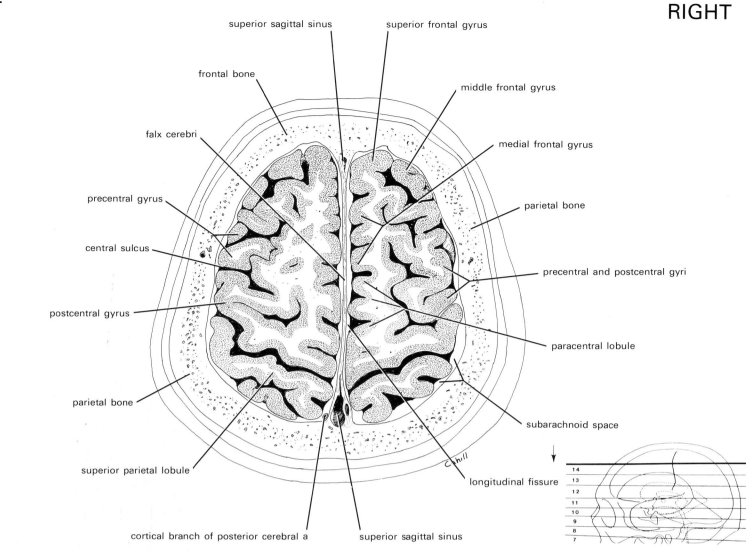

superior sagittal sinus superior frontal gyrus

frontal bone

middle frontal gyrus

falx cerebri

medial frontal gyrus

precentral gyrus

parietal bone

central sulcus

precentral and postcentral gyri

postcentral gyrus

paracentral lobule

parietal bone

subarachnoid space

superior parietal lobule

longitudinal fissure

cortical branch of posterior cerebral a   superior sagittal sinus

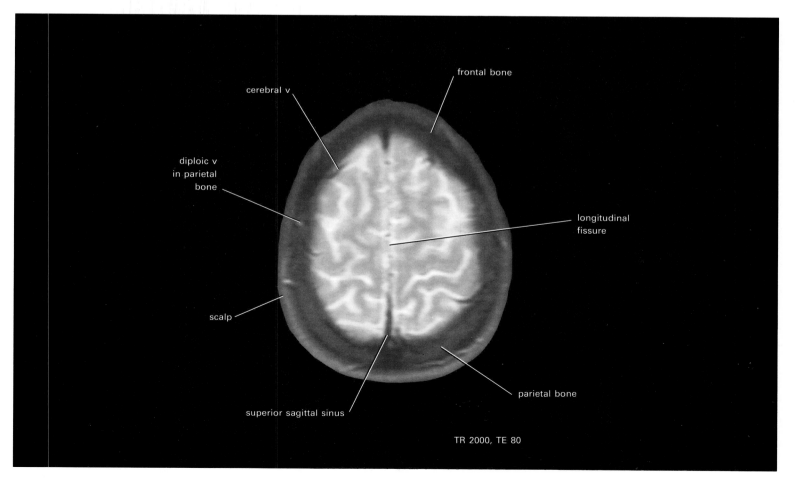

cerebral v   frontal bone

diploic v
in parietal
bone

longitudinal
fissure

scalp

parietal bone

superior sagittal sinus

TR 2000, TE 80

Section 14 from above.

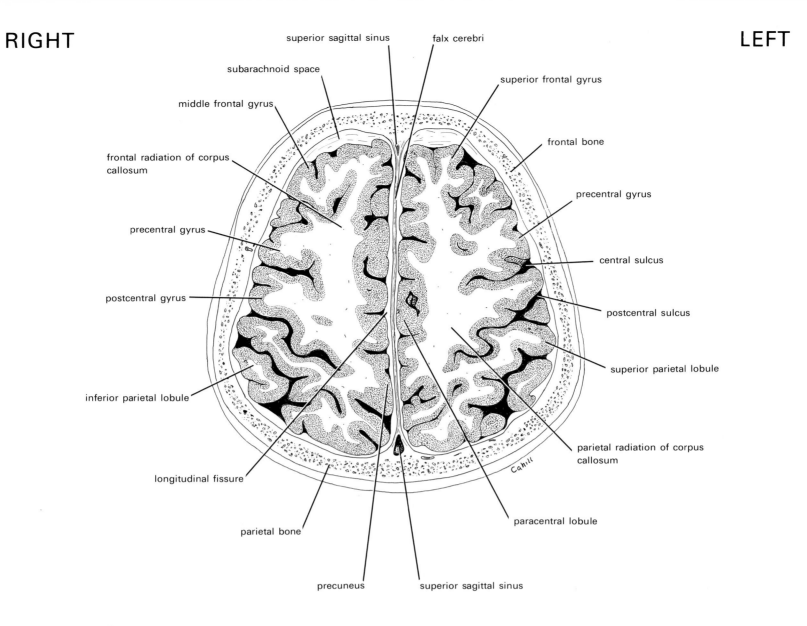

superior sagittal sinus
falx cerebri
subarachnoid space
superior frontal gyrus
middle frontal gyrus
frontal bone
frontal radiation of corpus callosum
precentral gyrus
precentral gyrus
central sulcus
postcentral gyrus
postcentral sulcus
superior parietal lobule
inferior parietal lobule
parietal radiation of corpus callosum
longitudinal fissure
parietal bone
paracentral lobule
Cahill
precuneus
superior sagittal sinus

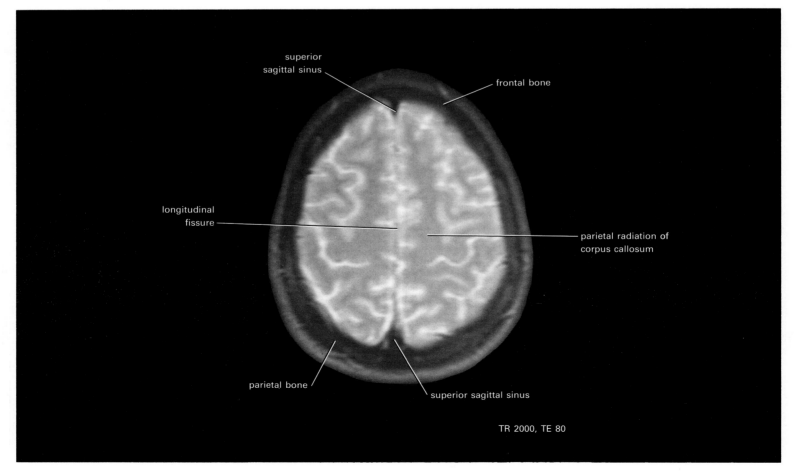

superior sagittal sinus
frontal bone
longitudinal fissure
parietal radiation of corpus callosum
parietal bone
superior sagittal sinus
TR 2000, TE 80

Section 14 from below.

LEFT   RIGHT

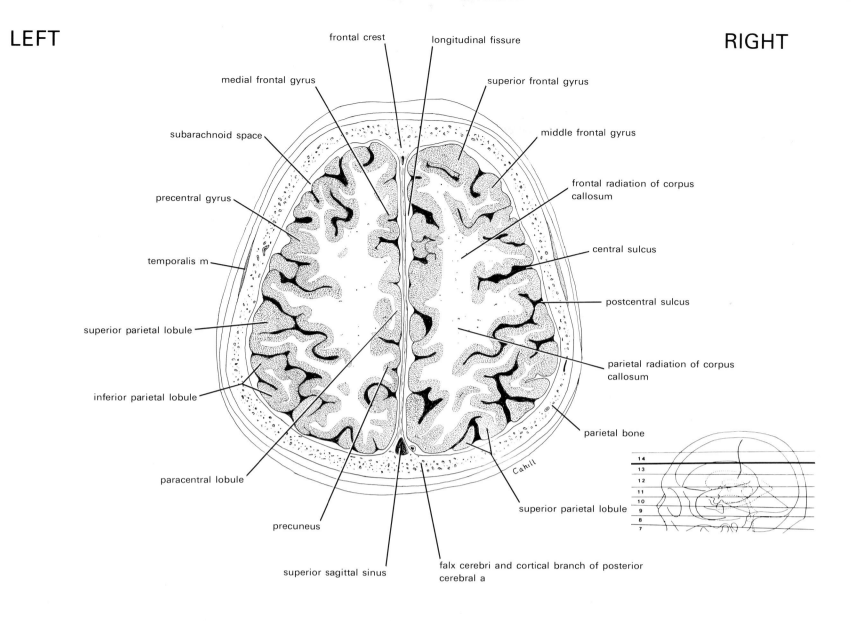

frontal crest
longitudinal fissure
medial frontal gyrus
superior frontal gyrus
subarachnoid space
middle frontal gyrus
frontal radiation of corpus callosum
precentral gyrus
central sulcus
temporalis m
postcentral sulcus
superior parietal lobule
inferior parietal lobule
parietal radiation of corpus callosum
parietal bone
paracentral lobule
superior parietal lobule
precuneus
superior sagittal sinus
falx cerebri and cortical branch of posterior cerebral a

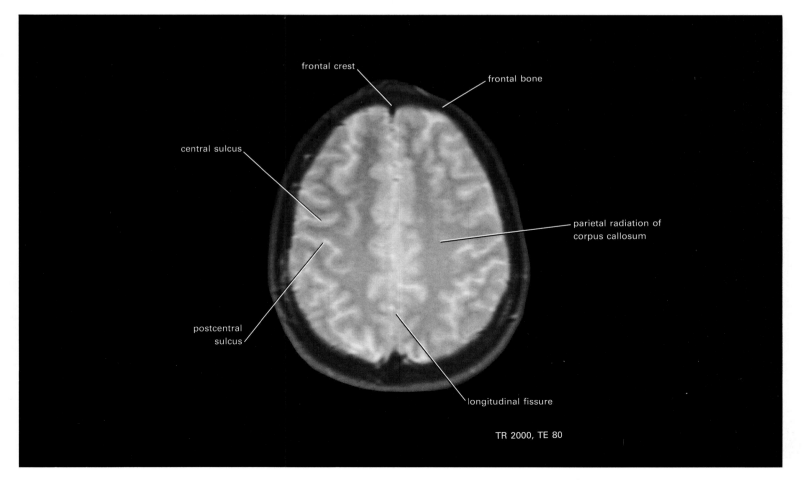

frontal crest
frontal bone
central sulcus
parietal radiation of corpus callosum
postcentral sulcus
longitudinal fissure
TR 2000, TE 80

Section 13 from above.

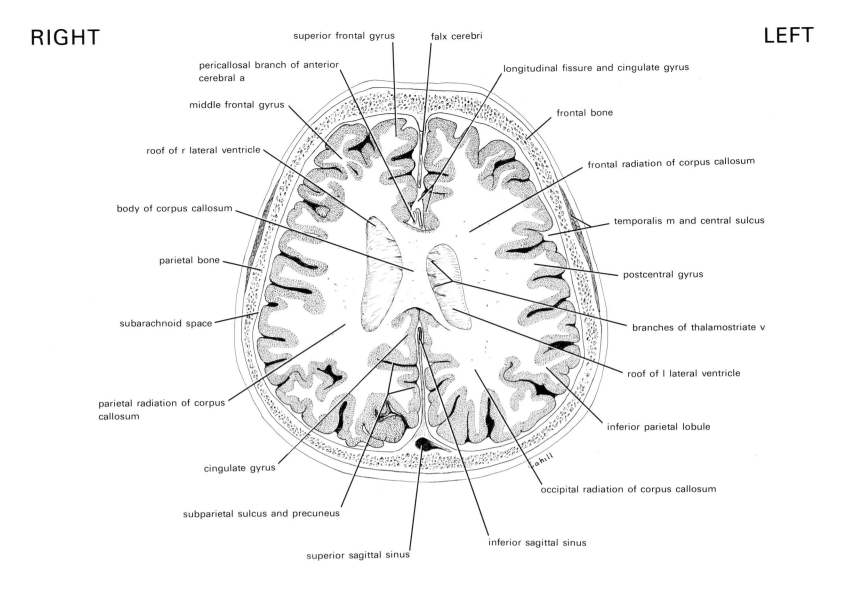

superior frontal gyrus
falx cerebri
pericallosal branch of anterior cerebral a
longitudinal fissure and cingulate gyrus
middle frontal gyrus
frontal bone
roof of r lateral ventricle
frontal radiation of corpus callosum
body of corpus callosum
temporalis m and central sulcus
parietal bone
postcentral gyrus
subarachnoid space
branches of thalamostriate v
roof of l lateral ventricle
parietal radiation of corpus callosum
inferior parietal lobule
cingulate gyrus
occipital radiation of corpus callosum
subparietal sulcus and precuneus
inferior sagittal sinus
superior sagittal sinus

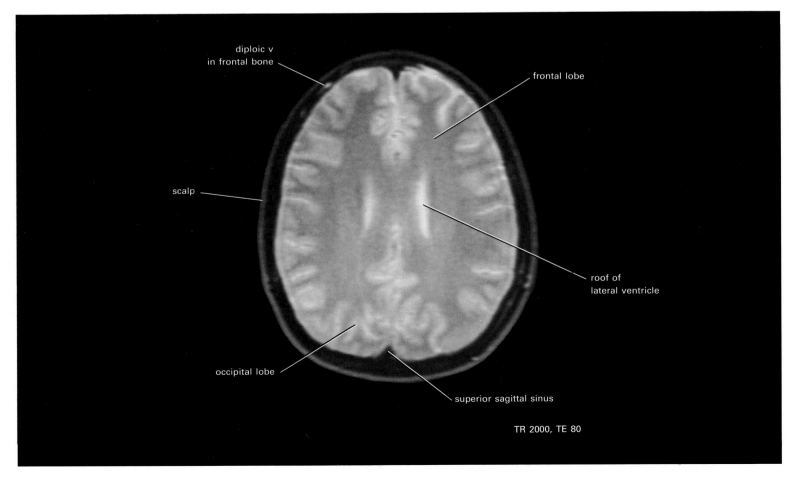

diploic v in frontal bone
frontal lobe
scalp
roof of lateral ventricle
occipital lobe
superior sagittal sinus
TR 2000, TE 80

Section 13 from below.

LEFT   RIGHT

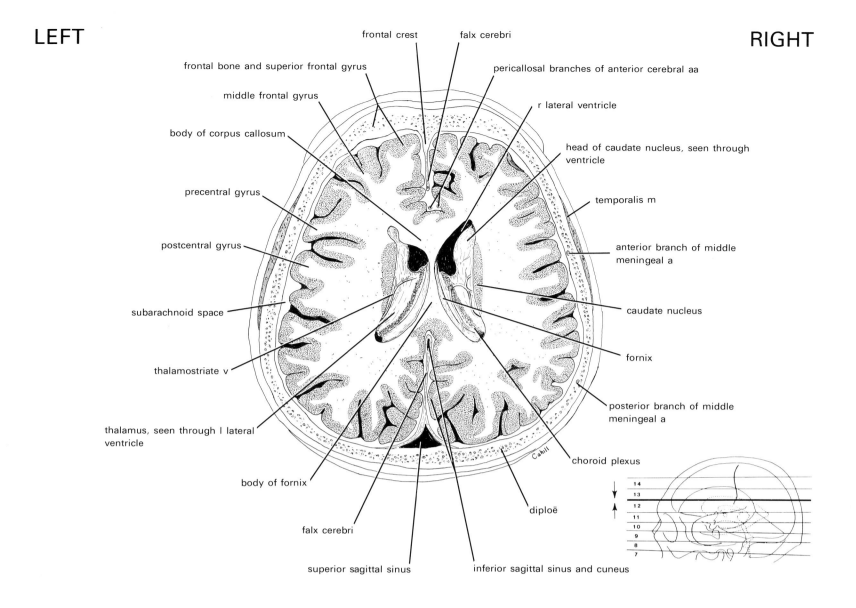

frontal crest   falx cerebri

frontal bone and superior frontal gyrus

pericallosal branches of anterior cerebral aa

middle frontal gyrus

r lateral ventricle

body of corpus callosum

head of caudate nucleus, seen through ventricle

precentral gyrus

temporalis m

postcentral gyrus

anterior branch of middle meningeal a

subarachnoid space

caudate nucleus

thalamostriate v

fornix

thalamus, seen through l lateral ventricle

posterior branch of middle meningeal a

body of fornix

choroid plexus

falx cerebri

diploë

superior sagittal sinus

inferior sagittal sinus and cuneus

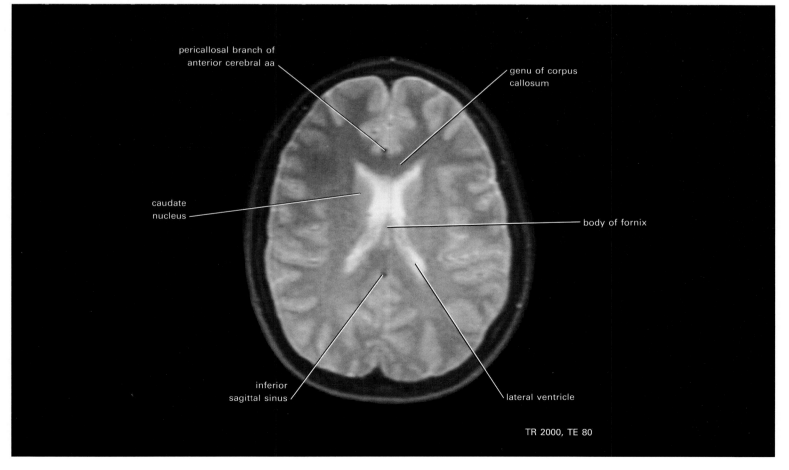

pericallosal branch of anterior cerebral aa

genu of corpus callosum

caudate nucleus

body of fornix

inferior sagittal sinus

lateral ventricle

TR 2000, TE 80

Section 12 from above.

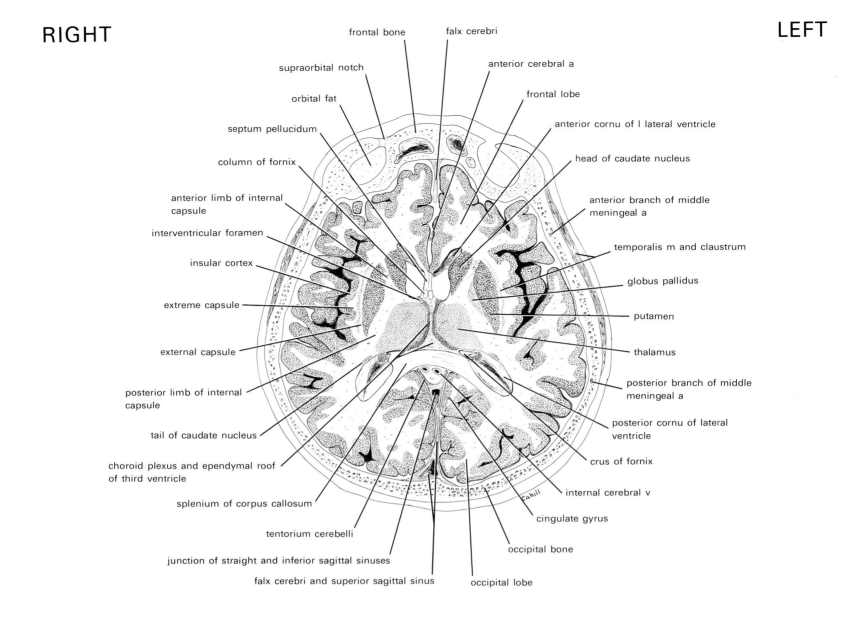

frontal bone · falx cerebri · supraorbital notch · anterior cerebral a · orbital fat · frontal lobe · septum pellucidum · anterior cornu of l lateral ventricle · column of fornix · head of caudate nucleus · anterior limb of internal capsule · anterior branch of middle meningeal a · interventricular foramen · temporalis m and claustrum · insular cortex · globus pallidus · extreme capsule · putamen · external capsule · thalamus · posterior limb of internal capsule · posterior branch of middle meningeal a · tail of caudate nucleus · posterior cornu of lateral ventricle · choroid plexus and ependymal roof of third ventricle · crus of fornix · splenium of corpus callosum · internal cerebral v · tentorium cerebelli · cingulate gyrus · junction of straight and inferior sagittal sinuses · occipital bone · falx cerebri and superior sagittal sinus · occipital lobe

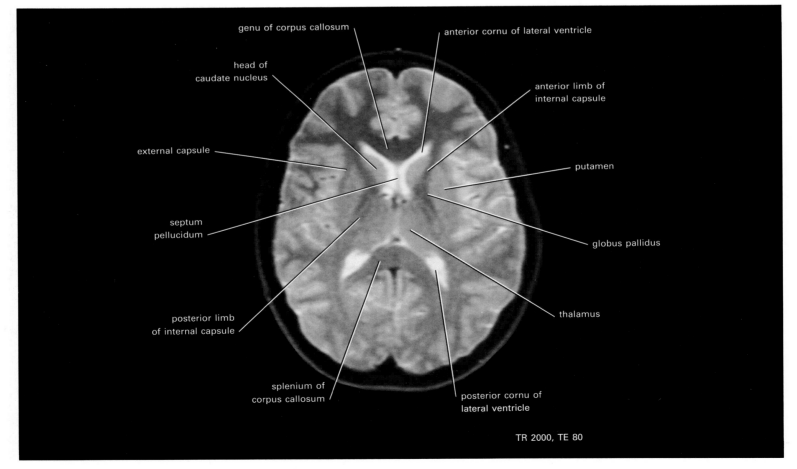

genu of corpus callosum · anterior cornu of lateral ventricle · head of caudate nucleus · anterior limb of internal capsule · external capsule · putamen · septum pellucidum · globus pallidus · posterior limb of internal capsule · thalamus · splenium of corpus callosum · posterior cornu of lateral ventricle

TR 2000, TE 80

Section 12 from below.

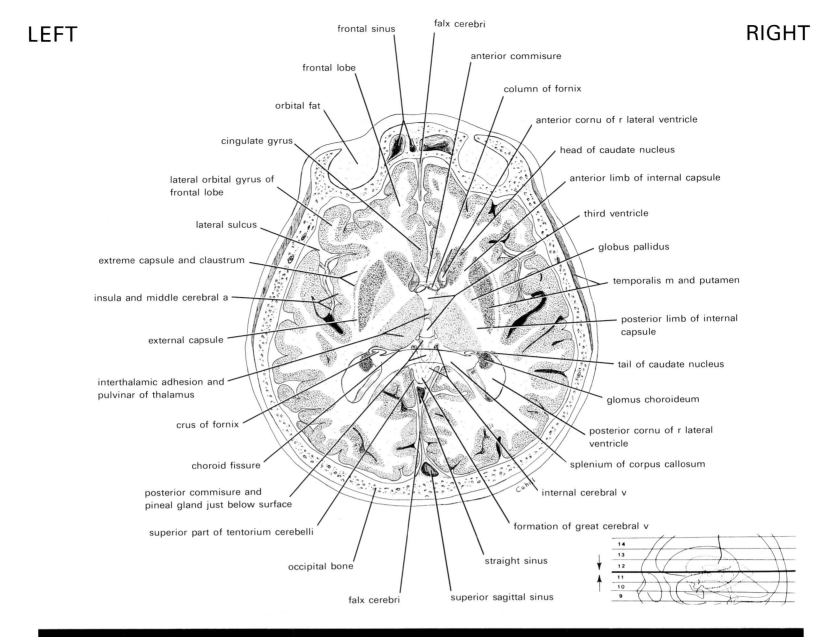

frontal sinus
falx cerebri
anterior commisure
frontal lobe
column of fornix
orbital fat
anterior cornu of r lateral ventricle
cingulate gyrus
head of caudate nucleus
lateral orbital gyrus of frontal lobe
anterior limb of internal capsule
lateral sulcus
third ventricle
extreme capsule and claustrum
globus pallidus
insula and middle cerebral a
temporalis m and putamen
external capsule
posterior limb of internal capsule
interthalamic adhesion and pulvinar of thalamus
tail of caudate nucleus
crus of fornix
glomus choroideum
choroid fissure
posterior cornu of r lateral ventricle
posterior commisure and pineal gland just below surface
splenium of corpus callosum
superior part of tentorium cerebelli
internal cerebral v
occipital bone
formation of great cerebral v
falx cerebri
straight sinus
superior sagittal sinus

14
13
12
11
10
9

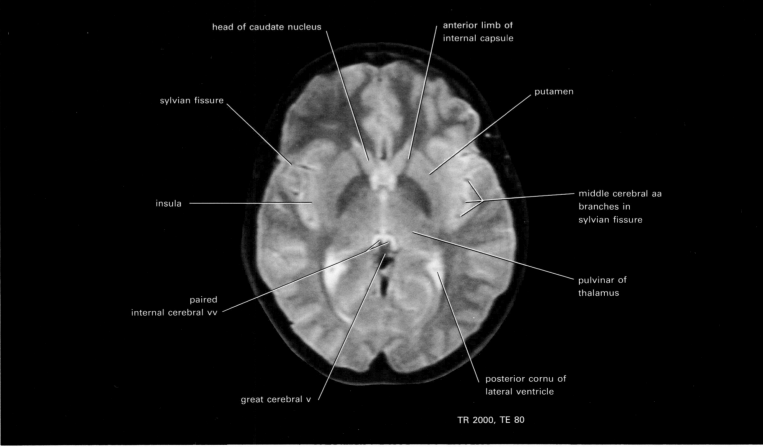

head of caudate nucleus
anterior limb of internal capsule
sylvian fissure
putamen
insula
middle cerebral aa branches in sylvian fissure
pulvinar of thalamus
paired internal cerebral vv
posterior cornu of lateral ventricle
great cerebral v

TR 2000, TE 80

Section 11 from above.

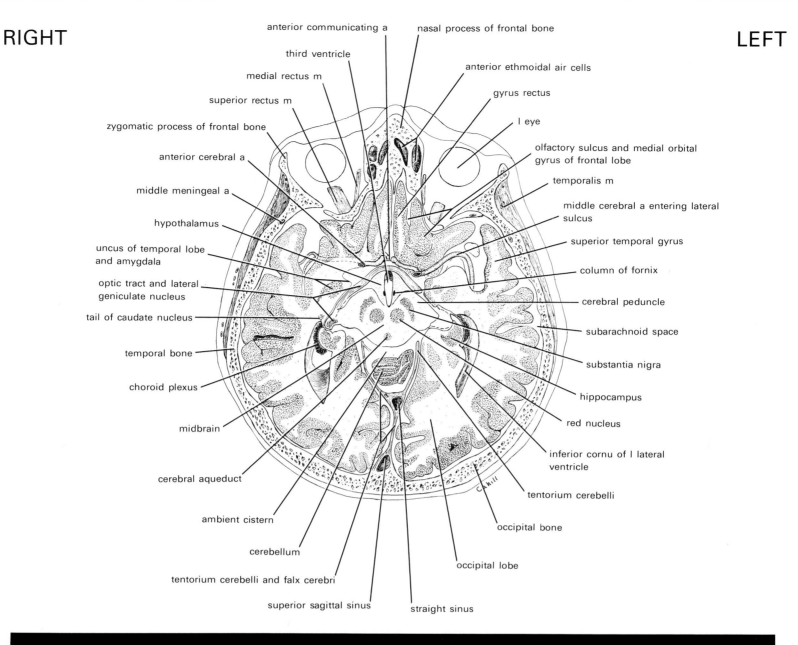

anterior communicating a
third ventricle
medial rectus m
superior rectus m
zygomatic process of frontal bone
anterior cerebral a
middle meningeal a
hypothalamus
uncus of temporal lobe and amygdala
optic tract and lateral geniculate nucleus
tail of caudate nucleus
temporal bone
choroid plexus
midbrain
cerebral aqueduct
ambient cistern
cerebellum
tentorium cerebelli and falx cerebri
superior sagittal sinus

nasal process of frontal bone
anterior ethmoidal air cells
gyrus rectus
l eye
olfactory sulcus and medial orbital gyrus of frontal lobe
temporalis m
middle cerebral a entering lateral sulcus
superior temporal gyrus
column of fornix
cerebral peduncle
subarachnoid space
substantia nigra
hippocampus
red nucleus
inferior cornu of l lateral ventricle
tentorium cerebelli
occipital bone
occipital lobe
straight sinus

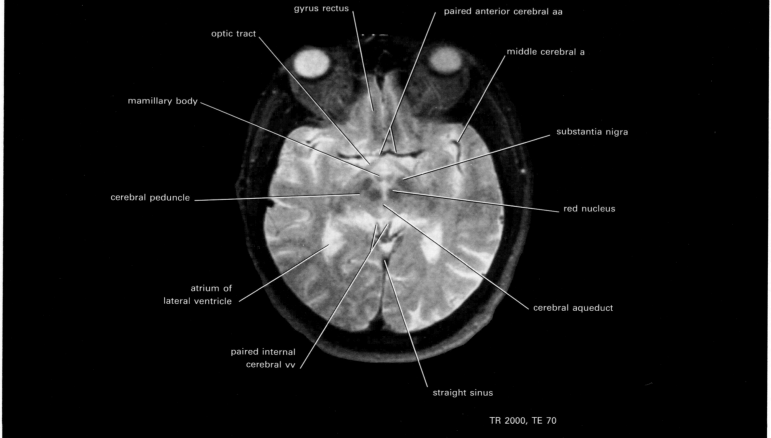

gyrus rectus
optic tract
mamillary body
cerebral peduncle
atrium of lateral ventricle
paired internal cerebral vv

paired anterior cerebral aa
middle cerebral a
substantia nigra
red nucleus
cerebral aqueduct
straight sinus

TR 2000, TE 70

Section 11 from below.

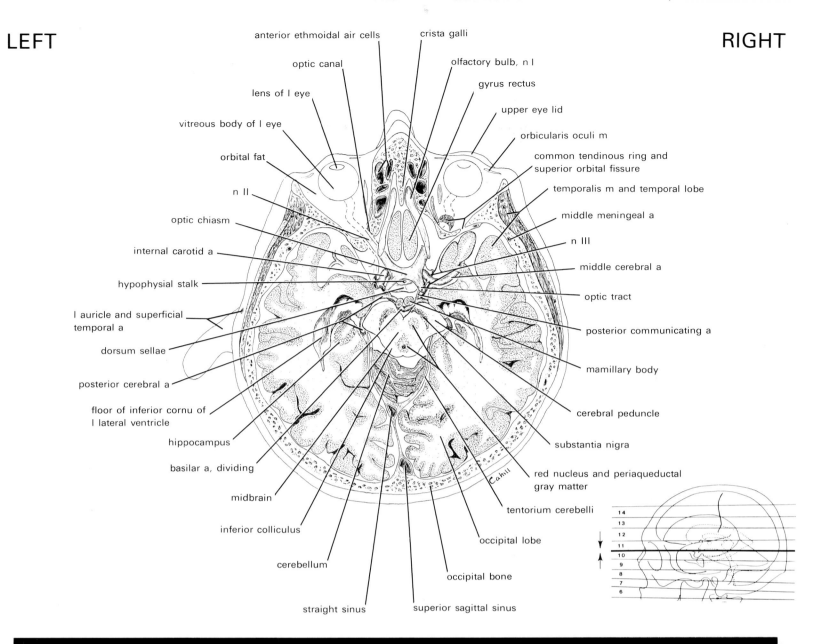

anterior ethmoidal air cells
crista galli
optic canal
olfactory bulb, n I
gyrus rectus
lens of l eye
upper eye lid
vitreous body of l eye
orbicularis oculi m
orbital fat
common tendinous ring and superior orbital fissure
n II
temporalis m and temporal lobe
optic chiasm
middle meningeal a
internal carotid a
n III
middle cerebral a
hypophysial stalk
optic tract
l auricle and superficial temporal a
posterior communicating a
dorsum sellae
posterior cerebral a
mamillary body
floor of inferior cornu of l lateral ventricle
cerebral peduncle
hippocampus
substantia nigra
basilar a, dividing
red nucleus and periaqueductal gray matter
midbrain
tentorium cerebelli
inferior colliculus
occipital lobe
cerebellum
occipital bone
straight sinus
superior sagittal sinus

14
13
12
11
10
9
8
7
6

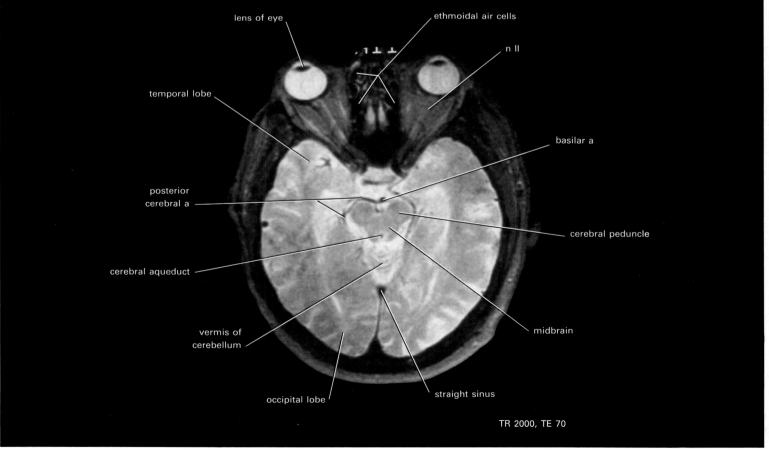

lens of eye
ethmoidal air cells
n II
temporal lobe
basilar a
posterior cerebral a
cerebral peduncle
cerebral aqueduct
midbrain
vermis of cerebellum
occipital lobe
straight sinus

TR 2000, TE 70

Section 10 from above.

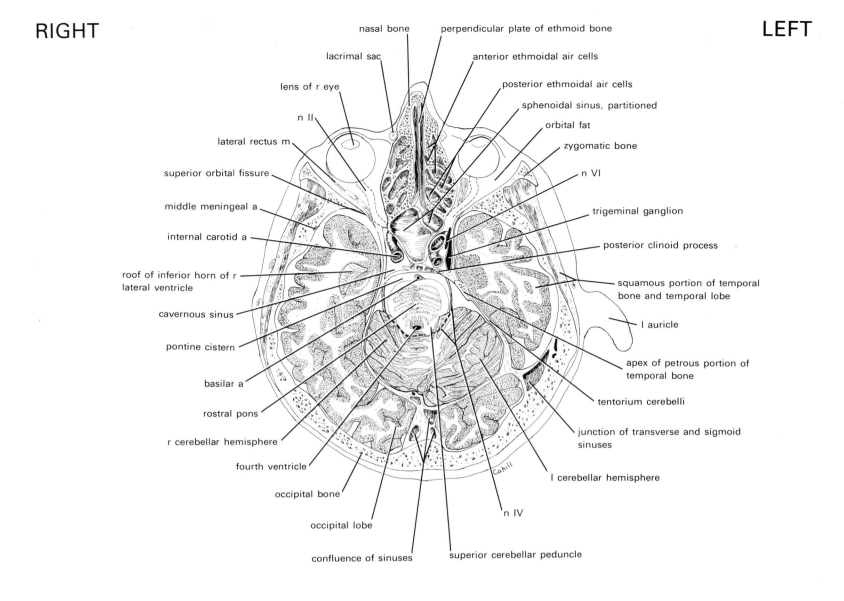

nasal bone
perpendicular plate of ethmoid bone
lacrimal sac
anterior ethmoidal air cells
lens of r eye
posterior ethmoidal air cells
n II
sphenoidal sinus, partitioned
lateral rectus m
orbital fat
superior orbital fissure
zygomatic bone
n VI
middle meningeal a
trigeminal ganglion
internal carotid a
posterior clinoid process
roof of inferior horn of r
lateral ventricle
squamous portion of temporal
bone and temporal lobe
cavernous sinus
l auricle
pontine cistern
apex of petrous portion of
temporal bone
basilar a
tentorium cerebelli
rostral pons
junction of transverse and sigmoid
sinuses
r cerebellar hemisphere
fourth ventricle
l cerebellar hemisphere
occipital bone
occipital lobe
n IV
confluence of sinuses
superior cerebellar peduncle

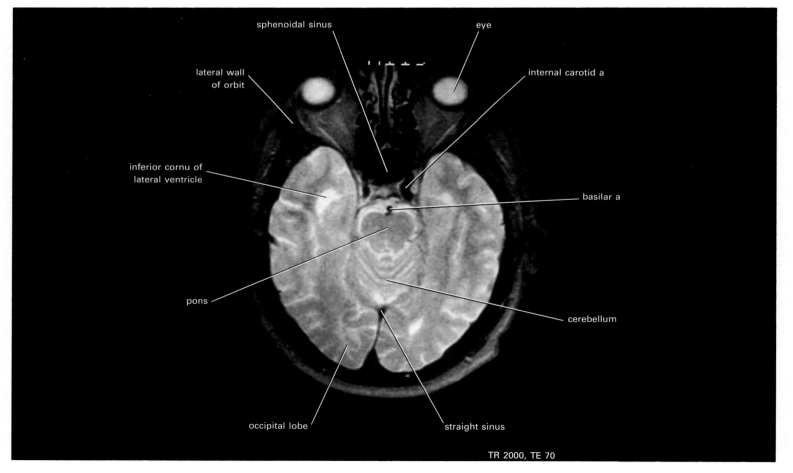

sphenoidal sinus
eye
lateral wall
of orbit
internal carotid a
inferior cornu of
lateral ventricle
basilar a
pons
cerebellum
occipital lobe
straight sinus
TR 2000, TE 70

Section 10 from below.

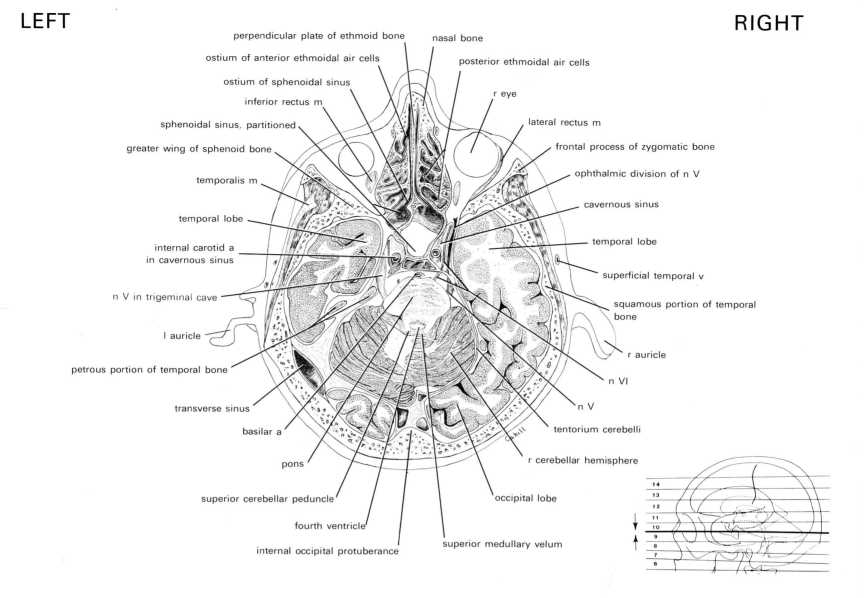

perpendicular plate of ethmoid bone

nasal bone

ostium of anterior ethmoidal air cells

posterior ethmoidal air cells

ostium of sphenoidal sinus

r eye

inferior rectus m

lateral rectus m

sphenoidal sinus, partitioned

frontal process of zygomatic bone

greater wing of sphenoid bone

ophthalmic division of n V

temporalis m

cavernous sinus

temporal lobe

temporal lobe

internal carotid a
in cavernous sinus

superficial temporal v

n V in trigeminal cave

squamous portion of temporal
bone

l auricle

r auricle

petrous portion of temporal bone

n VI

transverse sinus

n V

basilar a

tentorium cerebelli

pons

r cerebellar hemisphere

superior cerebellar peduncle

occipital lobe

fourth ventricle

internal occipital protuberance

superior medullary velum

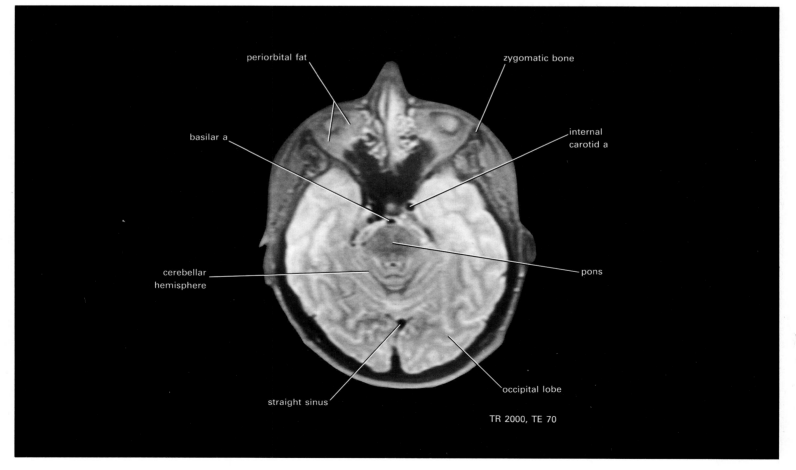

periorbital fat

zygomatic bone

basilar a

internal
carotid a

cerebellar
hemisphere

pons

straight sinus

occipital lobe

TR 2000, TE 70

Section 9 from above.

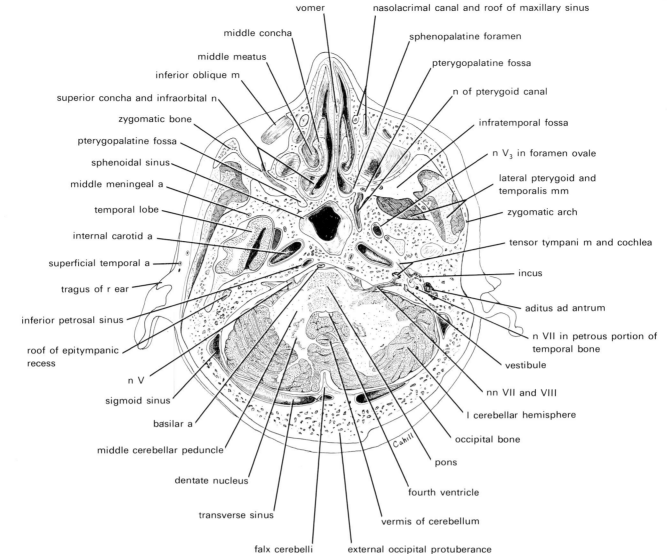

vomer
nasolacrimal canal and roof of maxillary sinus
middle concha
sphenopalatine foramen
middle meatus
pterygopalatine fossa
inferior oblique m
n of pterygoid canal
superior concha and infraorbital n
infratemporal fossa
zygomatic bone
pterygopalatine fossa
n V₃ in foramen ovale
sphenoidal sinus
lateral pterygoid and temporalis mm
middle meningeal a
zygomatic arch
temporal lobe
internal carotid a
tensor tympani m and cochlea
superficial temporal a
incus
tragus of r ear
aditus ad antrum
inferior petrosal sinus
n VII in petrous portion of temporal bone
roof of epitympanic recess
vestibule
n V
nn VII and VIII
sigmoid sinus
l cerebellar hemisphere
basilar a
occipital bone
middle cerebellar peduncle
pons
dentate nucleus
fourth ventricle
transverse sinus
vermis of cerebellum
falx cerebelli
external occipital protuberance

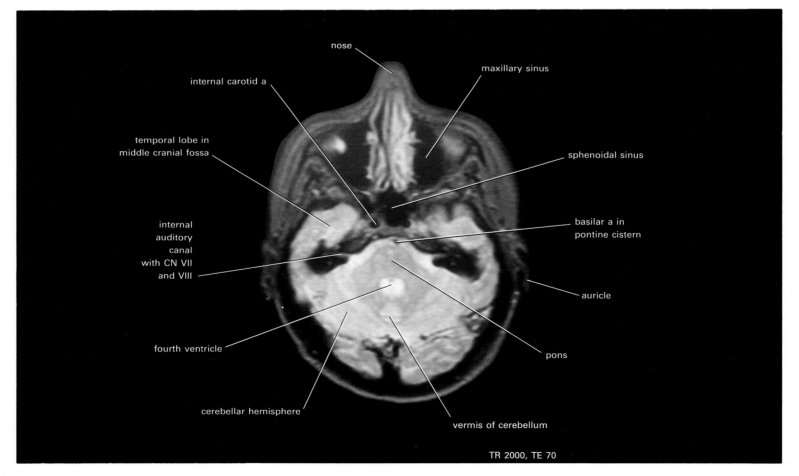

nose
internal carotid a
maxillary sinus
temporal lobe in middle cranial fossa
sphenoidal sinus
internal auditory canal with CN VII and VIII
basilar a in pontine cistern
auricle
fourth ventricle
pons
cerebellar hemisphere
vermis of cerebellum

TR 2000, TE 70

Section 9 from below.

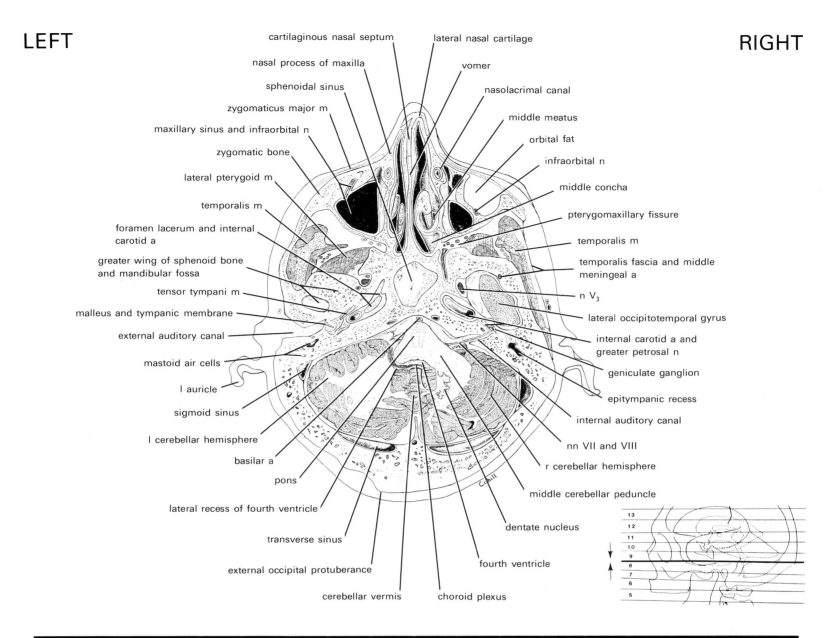

LEFT   RIGHT

cartilaginous nasal septum   lateral nasal cartilage
nasal process of maxilla   vomer
sphenoidal sinus   nasolacrimal canal
zygomaticus major m   middle meatus
maxillary sinus and infraorbital n   orbital fat
zygomatic bone   infraorbital n
lateral pterygoid m   middle concha
temporalis m   pterygomaxillary fissure
foramen lacerum and internal   temporalis m
carotid a   temporalis fascia and middle
greater wing of sphenoid bone   meningeal a
and mandibular fossa   n V₃
tensor tympani m   lateral occipitotemporal gyrus
malleus and tympanic membrane   internal carotid a and
external auditory canal   greater petrosal n
mastoid air cells   geniculate ganglion
l auricle   epitympanic recess
sigmoid sinus   internal auditory canal
l cerebellar hemisphere   nn VII and VIII
basilar a   r cerebellar hemisphere
pons   middle cerebellar peduncle
lateral recess of fourth ventricle   dentate nucleus
transverse sinus   fourth ventricle
external occipital protuberance
cerebellar vermis   choroid plexus

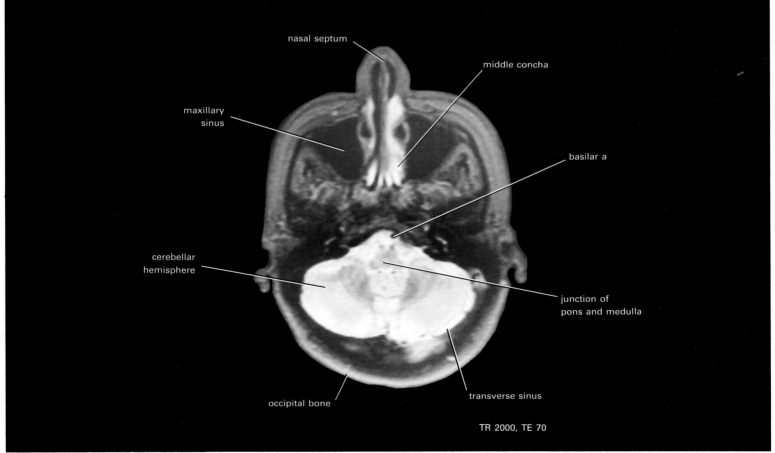

nasal septum
middle concha
maxillary
sinus
basilar a
cerebellar
hemisphere
junction of
pons and medulla
occipital bone   transverse sinus

TR 2000, TE 70

Section 8 from above.

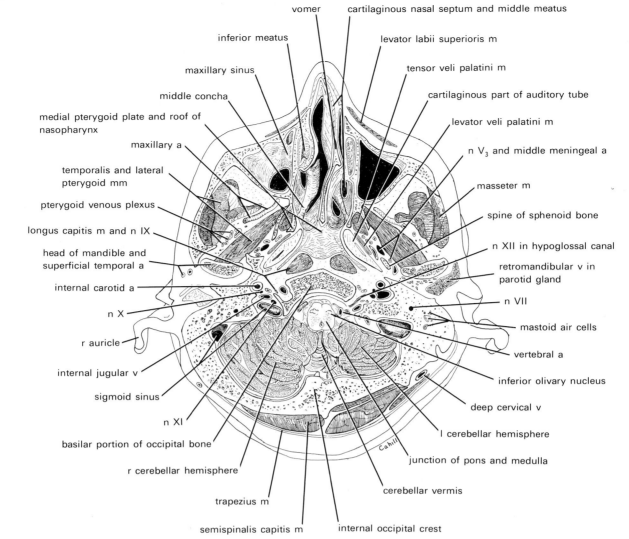

vomer
cartilaginous nasal septum and middle meatus
inferior meatus
levator labii superioris m
maxillary sinus
tensor veli palatini m
middle concha
cartilaginous part of auditory tube
medial pterygoid plate and roof of nasopharynx
levator veli palatini m
maxillary a
n V₃ and middle meningeal a
temporalis and lateral pterygoid mm
masseter m
pterygoid venous plexus
spine of sphenoid bone
longus capitis m and n IX
n XII in hypoglossal canal
head of mandible and superficial temporal a
retromandibular v in parotid gland
internal carotid a
n VII
n X
mastoid air cells
r auricle
vertebral a
internal jugular v
inferior olivary nucleus
sigmoid sinus
deep cervical v
n XI
l cerebellar hemisphere
basilar portion of occipital bone
junction of pons and medulla
r cerebellar hemisphere
cerebellar vermis
trapezius m
semispinalis capitis m
internal occipital crest

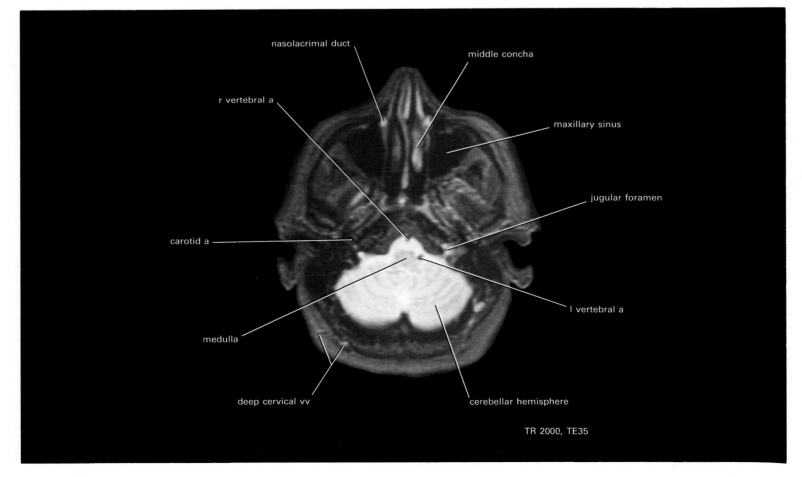

nasolacrimal duct
middle concha
r vertebral a
maxillary sinus
carotid a
jugular foramen
l vertebral a
medulla
deep cervical vv
cerebellar hemisphere
TR 2000, TE35

Section 8 from below.

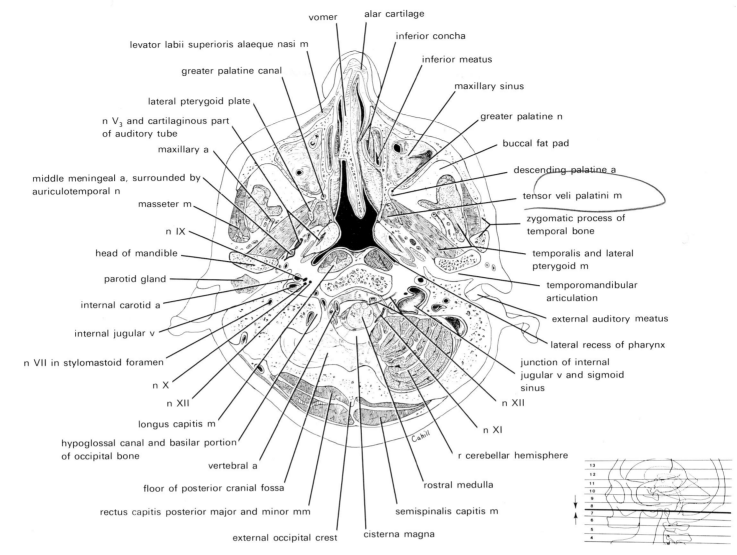

vomer
alar cartilage
levator labii superioris alaeque nasi m
inferior concha
greater palatine canal
inferior meatus
lateral pterygoid plate
maxillary sinus
n V₃ and cartilaginous part
of auditory tube
greater palatine n
maxillary a
buccal fat pad
middle meningeal a, surrounded by
auriculotemporal n
descending palatine a
tensor veli palatini m
masseter m
zygomatic process of
temporal bone
n IX
temporalis and lateral
pterygoid m
head of mandible
temporomandibular
articulation
parotid gland
internal carotid a
external auditory meatus
internal jugular v
lateral recess of pharynx
n VII in stylomastoid foramen
junction of internal
jugular v and sigmoid
n X
sinus
n XII
n XII
longus capitis m
n XI
hypoglossal canal and basilar portion
of occipital bone
r cerebellar hemisphere
vertebral a
rostral medulla
floor of posterior cranial fossa
rectus capitis posterior major and minor mm
semispinalis capitis m
external occipital crest
cisterna magna

Cahill

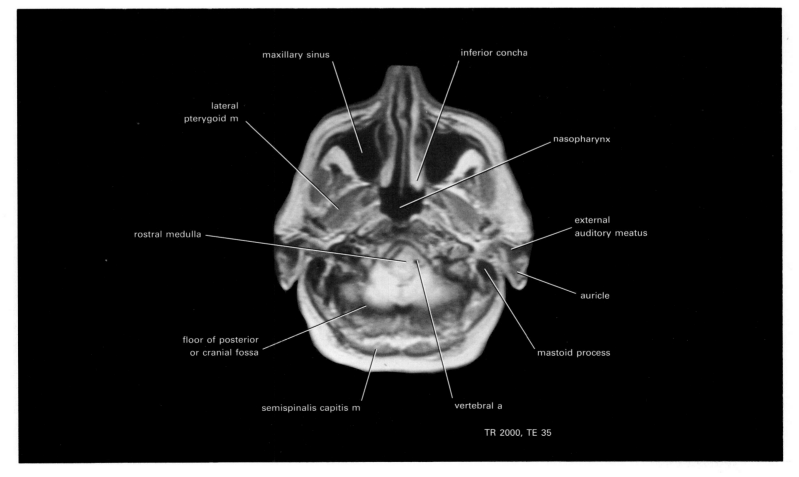

maxillary sinus
inferior concha
lateral
pterygoid m
nasopharynx
rostral medulla
external
auditory meatus
auricle
floor of posterior
or cranial fossa
mastoid process
semispinalis capitis m
vertebral a

TR 2000, TE 35

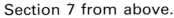

Section 7 from above.

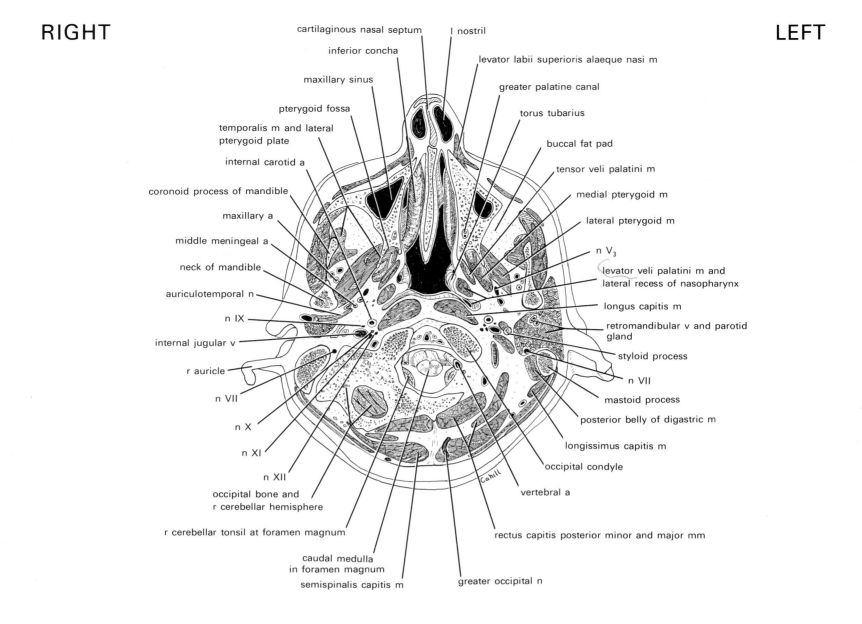

cartilaginous nasal septum
l nostril
inferior concha
levator labii superioris alaeque nasi m
maxillary sinus
greater palatine canal
pterygoid fossa
torus tubarius
temporalis m and lateral pterygoid plate
buccal fat pad
internal carotid a
tensor veli palatini m
coronoid process of mandible
medial pterygoid m
maxillary a
lateral pterygoid m
middle meningeal a
n V₃
neck of mandible
levator veli palatini m and lateral recess of nasopharynx
auriculotemporal n
longus capitis m
n IX
retromandibular v and parotid gland
internal jugular v
styloid process
r auricle
n VII
n VII
mastoid process
n X
posterior belly of digastric m
n XI
longissimus capitis m
n XII
occipital condyle
occipital bone and r cerebellar hemisphere
vertebral a
r cerebellar tonsil at foramen magnum
rectus capitis posterior minor and major mm
caudal medulla in foramen magnum
greater occipital n
semispinalis capitis m

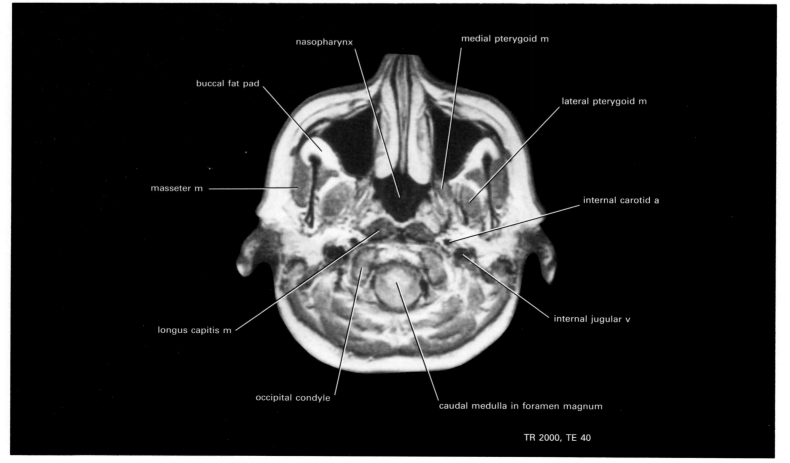

nasopharynx
medial pterygoid m
buccal fat pad
lateral pterygoid m
masseter m
internal carotid a
internal jugular v
longus capitis m
occipital condyle
caudal medulla in foramen magnum
TR 2000, TE 40

Section 7 from below.

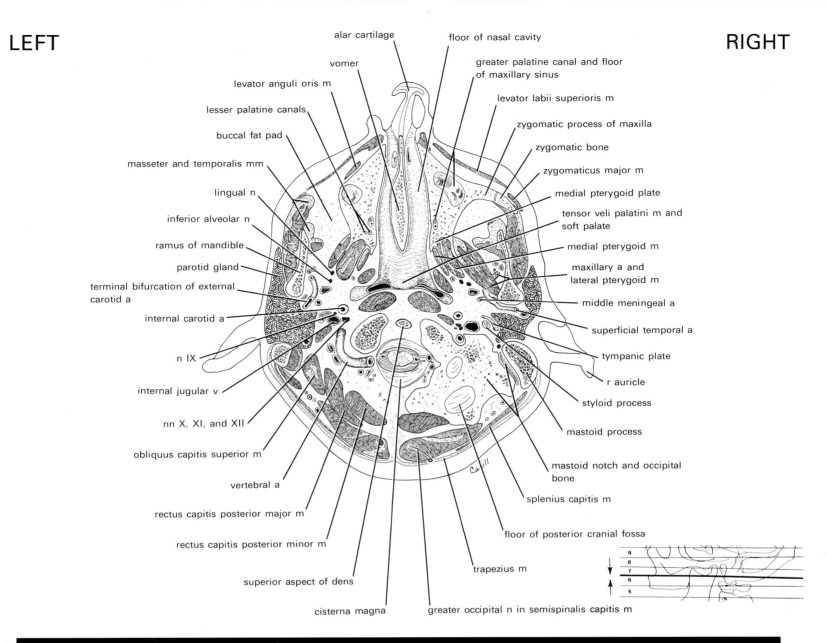

alar cartilage
floor of nasal cavity
vomer
greater palatine canal and floor of maxillary sinus
levator anguli oris m
levator labii superioris m
lesser palatine canals
zygomatic process of maxilla
buccal fat pad
zygomatic bone
masseter and temporalis mm
zygomaticus major m
lingual n
medial pterygoid plate
inferior alveolar n
tensor veli palatini m and soft palate
ramus of mandible
medial pterygoid m
parotid gland
maxillary a and lateral pterygoid m
terminal bifurcation of external carotid a
middle meningeal a
internal carotid a
superficial temporal a
n IX
tympanic plate
internal jugular v
r auricle
nn X, XI, and XII
styloid process
obliquus capitis superior m
mastoid process
vertebral a
mastoid notch and occipital bone
rectus capitis posterior major m
splenius capitis m
rectus capitis posterior minor m
floor of posterior cranial fossa
superior aspect of dens
trapezius m
cisterna magna
greater occipital n in semispinalis capitis m

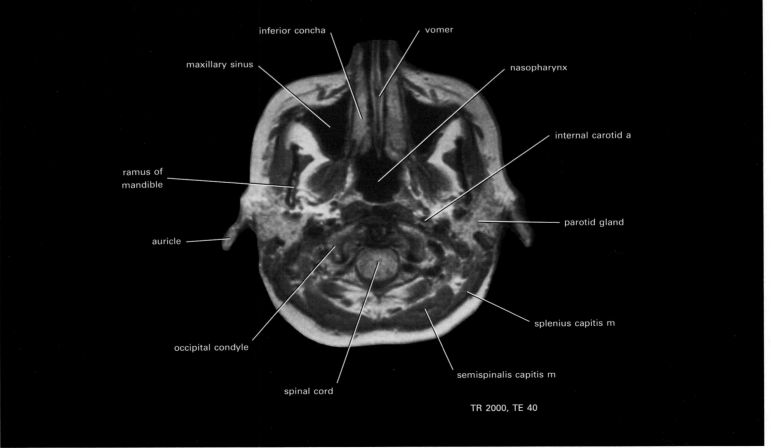

inferior concha
vomer
maxillary sinus
nasopharynx
internal carotid a
ramus of mandible
parotid gland
auricle
splenius capitis m
occipital condyle
semispinalis capitis m
spinal cord
TR 2000, TE 40

Section 6 from above.

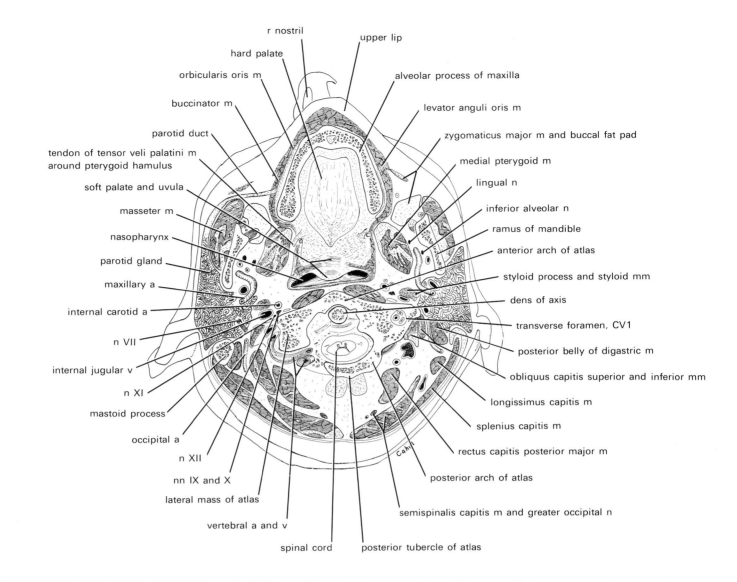

r nostril
hard palate
orbicularis oris m
buccinator m
parotid duct
tendon of tensor veli palatini m
around pterygoid hamulus
soft palate and uvula
masseter m
nasopharynx
parotid gland
maxillary a
internal carotid a
n VII
internal jugular v
n XI
mastoid process
occipital a
n XII
nn IX and X
lateral mass of atlas
vertebral a and v
spinal cord

upper lip
alveolar process of maxilla
levator anguli oris m
zygomaticus major m and buccal fat pad
medial pterygoid m
lingual n
inferior alveolar n
ramus of mandible
anterior arch of atlas
styloid process and styloid mm
dens of axis
transverse foramen, CV1
posterior belly of digastric m
obliquus capitis superior and inferior mm
longissimus capitis m
splenius capitis m
rectus capitis posterior major m
posterior arch of atlas
semispinalis capitis m and greater occipital n
posterior tubercle of atlas

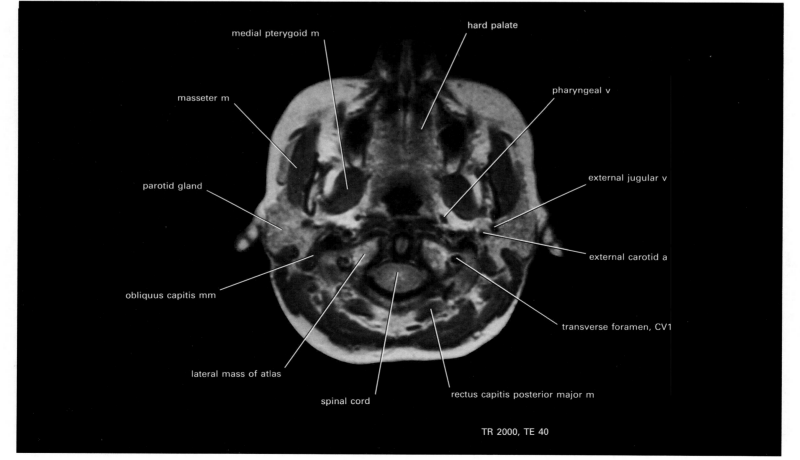

medial pterygoid m
masseter m
parotid gland
obliquus capitis mm
lateral mass of atlas
spinal cord

hard palate
pharyngeal v
external jugular v
external carotid a
transverse foramen, CV1
rectus capitis posterior major m

TR 2000, TE 40

Section 6 from below.

LEFT   RIGHT

incisive fossa   upper lip

oral cavity

mucosa of hard palate

orbicularis oris m

alveolar process of edentulous maxilla

vestibule of mouth

levator anguli oris m

pterygomandibular raphe

buccinator m

medial pterygoid m

parotid duct

soft palate and nasopharynx

insertions of temporalis m

masseter m and ramus of
mandible

superior constrictor m and lingual n

longus capitis m and anterior
tubercle of atlas

inferior alveolar n and a

internal carotid a

pterygoid venous plexus

external carotid a

styloglossus m

n IX

stylopharyngeus m

internal jugular v

stylohyoid m

n XI

lobule of r ear

n XII

parotid gland and occipital a

vertebral a in transverse
foramen, CV1

mastoid process

lateral mass of atlas

obliquus capitis inferior m

transverse ligament of atlas

longissimus capitis m

greater occipital n

vertebral a

rectus capitis posterior minor m

splenius capitis m

posterior tubercle of atlas   spinal cord

semispinalis capitis m

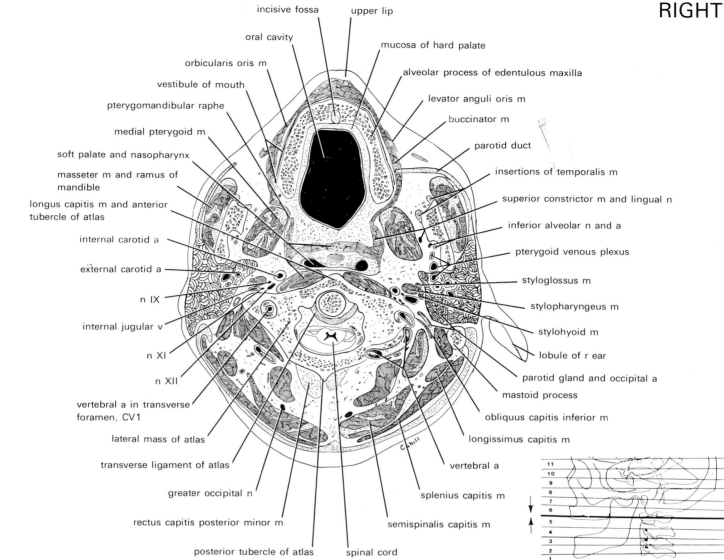

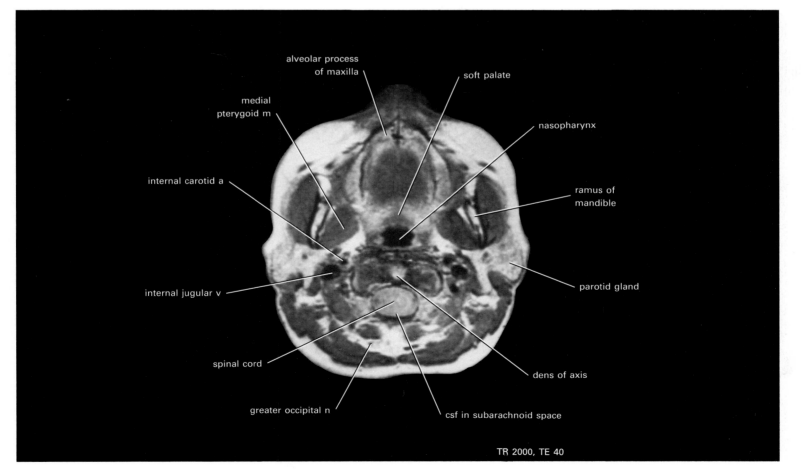

alveolar process
of maxilla   soft palate

medial
pterygoid m

nasopharynx

internal carotid a

ramus of
mandible

internal jugular v

parotid gland

spinal cord

dens of axis

greater occipital n   csf in subarachnoid space

TR 2000, TE 40

Section 5 from above.

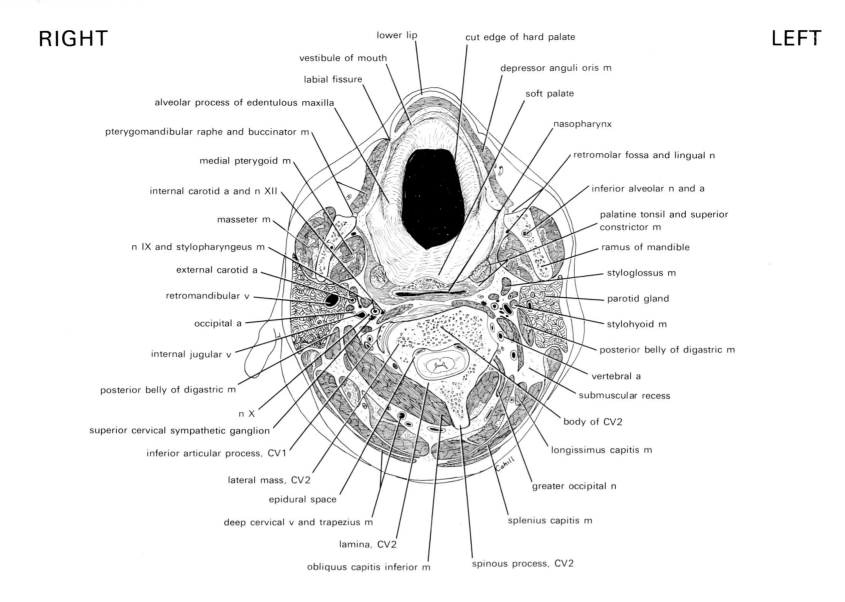

lower lip
cut edge of hard palate
vestibule of mouth
depressor anguli oris m
labial fissure
soft palate
alveolar process of edentulous maxilla
nasopharynx
pterygomandibular raphe and buccinator m
retromolar fossa and lingual n
medial pterygoid m
inferior alveolar n and a
internal carotid a and n XII
palatine tonsil and superior constrictor m
masseter m
ramus of mandible
n IX and stylopharyngeus m
styloglossus m
external carotid a
parotid gland
retromandibular v
stylohyoid m
occipital a
posterior belly of digastric m
internal jugular v
vertebral a
posterior belly of digastric m
submuscular recess
n X
body of CV2
superior cervical sympathetic ganglion
longissimus capitis m
inferior articular process, CV1
greater occipital n
lateral mass, CV2
epidural space
splenius capitis m
deep cervical v and trapezius m
lamina, CV2
obliquus capitis inferior m
spinous process, CV2

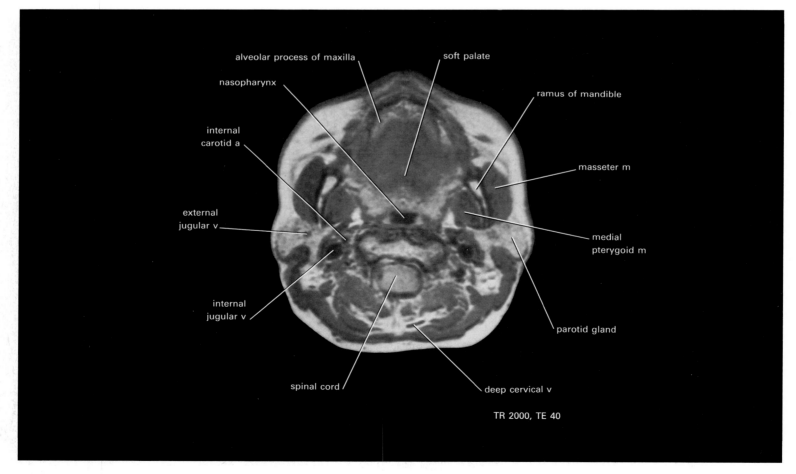

alveolar process of maxilla
soft palate
nasopharynx
ramus of mandible
internal carotid a
masseter m
external jugular v
medial pterygoid m
internal jugular v
parotid gland
spinal cord
deep cervical v

TR 2000, TE 40

Section 5 from below.

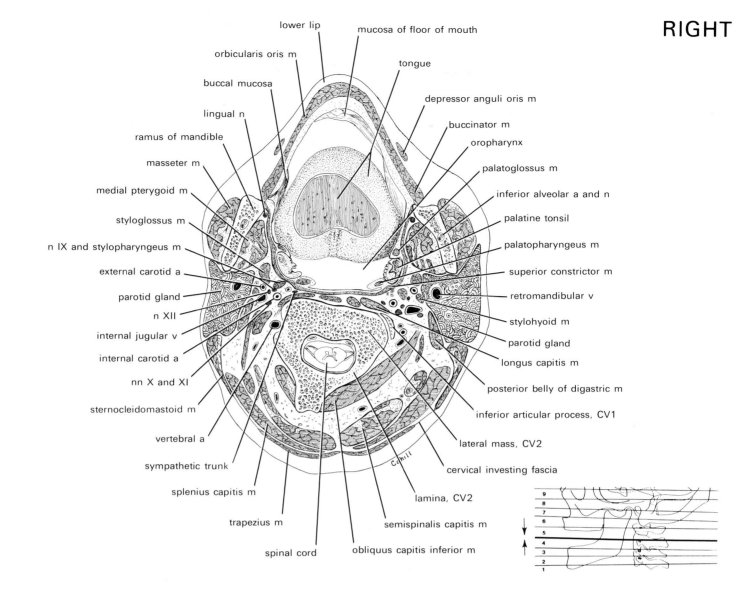

lower lip
mucosa of floor of mouth
orbicularis oris m
tongue
buccal mucosa
depressor anguli oris m
lingual n
buccinator m
ramus of mandible
oropharynx
masseter m
palatoglossus m
medial pterygoid m
inferior alveolar a and n
styloglossus m
palatine tonsil
n IX and stylopharyngeus m
palatopharyngeus m
external carotid a
superior constrictor m
parotid gland
retromandibular v
n XII
stylohyoid m
internal jugular v
parotid gland
internal carotid a
longus capitis m
nn X and XI
posterior belly of digastric m
sternocleidomastoid m
inferior articular process, CV1
vertebral a
lateral mass, CV2
sympathetic trunk
cervical investing fascia
splenius capitis m
lamina, CV2
trapezius m
semispinalis capitis m
spinal cord
obliquus capitis inferior m

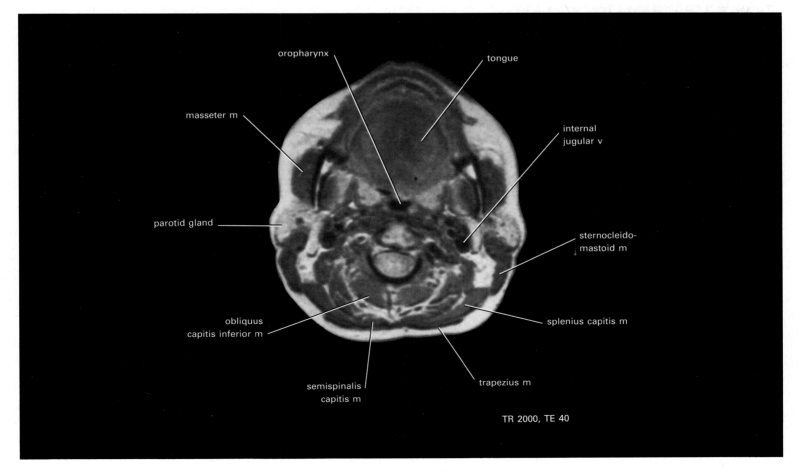

oropharynx
tongue
masseter m
internal jugular v
parotid gland
sternocleido-mastoid m
obliquus capitis inferior m
splenius capitis m
semispinalis capitis m
trapezius m

TR 2000, TE 40

Section 4 from above.

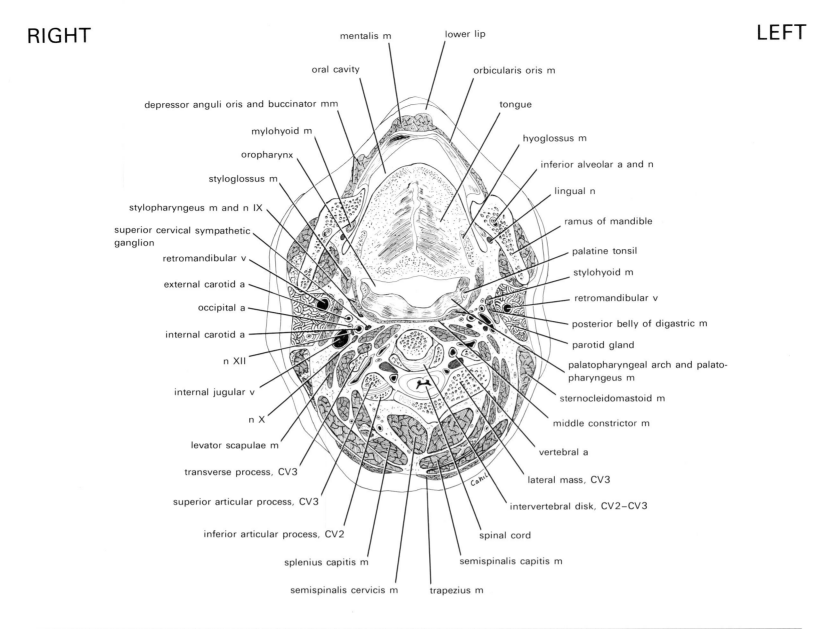

mentalis m
lower lip
oral cavity
orbicularis oris m
depressor anguli oris and buccinator mm
tongue
mylohyoid m
hyoglossus m
oropharynx
inferior alveolar a and n
styloglossus m
lingual n
stylopharyngeus m and n IX
ramus of mandible
superior cervical sympathetic ganglion
palatine tonsil
retromandibular v
stylohyoid m
external carotid a
retromandibular v
occipital a
posterior belly of digastric m
internal carotid a
parotid gland
n XII
palatopharyngeal arch and palato-pharyngeus m
internal jugular v
sternocleidomastoid m
n X
middle constrictor m
levator scapulae m
vertebral a
transverse process, CV3
lateral mass, CV3
superior articular process, CV3
intervertebral disk, CV2–CV3
inferior articular process, CV2
spinal cord
splenius capitis m
semispinalis capitis m
semispinalis cervicis m
trapezius m

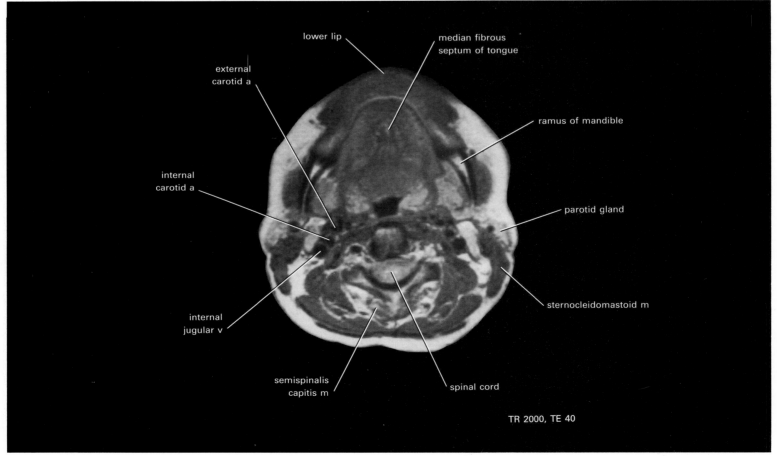

lower lip
median fibrous septum of tongue
external carotid a
ramus of mandible
internal carotid a
parotid gland
internal jugular v
sternocleidomastoid m
semispinalis capitis m
spinal cord
TR 2000, TE 40

Section 4 from below.

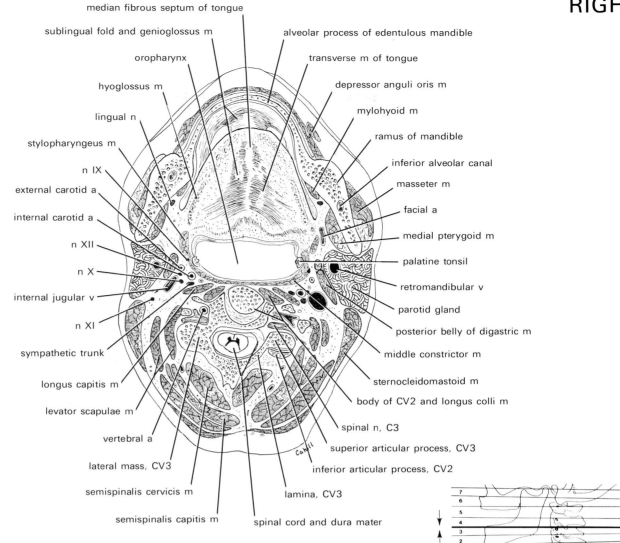

median fibrous septum of tongue

sublingual fold and genioglossus m

oropharynx

hyoglossus m

lingual n

stylopharyngeus m

n IX

external carotid a

internal carotid a

n XII

n X

internal jugular v

n XI

sympathetic trunk

longus capitis m

levator scapulae m

vertebral a

lateral mass, CV3

semispinalis cervicis m

semispinalis capitis m

alveolar process of edentulous mandible

transverse m of tongue

depressor anguli oris m

mylohyoid m

ramus of mandible

inferior alveolar canal

masseter m

facial a

medial pterygoid m

palatine tonsil

retromandibular v

parotid gland

posterior belly of digastric m

middle constrictor m

sternocleidomastoid m

body of CV2 and longus colli m

spinal n, C3

superior articular process, CV3

inferior articular process, CV2

lamina, CV3

spinal cord and dura mater

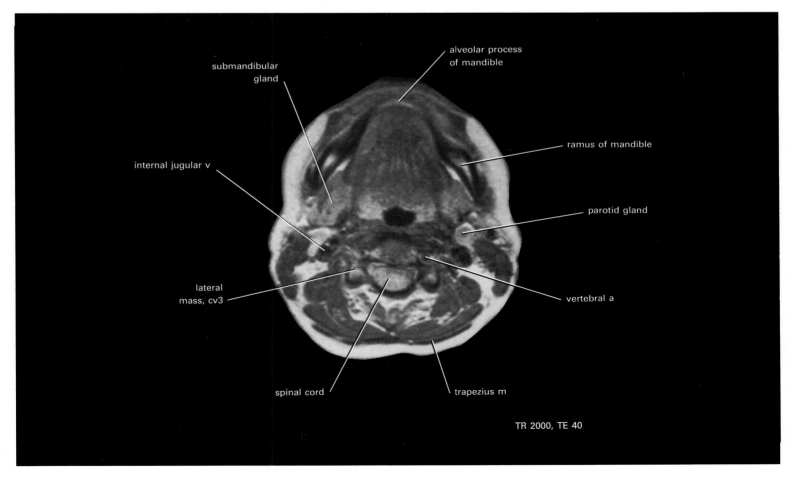

alveolar process
of mandible

submandibular
gland

internal jugular v

lateral
mass, cv3

spinal cord

ramus of mandible

parotid gland

vertebral a

trapezius m

TR 2000, TE 40

Section 3 from above.

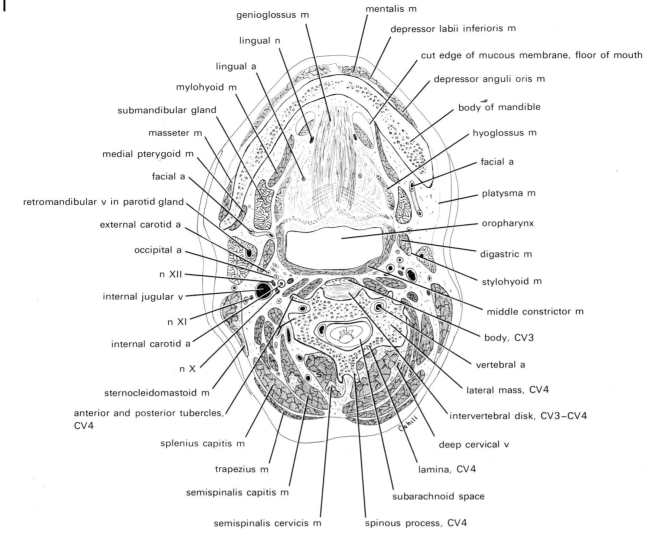

genioglossus m
mentalis m
lingual n
depressor labii inferioris m
lingual a
cut edge of mucous membrane, floor of mouth
mylohyoid m
depressor anguli oris m
submandibular gland
body of mandible
masseter m
hyoglossus m
medial pterygoid m
facial a
facial a
platysma m
retromandibular v in parotid gland
oropharynx
external carotid a
occipital a
digastric m
n XII
stylohyoid m
internal jugular v
middle constrictor m
n XI
body, CV3
internal carotid a
vertebral a
n X
lateral mass, CV4
sternocleidomastoid m
intervertebral disk, CV3–CV4
anterior and posterior tubercles, CV4
deep cervical v
splenius capitis m
lamina, CV4
trapezius m
subarachnoid space
semispinalis capitis m
spinous process, CV4
semispinalis cervicis m

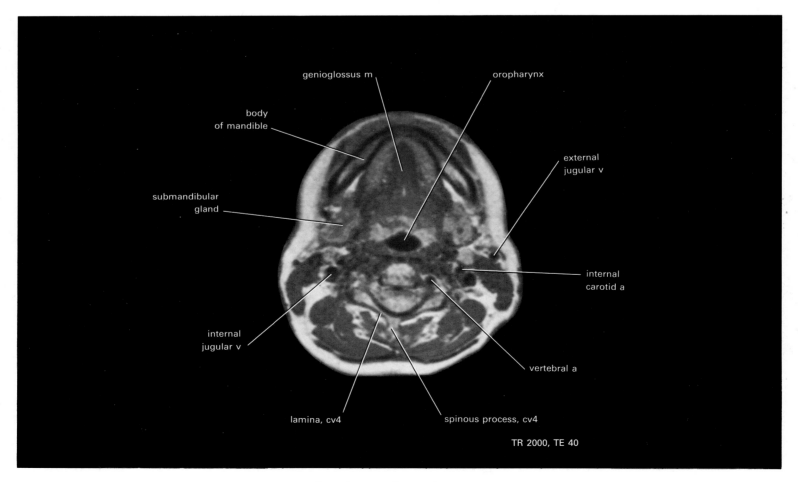

genioglossus m
oropharynx
body of mandible
external jugular v
submandibular gland
internal carotid a
internal jugular v
vertebral a
lamina, cv4
spinous process, cv4

TR 2000, TE 40

Section 3 from below.

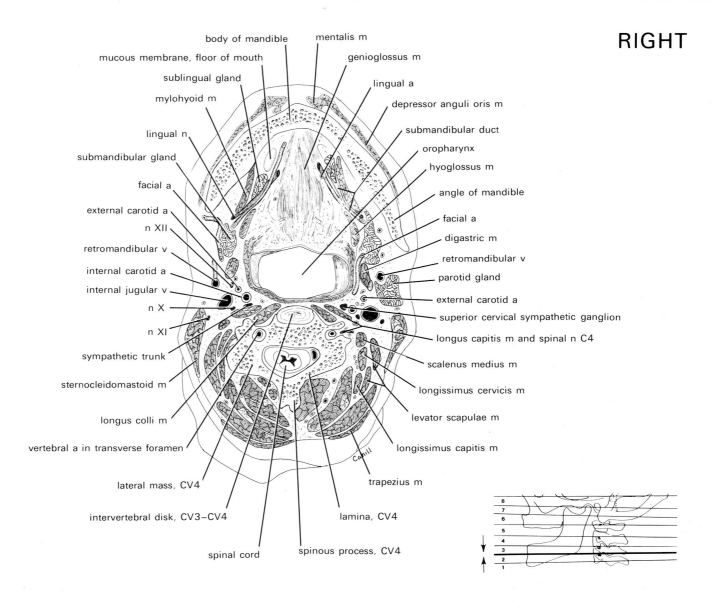

body of mandible
mentalis m
mucous membrane, floor of mouth
genioglossus m
sublingual gland
lingual a
mylohyoid m
depressor anguli oris m
lingual n
submandibular duct
submandibular gland
oropharynx
facial a
hyoglossus m
external carotid a
angle of mandible
n XII
facial a
retromandibular v
digastric m
internal carotid a
retromandibular v
internal jugular v
parotid gland
n X
external carotid a
n XI
superior cervical sympathetic ganglion
sympathetic trunk
longus capitis m and spinal n C4
sternocleidomastoid m
scalenus medius m
longus colli m
longissimus cervicis m
vertebral a in transverse foramen
levator scapulae m
longissimus capitis m
lateral mass, CV4
trapezius m
intervertebral disk, CV3–CV4
lamina, CV4
spinal cord
spinous process, CV4

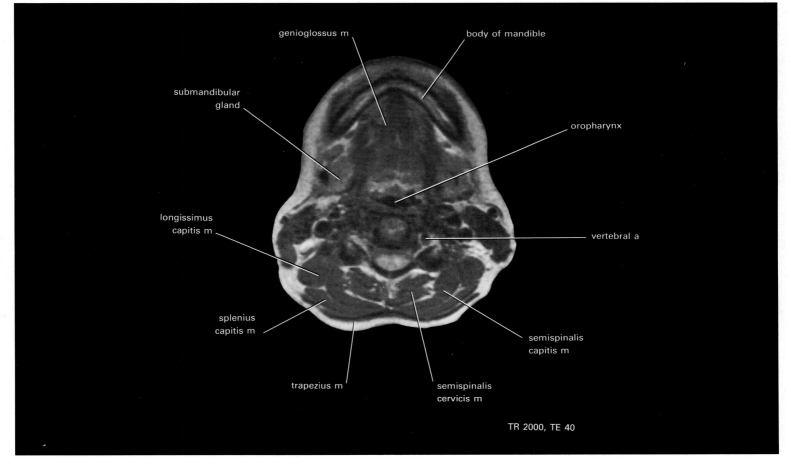

genioglossus m
body of mandible
submandibular gland
oropharynx
longissimus capitis m
vertebral a
splenius capitis m
semispinalis capitis m
trapezius m
semispinalis cervicis m

TR 2000, TE 40

Section 2 from above.

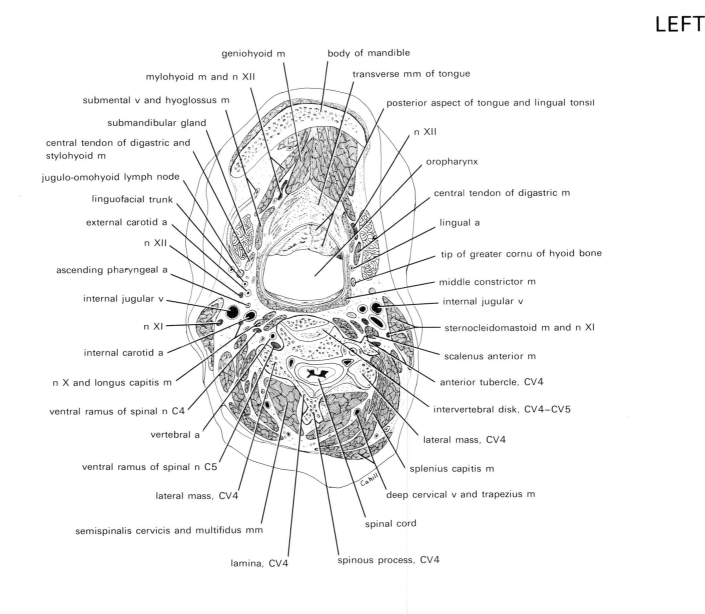

geniohyoid m
body of mandible
mylohyoid m and n XII
transverse mm of tongue
submental v and hyoglossus m
posterior aspect of tongue and lingual tonsil
submandibular gland
n XII
central tendon of digastric and stylohyoid m
oropharynx
jugulo-omohyoid lymph node
central tendon of digastric m
linguofacial trunk
lingual a
external carotid a
tip of greater cornu of hyoid bone
n XII
ascending pharyngeal a
middle constrictor m
internal jugular v
internal jugular v
n XI
sternocleidomastoid m and n XI
internal carotid a
scalenus anterior m
n X and longus capitis m
anterior tubercle, CV4
ventral ramus of spinal n C4
intervertebral disk, CV4–CV5
vertebral a
lateral mass, CV4
ventral ramus of spinal n C5
splenius capitis m
lateral mass, CV4
deep cervical v and trapezius m
semispinalis cervicis and multifidus mm
spinal cord
lamina, CV4
spinous process, CV4

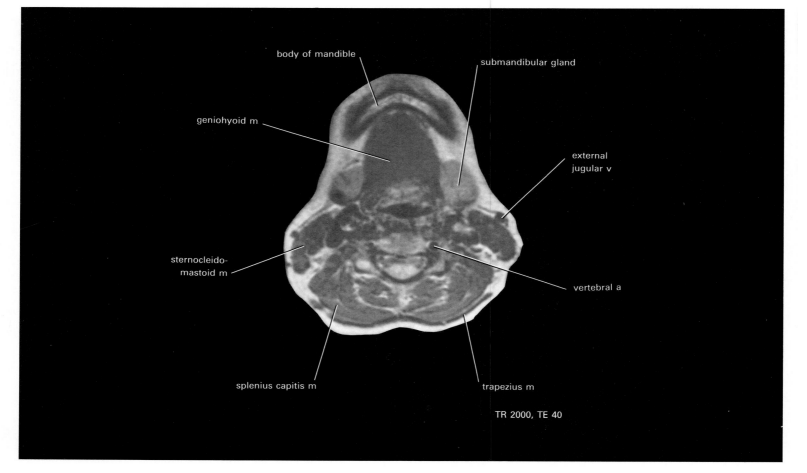

body of mandible
submandibular gland
geniohyoid m
external jugular v
sternocleido-mastoid m
vertebral a
splenius capitis m
trapezius m
TR 2000, TE 40

Section 2 from below.

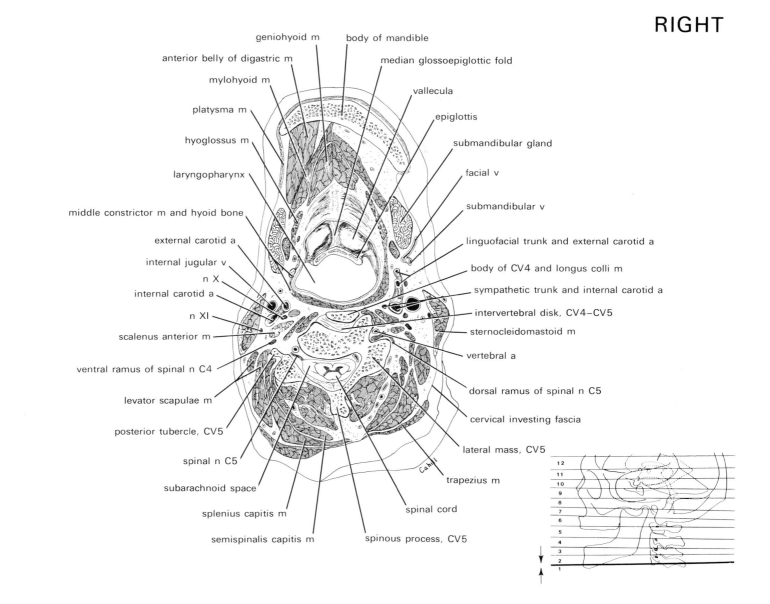

geniohyoid m
anterior belly of digastric m
mylohyoid m
platysma m
hyoglossus m
laryngopharynx
middle constrictor m and hyoid bone
external carotid a
internal jugular v
n X
internal carotid a
n XI
scalenus anterior m
ventral ramus of spinal n C4
levator scapulae m
posterior tubercle, CV5
spinal n C5
subarachnoid space
splenius capitis m
semispinalis capitis m

body of mandible
median glossoepiglottic fold
vallecula
epiglottis
submandibular gland
facial v
submandibular v
linguofacial trunk and external carotid a
body of CV4 and longus colli m
sympathetic trunk and internal carotid a
intervertebral disk, CV4–CV5
sternocleidomastoid m
vertebral a
dorsal ramus of spinal n C5
cervical investing fascia
lateral mass, CV5
trapezius m
spinal cord
spinous process, CV5

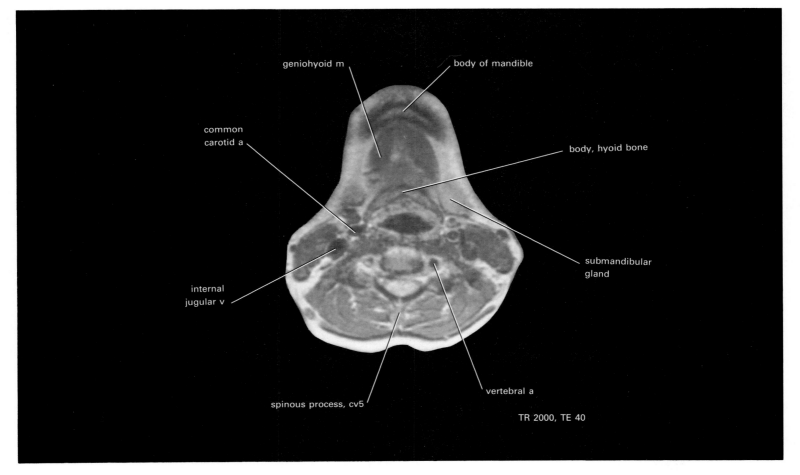

geniohyoid m
body of mandible
common carotid a
body, hyoid bone
internal jugular v
submandibular gland
vertebral a
spinous process, cv5

TR 2000, TE 40

Section 1 from above.

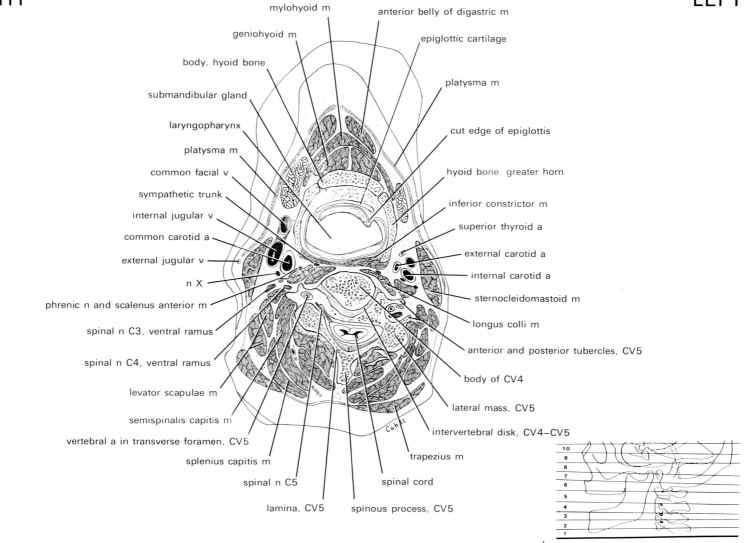

mylohyoid m
geniohyoid m
body, hyoid bone
submandibular gland
laryngopharynx
platysma m
common facial v
sympathetic trunk
internal jugular v
common carotid a
external jugular v
n X
phrenic n and scalenus anterior m
spinal n C3, ventral ramus
spinal n C4, ventral ramus
levator scapulae m
semispinalis capitis m
vertebral a in transverse foramen, CV5
splenius capitis m
spinal n C5
lamina, CV5

anterior belly of digastric m
epiglottic cartilage
platysma m
cut edge of epiglottis
hyoid bone, greater horn
inferior constrictor m
superior thyroid a
external carotid a
internal carotid a
sternocleidomastoid m
longus colli m
anterior and posterior tubercles, CV5
body of CV4
lateral mass, CV5
intervertebral disk, CV4–CV5
trapezius m
spinal cord
spinous process, CV5

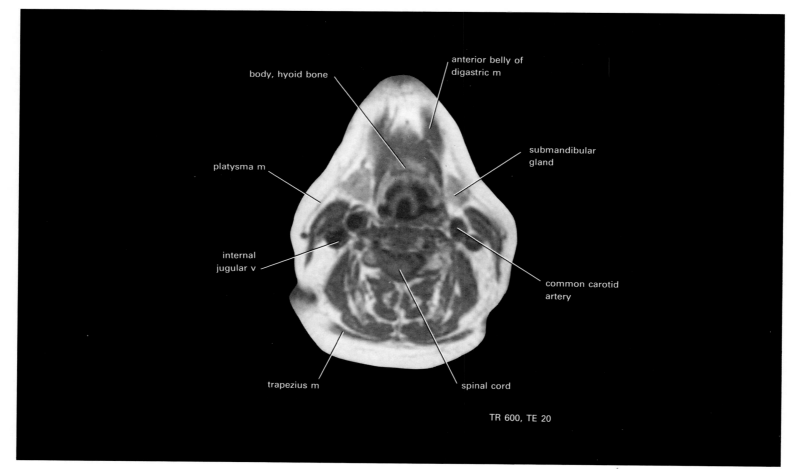

body, hyoid bone
anterior belly of digastric m
platysma m
submandibular gland
internal jugular v
common carotid artery
trapezius m
spinal cord

TR 600, TE 20

Section 1 from below.

Section 1 from below.

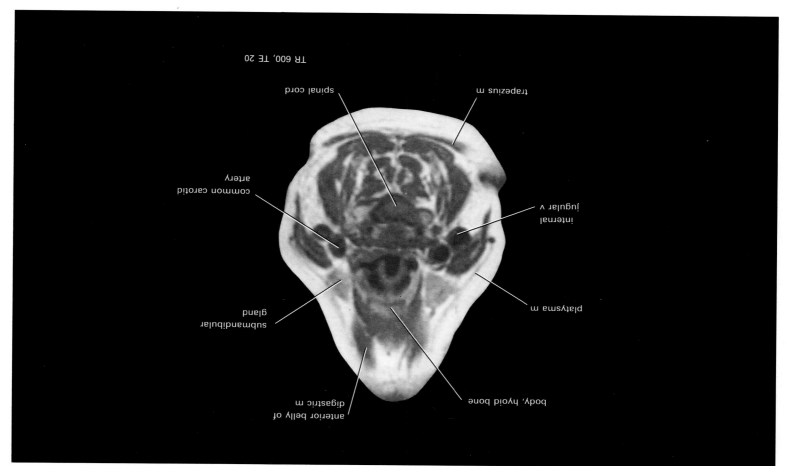

TR 600, TE 20

spinal cord

trapezius m

common carotid
artery

internal
jugular v

platysma m

submandibular
gland

body, hyoid bone

anterior belly of
digastric m

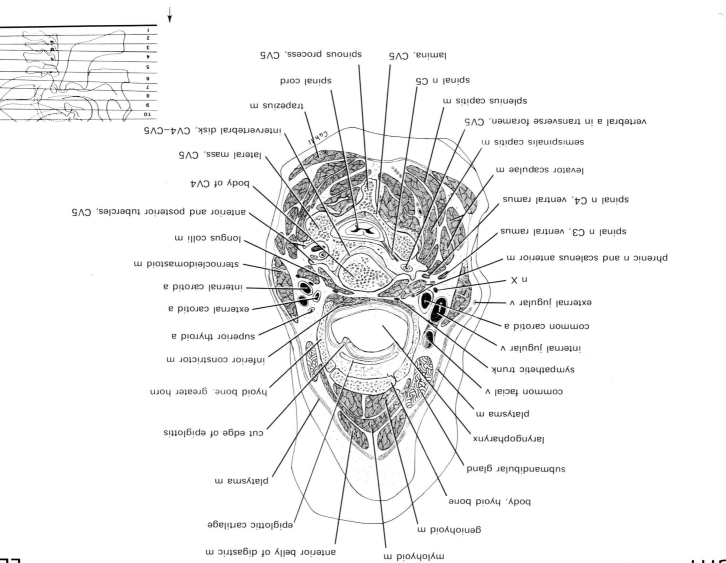

spinous process, CV5

lamina, CV5

spinal cord

spinal n C5

trapezius m

splenius capitis m

intervertebral disk, CV4–CV5

vertebral a in transverse foramen, CV5

lateral mass, CV5

semispinalis capitis m

body of CV4

levator scapulae m

anterior and posterior tubercles, CV5

spinal n C4, ventral ramus

longus colli m

spinal n C3, ventral ramus

sternocleidomastoid m

phrenic n and scalenus anterior m

internal carotid a

n X

external carotid a

external jugular v

superior thyroid a

common carotid a

inferior constrictor m

internal jugular v

hyoid bone, greater horn

sympathetic trunk

cut edge of epiglottis

common facial v

platysma m

platysma m

laryngopharynx

submandibular gland

body, hyoid bone

epiglottic cartilage

geniohyoid m

anterior belly of digastric m

mylohyoid m

LEFT

RIGHT

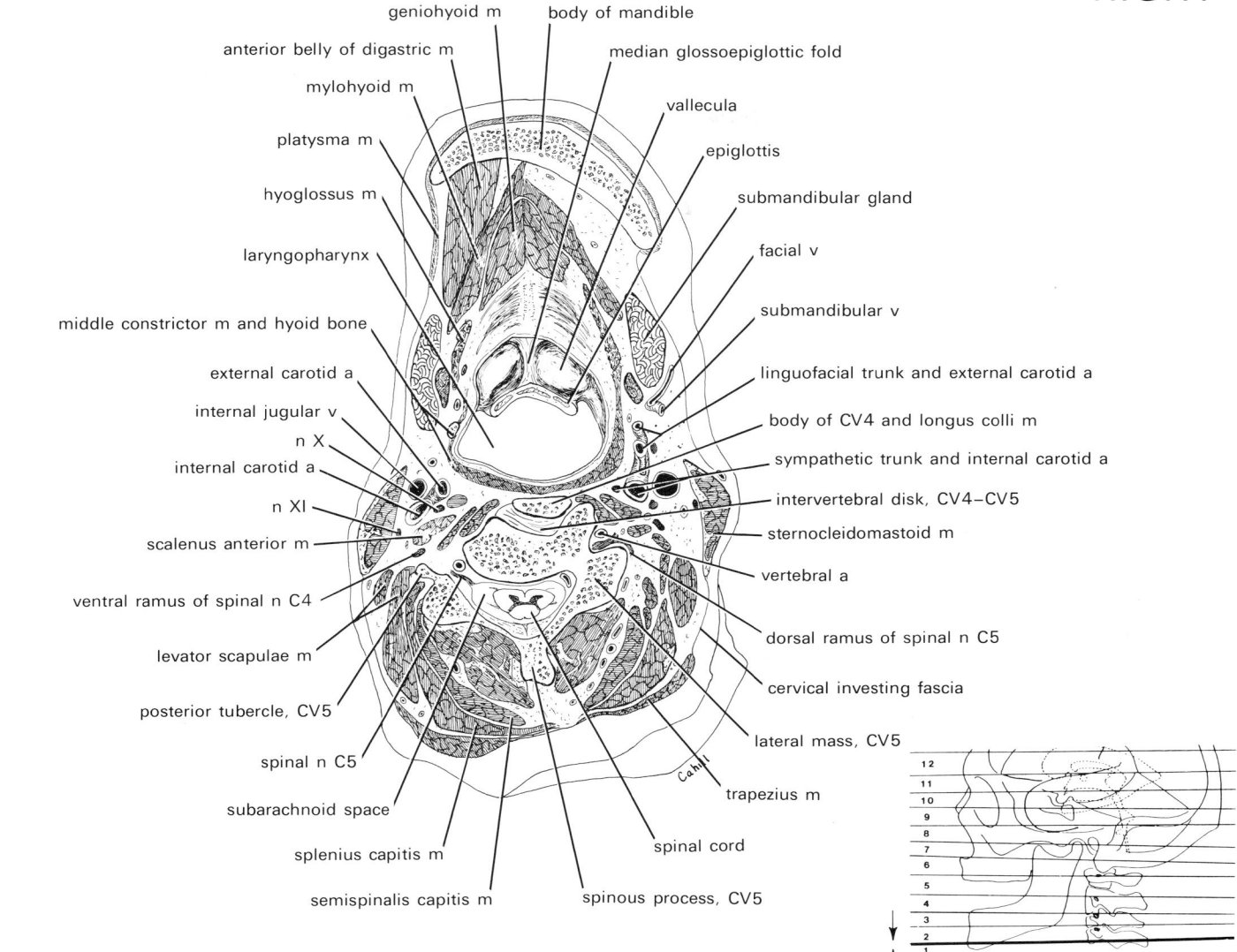

geniohyoid m

body of mandible

anterior belly of digastric m

median glossoepiglottic fold

mylohyoid m

vallecula

platysma m

epiglottis

hyoglossus m

submandibular gland

laryngopharynx

facial v

middle constrictor m and hyoid bone

submandibular v

external carotid a

linguofacial trunk and external carotid a

internal jugular v

body of CV4 and longus colli m

n X

sympathetic trunk and internal carotid a

internal carotid a

intervertebral disk, CV4–CV5

n XI

sternocleidomastoid m

scalenus anterior m

vertebral a

ventral ramus of spinal n C4

dorsal ramus of spinal n C5

levator scapulae m

cervical investing fascia

posterior tubercle, CV5

lateral mass, CV5

spinal n C5

trapezius m

subarachnoid space

spinal cord

splenius capitis m

semispinalis capitis m

spinous process, CV5

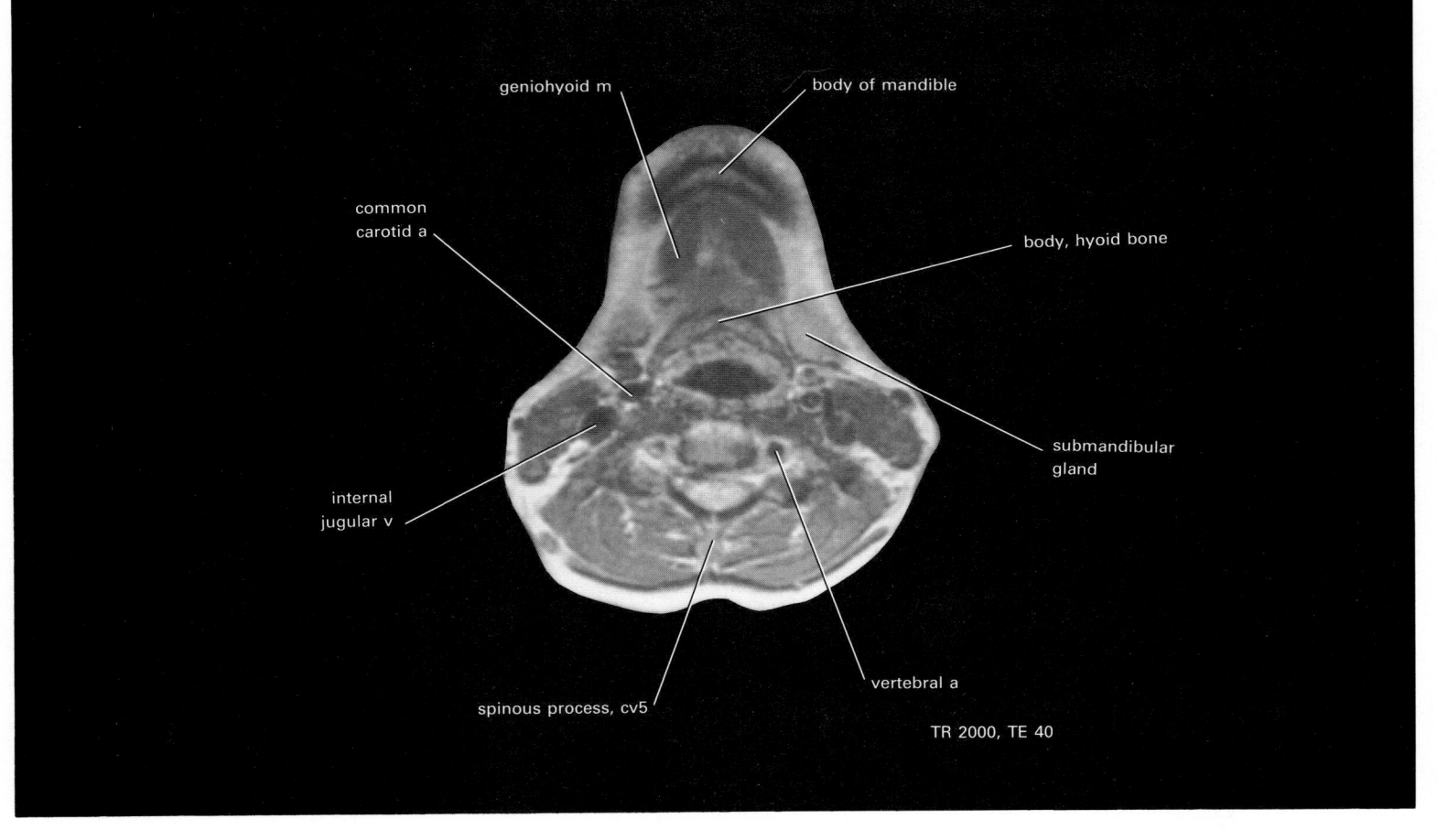

geniohyoid m

body of mandible

common
carotid a

body, hyoid bone

submandibular
gland

internal
jugular v

vertebral a

spinous process, cv5

TR 2000, TE 40

Section 1 from above.

The Head and Neck in Sagittal Planes

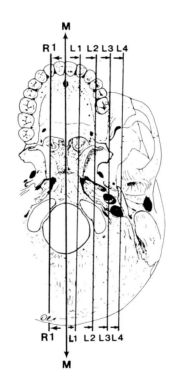

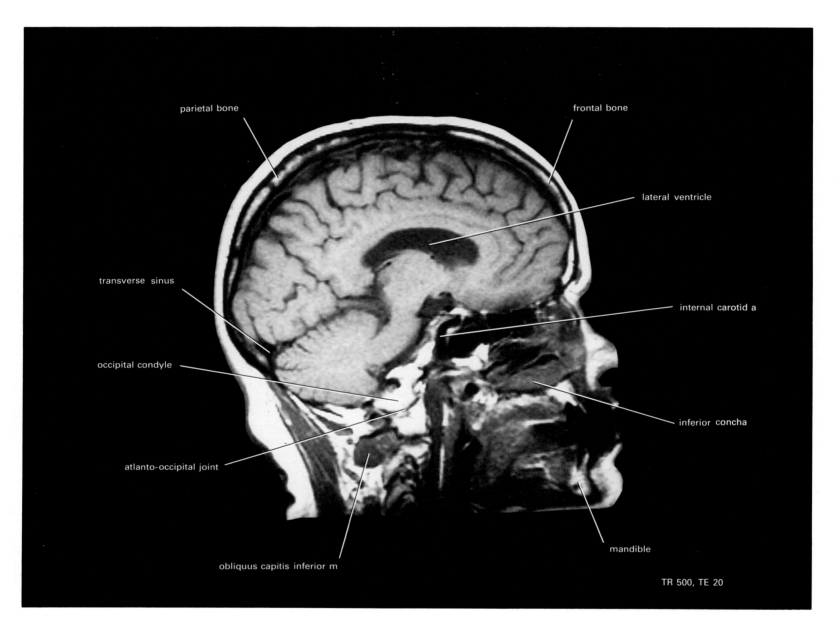

Section R1 from midline.

BONES, MUSCLES, VESSELS, AND VISCERA

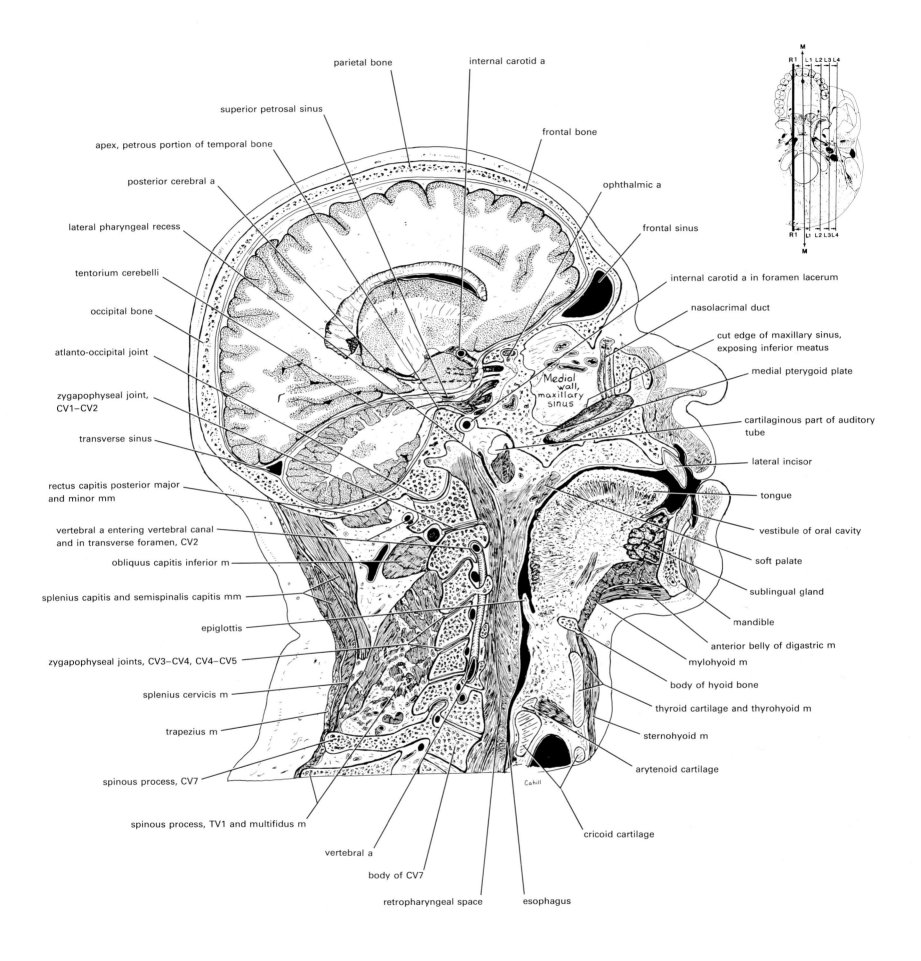

parietal bone

internal carotid a

superior petrosal sinus

frontal bone

apex, petrous portion of temporal bone

ophthalmic a

posterior cerebral a

frontal sinus

lateral pharyngeal recess

internal carotid a in foramen lacerum

tentorium cerebelli

nasolacrimal duct

occipital bone

cut edge of maxillary sinus, exposing inferior meatus

atlanto-occipital joint

medial pterygoid plate

zygapophyseal joint, CV1–CV2

cartilaginous part of auditory tube

transverse sinus

lateral incisor

rectus capitis posterior major and minor mm

tongue

vertebral a entering vertebral canal and in transverse foramen, CV2

vestibule of oral cavity

obliquus capitis inferior m

soft palate

splenius capitis and semispinalis capitis mm

sublingual gland

epiglottis

mandible

zygapophyseal joints, CV3–CV4, CV4–CV5

anterior belly of digastric m

splenius cervicis m

mylohyoid m

trapezius m

body of hyoid bone

spinous process, CV7

thyroid cartilage and thyrohyoid m

spinous process, TV1 and multifidus m

sternohyoid m

vertebral a

arytenoid cartilage

body of CV7

cricoid cartilage

retropharyngeal space

esophagus

Medial wall, maxillary sinus

Cahill

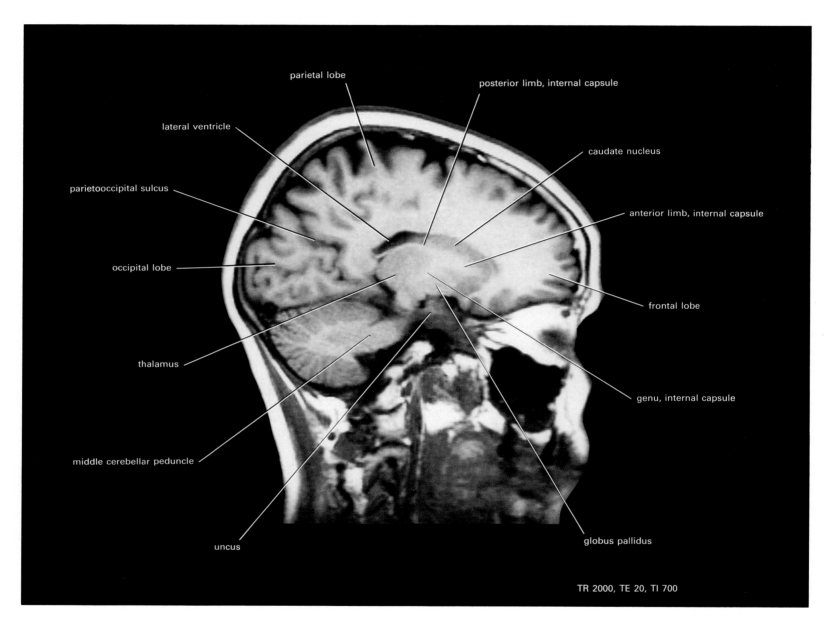

parietal lobe

posterior limb, internal capsule

lateral ventricle

caudate nucleus

parietooccipital sulcus

anterior limb, internal capsule

occipital lobe

frontal lobe

thalamus

genu, internal capsule

middle cerebellar peduncle

uncus

globus pallidus

TR 2000, TE 20, TI 700

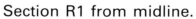

Section R1 from midline.

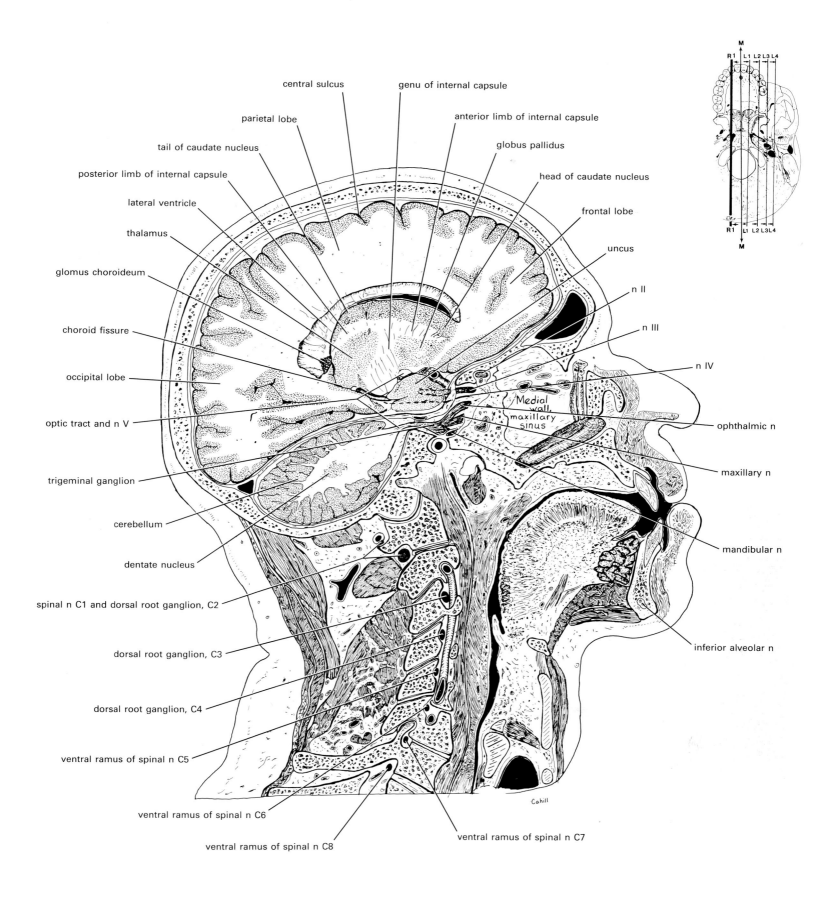

central sulcus

genu of internal capsule

parietal lobe

anterior limb of internal capsule

tail of caudate nucleus

globus pallidus

posterior limb of internal capsule

head of caudate nucleus

lateral ventricle

frontal lobe

thalamus

uncus

glomus choroideum

n II

choroid fissure

n III

occipital lobe

n IV

Medial wall, maxillary sinus

optic tract and n V

ophthalmic n

trigeminal ganglion

maxillary n

cerebellum

dentate nucleus

mandibular n

spinal n C1 and dorsal root ganglion, C2

dorsal root ganglion, C3

dorsal root ganglion, C4

inferior alveolar n

ventral ramus of spinal n C5

ventral ramus of spinal n C6

ventral ramus of spinal n C7

ventral ramus of spinal n C8

Cahill

Section R1 from midline.

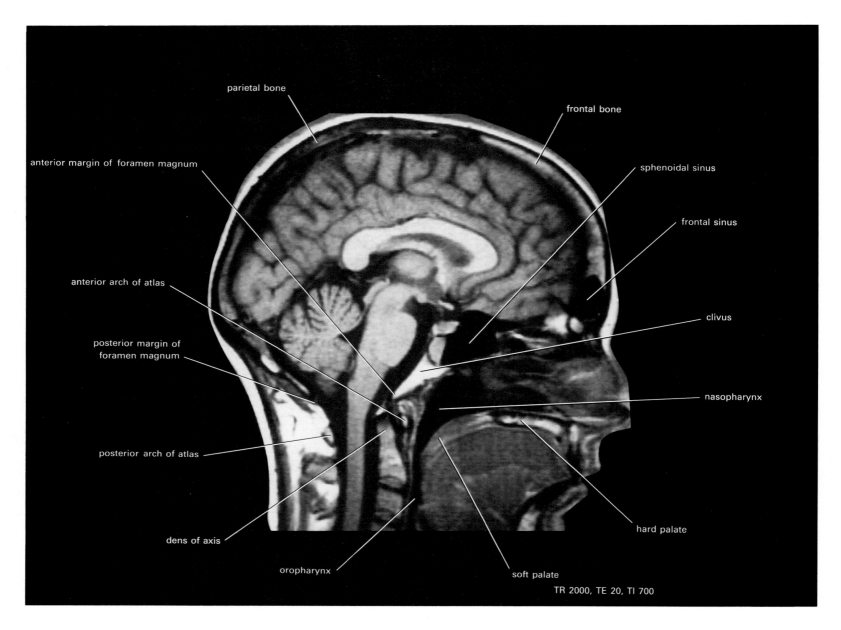

parietal bone

frontal bone

anterior margin of foramen magnum

sphenoidal sinus

frontal sinus

anterior arch of atlas

clivus

posterior margin of
foramen magnum

nasopharynx

posterior arch of atlas

hard palate

dens of axis

soft palate

oropharynx

TR 2000, TE 20, TI 700

Median Section

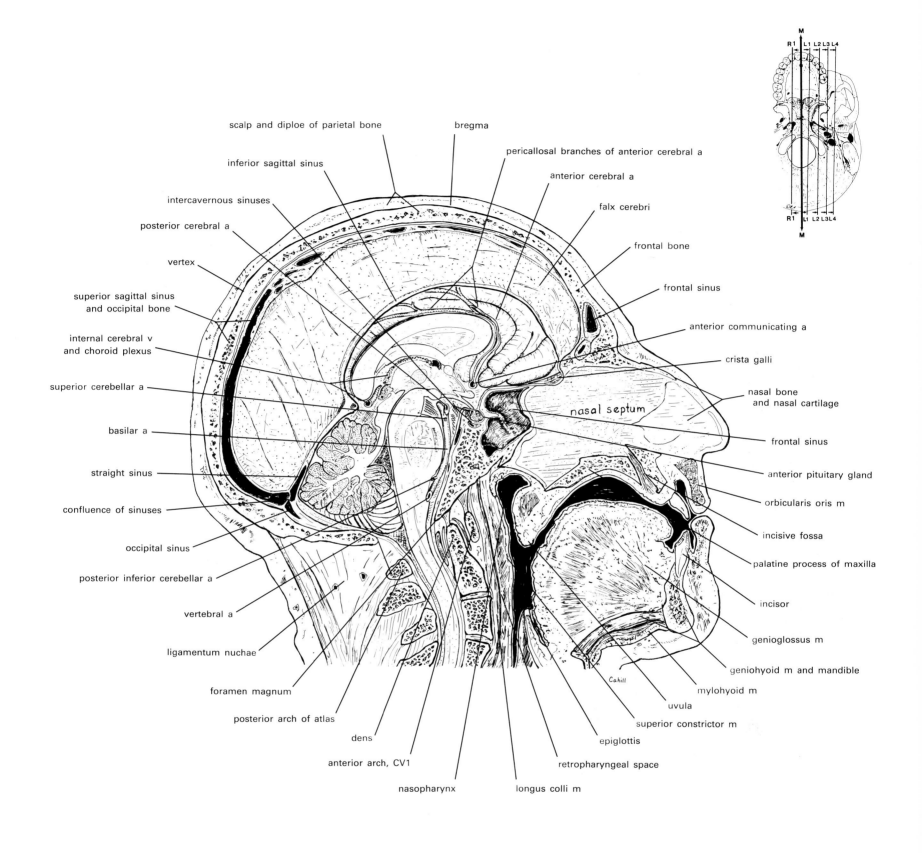

scalp and diploe of parietal bone

bregma

pericallosal branches of anterior cerebral a

inferior sagittal sinus

anterior cerebral a

intercavernous sinuses

falx cerebri

posterior cerebral a

frontal bone

vertex

frontal sinus

superior sagittal sinus
and occipital bone

anterior communicating a

internal cerebral v
and choroid plexus

crista galli

superior cerebellar a

nasal bone
and nasal cartilage

nasal septum

basilar a

frontal sinus

straight sinus

anterior pituitary gland

confluence of sinuses

orbicularis oris m

occipital sinus

incisive fossa

posterior inferior cerebellar a

palatine process of maxilla

vertebral a

incisor

ligamentum nuchae

genioglossus m

Cahill

geniohyoid m and mandible

foramen magnum

mylohyoid m

posterior arch of atlas

uvula

dens

superior constrictor m

anterior arch, CV1

epiglottis

retropharyngeal space

nasopharynx

longus colli m

Median Section

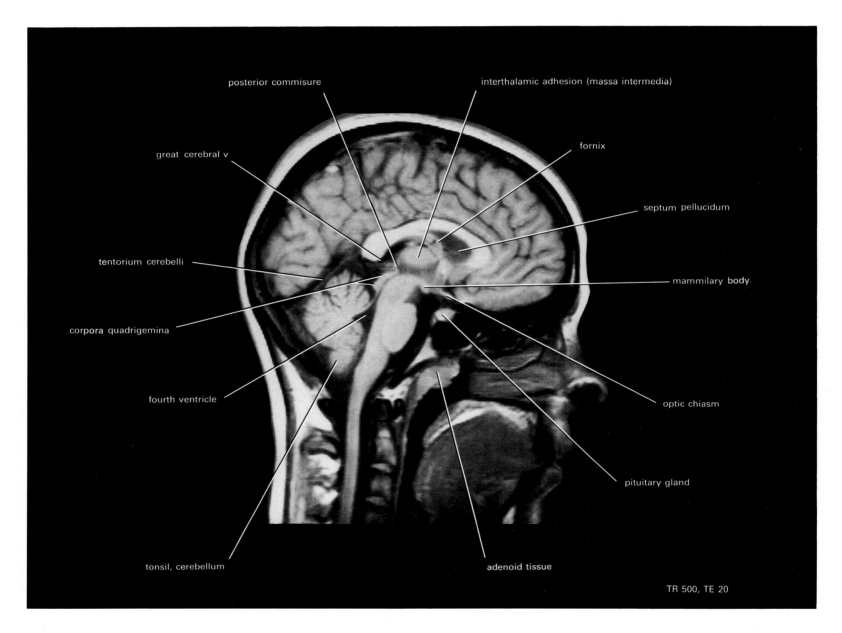

posterior commisure

interthalamic adhesion (massa intermedia)

great cerebral v

fornix

septum pellucidum

tentorium cerebelli

mammilary body

corpora quadrigemina

optic chiasm

fourth ventricle

pituitary gland

tonsil, cerebellum

adenoid tissue

TR 500, TE 20

Median Section

NERVOUS SYSTEM

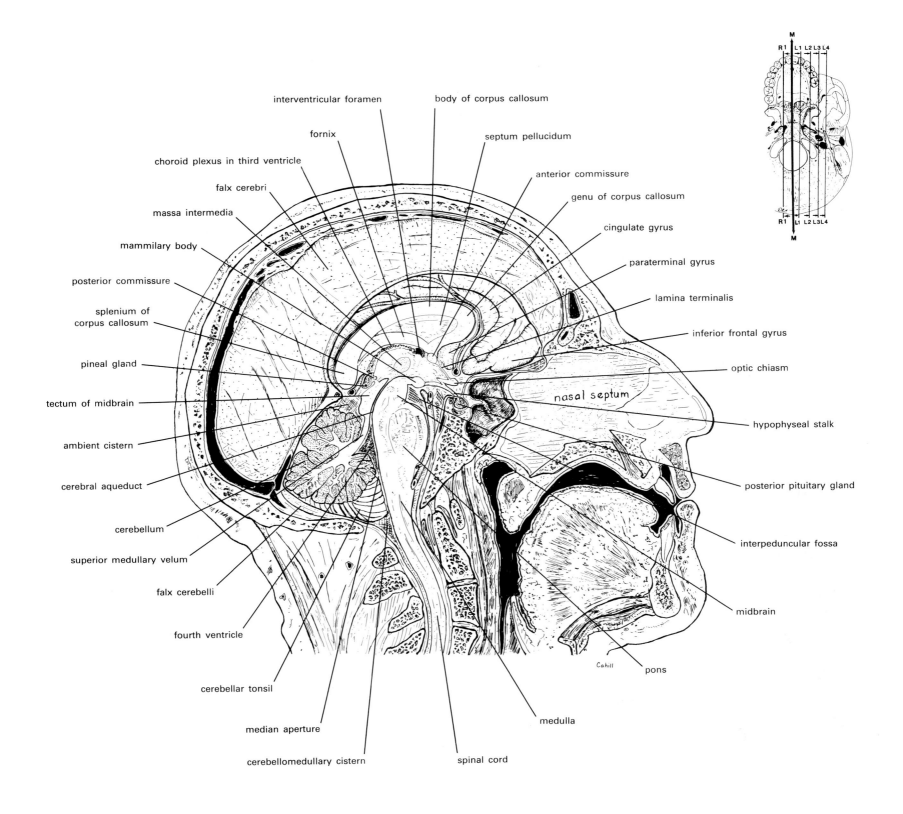

interventricular foramen

body of corpus callosum

fornix

septum pellucidum

choroid plexus in third ventricle

anterior commissure

falx cerebri

genu of corpus callosum

massa intermedia

cingulate gyrus

mammilary body

paraterminal gyrus

posterior commissure

lamina terminalis

splenium of
corpus callosum

inferior frontal gyrus

pineal gland

optic chiasm

tectum of midbrain

nasal septum

hypophyseal stalk

ambient cistern

cerebral aqueduct

posterior pituitary gland

cerebellum

superior medullary velum

interpeduncular fossa

falx cerebelli

midbrain

fourth ventricle

pons

cerebellar tonsil

medulla

median aperture

cerebellomedullary cistern

spinal cord

Cahill

Median Section

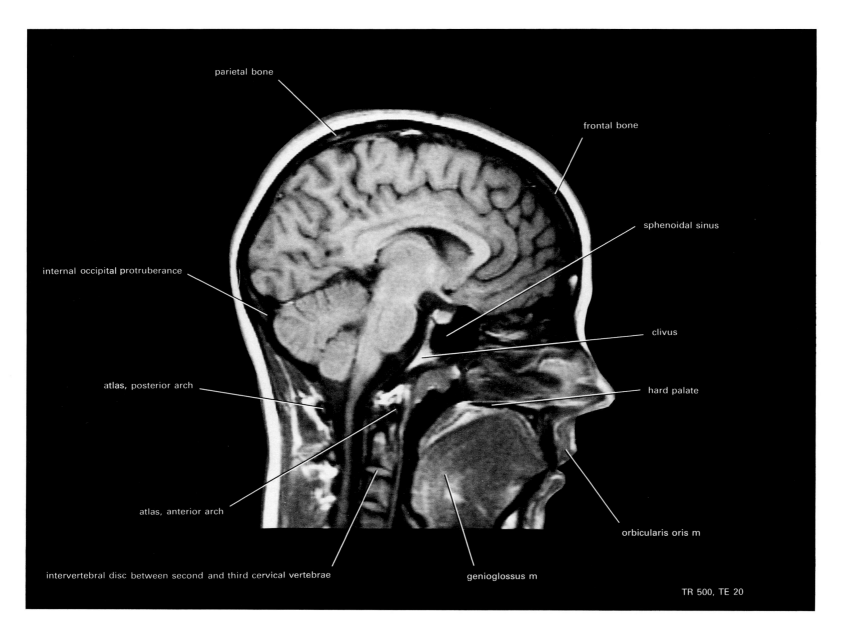

Section L1 from midline.

BONES AND MUSCLES

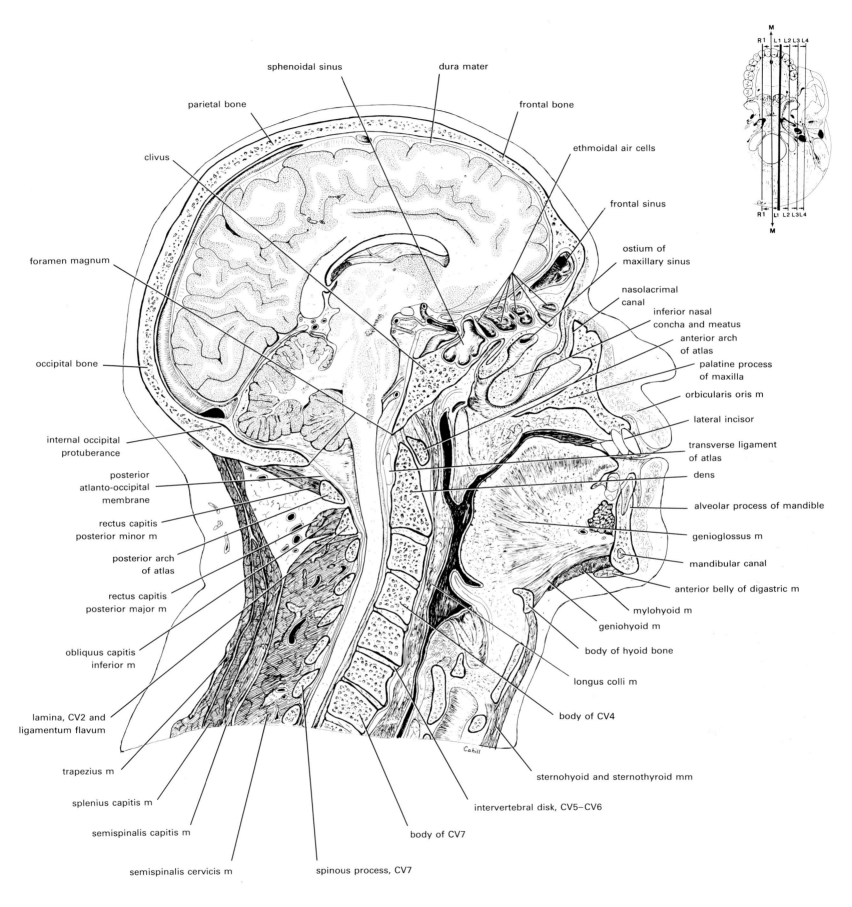

sphenoidal sinus

dura mater

parietal bone

frontal bone

clivus

ethmoidal air cells

frontal sinus

foramen magnum

ostium of
maxillary sinus

nasolacrimal
canal

inferior nasal
concha and meatus

anterior arch
of atlas

occipital bone

palatine process
of maxilla

orbicularis oris m

lateral incisor

transverse ligament
of atlas

internal occipital
protuberance

dens

posterior
atlanto-occipital
membrane

alveolar process of mandible

rectus capitis
posterior minor m

genioglossus m

posterior arch
of atlas

mandibular canal

anterior belly of digastric m

rectus capitis
posterior major m

mylohyoid m

geniohyoid m

body of hyoid bone

obliquus capitis
inferior m

longus colli m

body of CV4

lamina, CV2 and
ligamentum flavum

Cahill

trapezius m

sternohyoid and sternothyroid mm

splenius capitis m

intervertebral disk, CV5–CV6

semispinalis capitis m

body of CV7

semispinalis cervicis m

spinous process, CV7

Section L1 from midline.

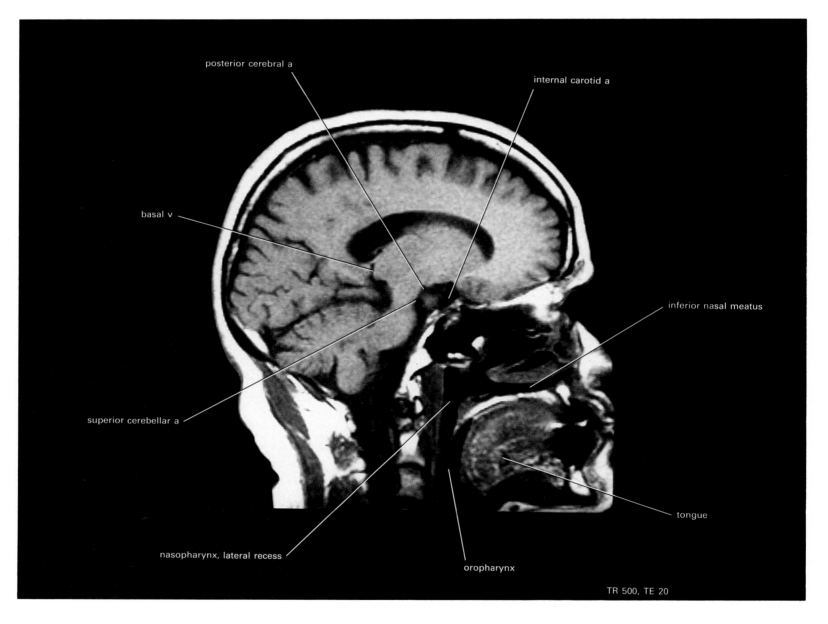

Section L1 from midline.

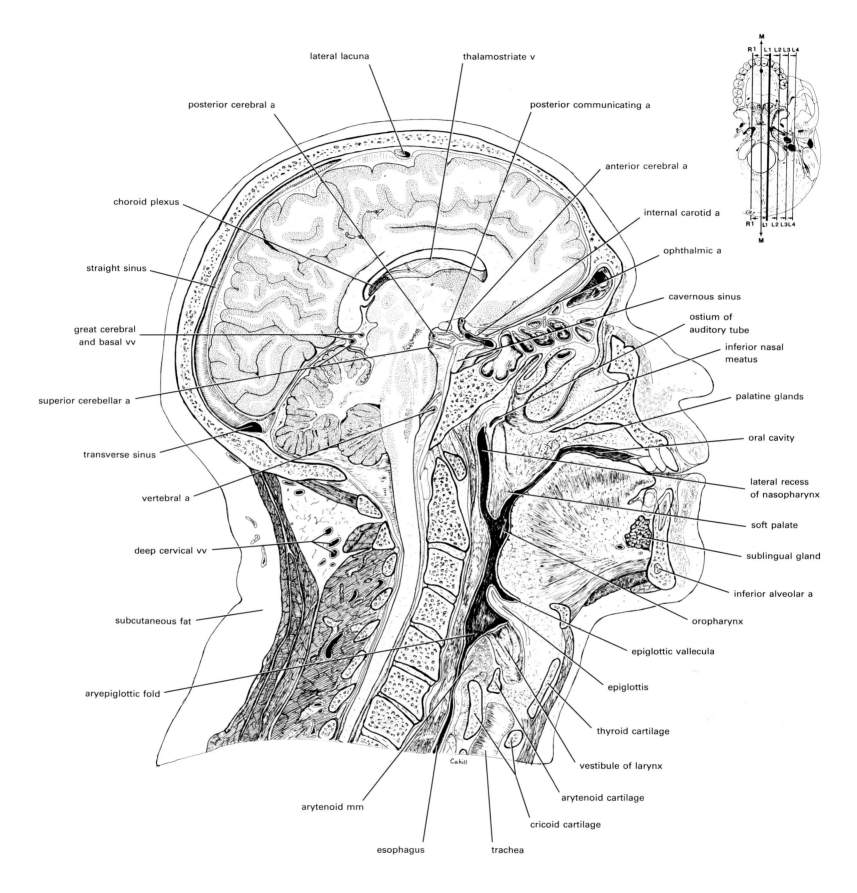

lateral lacuna

thalamostriate v

posterior cerebral a

posterior communicating a

anterior cerebral a

internal carotid a

choroid plexus

ophthalmic a

straight sinus

cavernous sinus

ostium of
auditory tube

great cerebral
and basal vv

inferior nasal
meatus

superior cerebellar a

palatine glands

oral cavity

transverse sinus

lateral recess
of nasopharynx

vertebral a

soft palate

deep cervical vv

sublingual gland

inferior alveolar a

subcutaneous fat

oropharynx

epiglottic vallecula

aryepiglottic fold

epiglottis

thyroid cartilage

vestibule of larynx

arytenoid cartilage

arytenoid mm

cricoid cartilage

esophagus

trachea

Cahill

Section L1 from midline.

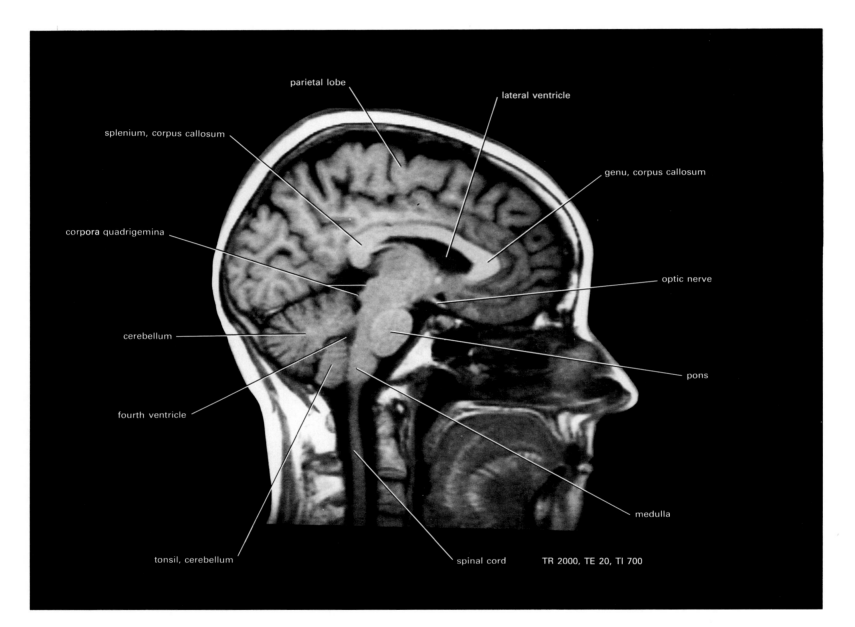

Section L1 from midline.

NERVOUS SYSTEM

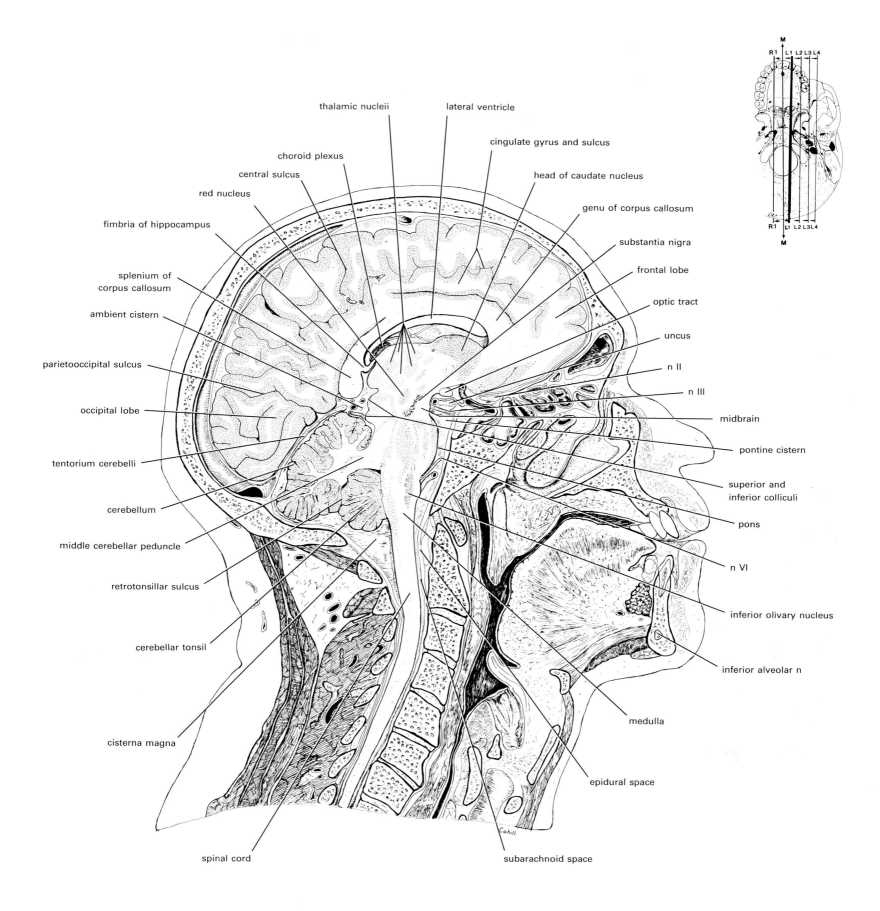

thalamic nucleii

lateral ventricle

cingulate gyrus and sulcus

choroid plexus

central sulcus

head of caudate nucleus

red nucleus

genu of corpus callosum

fimbria of hippocampus

substantia nigra

splenium of corpus callosum

frontal lobe

ambient cistern

optic tract

parietooccipital sulcus

uncus

n II

occipital lobe

n III

midbrain

tentorium cerebelli

pontine cistern

cerebellum

superior and inferior colliculi

middle cerebellar peduncle

pons

retrotonsillar sulcus

n VI

inferior olivary nucleus

cerebellar tonsil

inferior alveolar n

medulla

cisterna magna

epidural space

spinal cord

subarachnoid space

Section L1 from midline.

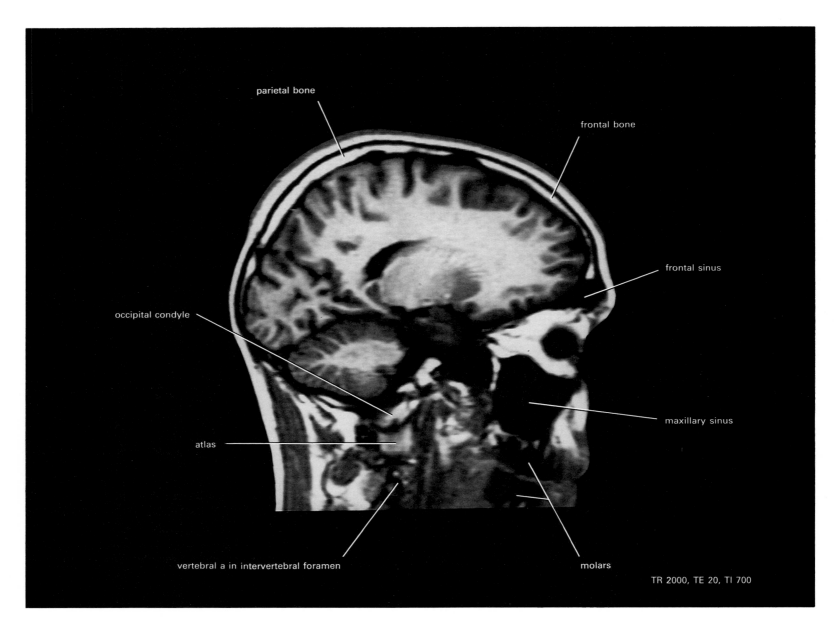

parietal bone

frontal bone

frontal sinus

occipital condyle

maxillary sinus

atlas

vertebral a in intervertebral foramen

molars

TR 2000, TE 20, TI 700

Section L2 from midline.

BONES AND JOINTS

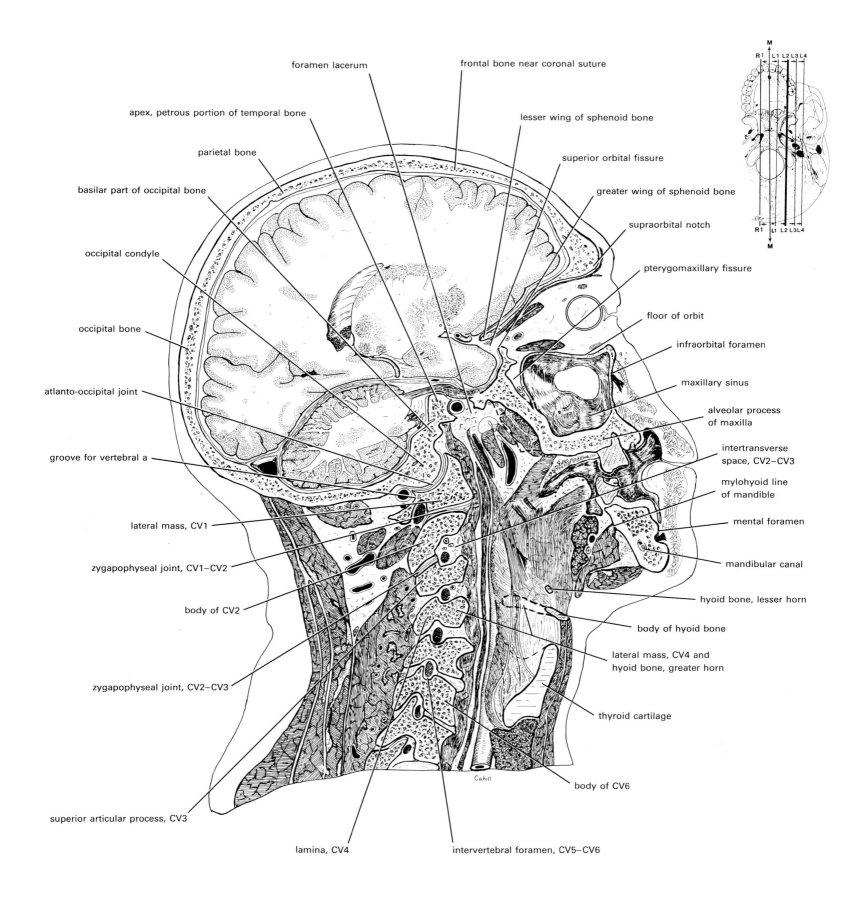

foramen lacerum

frontal bone near coronal suture

apex, petrous portion of temporal bone

lesser wing of sphenoid bone

parietal bone

superior orbital fissure

basilar part of occipital bone

greater wing of sphenoid bone

supraorbital notch

occipital condyle

pterygomaxillary fissure

floor of orbit

occipital bone

infraorbital foramen

maxillary sinus

atlanto-occipital joint

alveolar process
of maxilla

intertransverse
space, CV2–CV3

groove for vertebral a

mylohyoid line
of mandible

mental foramen

lateral mass, CV1

mandibular canal

zygapophyseal joint, CV1–CV2

hyoid bone, lesser horn

body of CV2

body of hyoid bone

lateral mass, CV4 and
hyoid bone, greater horn

zygapophyseal joint, CV2–CV3

thyroid cartilage

Cahill

body of CV6

superior articular process, CV3

lamina, CV4

intervertebral foramen, CV5–CV6

M

R1   L1 L2 L3 L4

R1   L1 L2 L3 L4

M

Section L2 from midline.

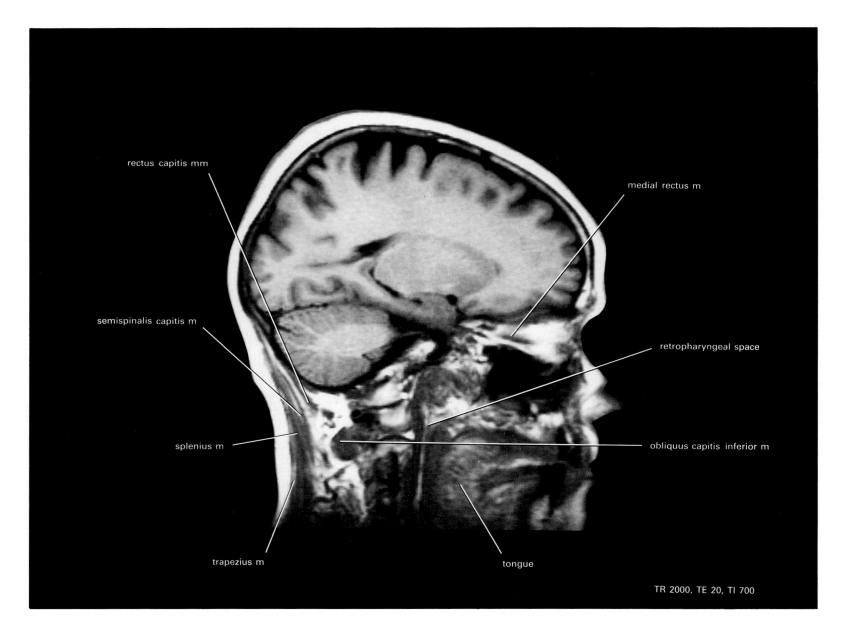

rectus capitis mm

medial rectus m

semispinalis capitis m

retropharyngeal space

splenius m

obliquus capitis inferior m

trapezius m

tongue

TR 2000, TE 20, TI 700

Section L2 from midline.

MUSCLES

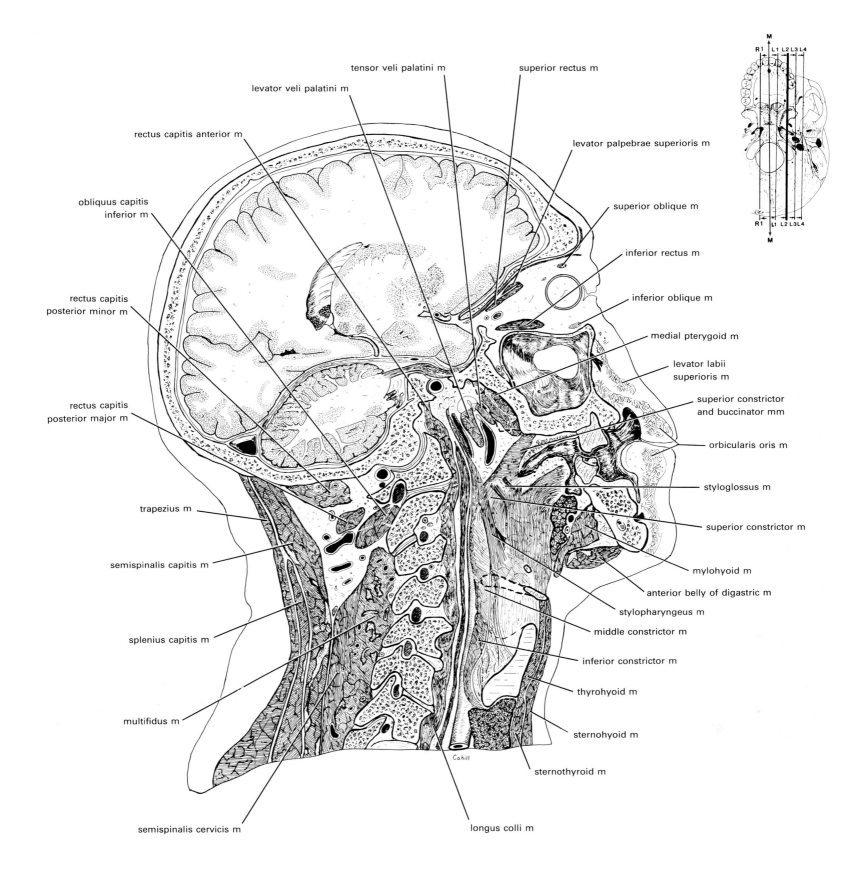

tensor veli palatini m

levator veli palatini m

superior rectus m

rectus capitis anterior m

levator palpebrae superioris m

obliquus capitis inferior m

superior oblique m

inferior rectus m

inferior oblique m

rectus capitis posterior minor m

medial pterygoid m

levator labii superioris m

rectus capitis posterior major m

superior constrictor and buccinator mm

orbicularis oris m

styloglossus m

trapezius m

superior constrictor m

semispinalis capitis m

mylohyoid m

anterior belly of digastric m

splenius capitis m

stylopharyngeus m

middle constrictor m

inferior constrictor m

thyrohyoid m

multifidus m

sternohyoid m

sternothyroid m

semispinalis cervicis m

longus colli m

Cahill

Section L2 from midline.

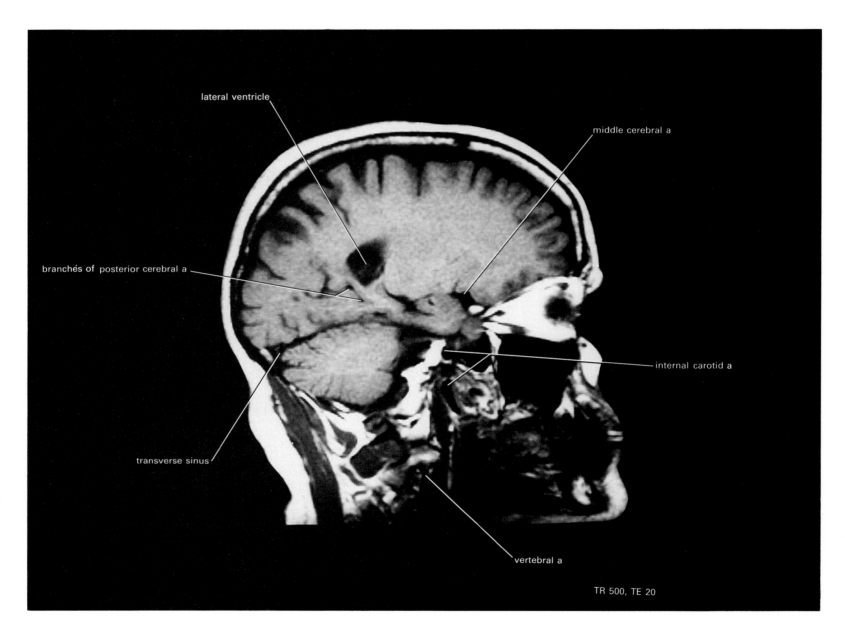

Section L2 from midline.

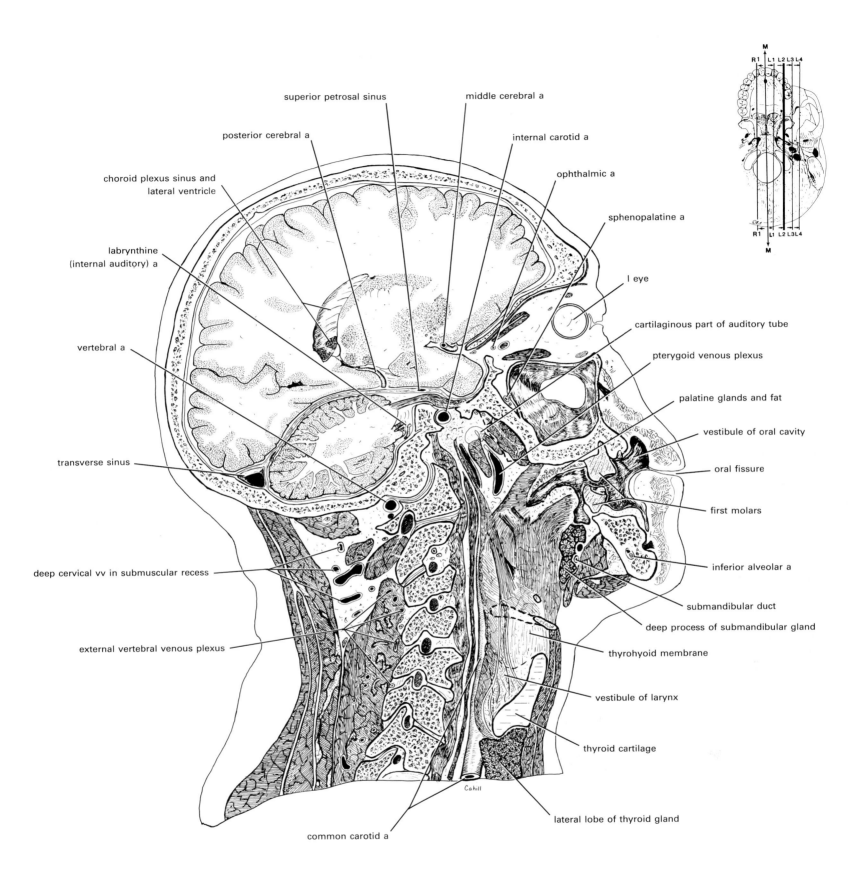

superior petrosal sinus

middle cerebral a

posterior cerebral a

internal carotid a

choroid plexus sinus and
lateral ventricle

ophthalmic a

sphenopalatine a

labrynthine
(internal auditory) a

l eye

cartilaginous part of auditory tube

vertebral a

pterygoid venous plexus

palatine glands and fat

vestibule of oral cavity

transverse sinus

oral fissure

first molars

deep cervical vv in submuscular recess

inferior alveolar a

submandibular duct

external vertebral venous plexus

deep process of submandibular gland

thyrohyoid membrane

vestibule of larynx

thyroid cartilage

lateral lobe of thyroid gland

common carotid a

Cahill

Section L2 from midline.

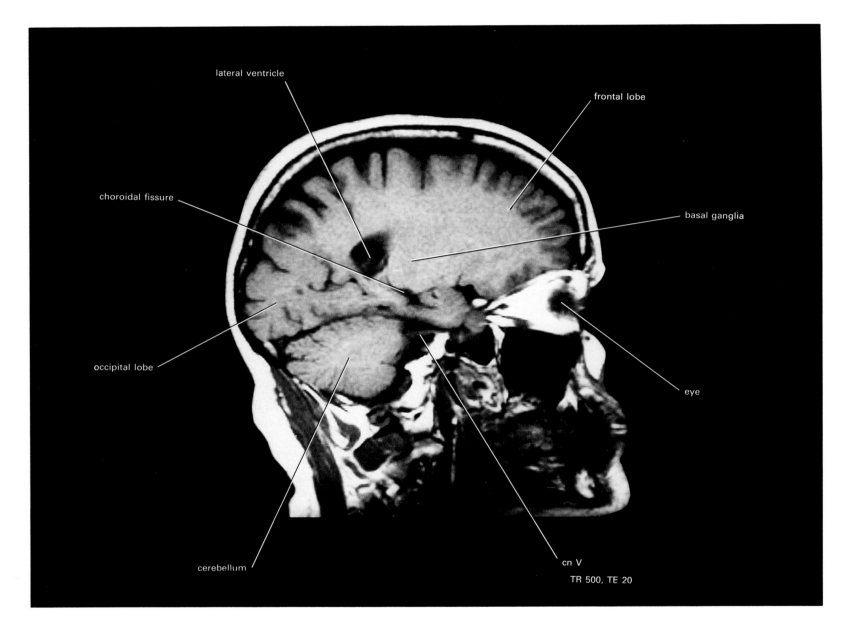

Section L2 from midline.

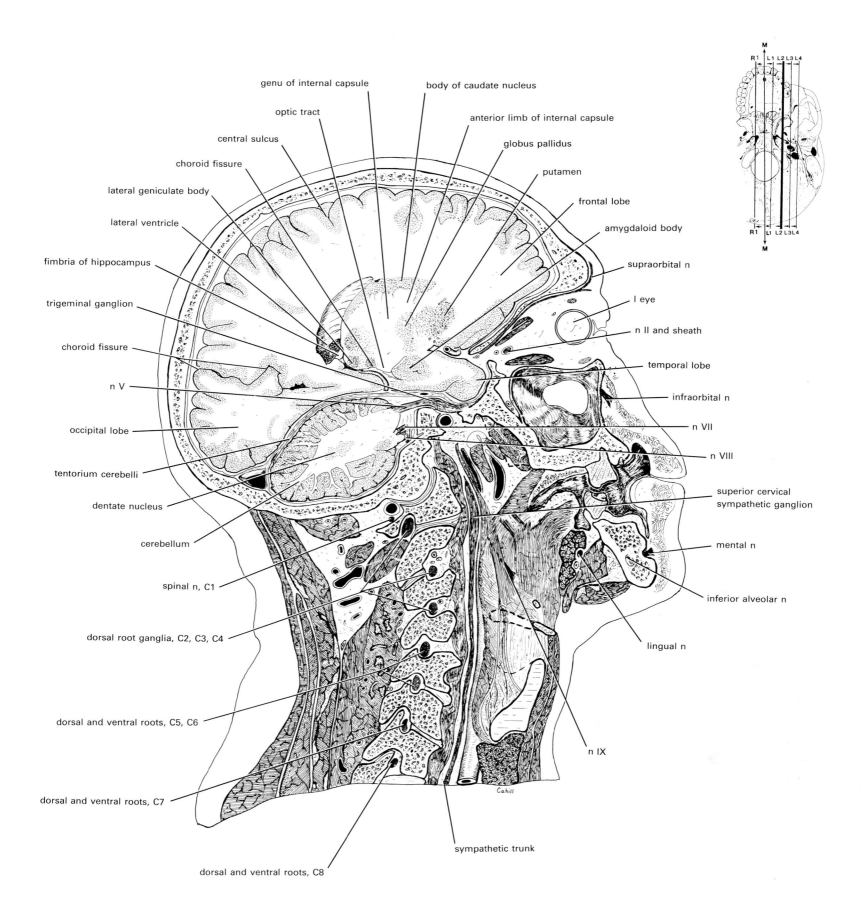

genu of internal capsule

body of caudate nucleus

optic tract

anterior limb of internal capsule

central sulcus

globus pallidus

choroid fissure

putamen

lateral geniculate body

frontal lobe

lateral ventricle

amygdaloid body

fimbria of hippocampus

supraorbital n

trigeminal ganglion

l eye

n II and sheath

choroid fissure

temporal lobe

n V

infraorbital n

occipital lobe

n VII

tentorium cerebelli

n VIII

dentate nucleus

superior cervical
sympathetic ganglion

cerebellum

mental n

spinal n, C1

inferior alveolar n

dorsal root ganglia, C2, C3, C4

lingual n

dorsal and ventral roots, C5, C6

n IX

dorsal and ventral roots, C7

Cahill

dorsal and ventral roots, C8

sympathetic trunk

Section L2 from midline.

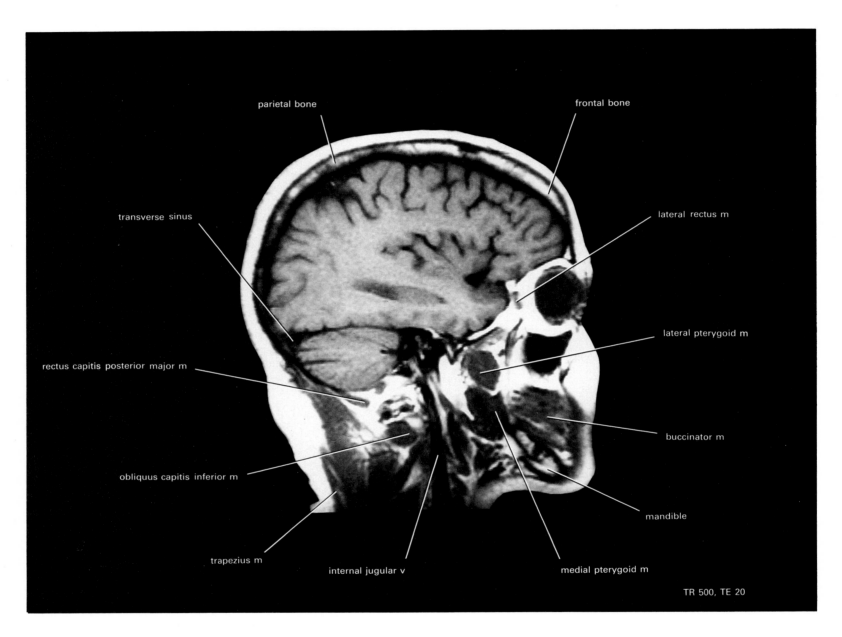

Section L3 from midline.

BONES AND MUSCLES

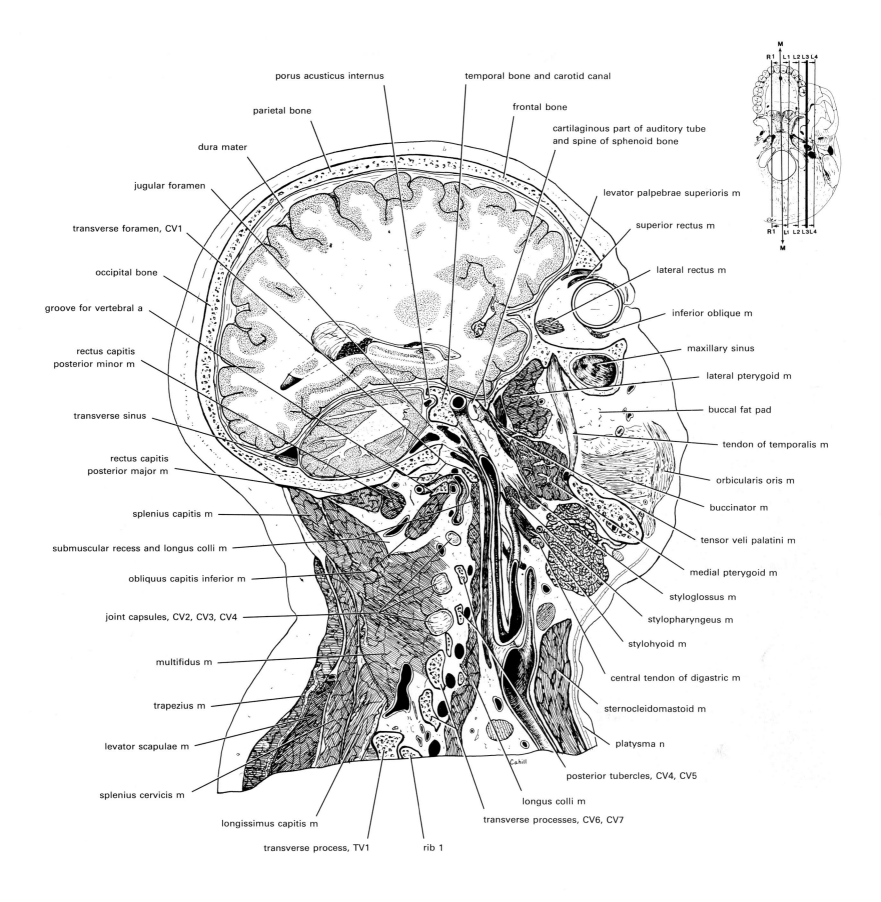

porus acusticus internus

temporal bone and carotid canal

parietal bone

frontal bone

dura mater

cartilaginous part of auditory tube
and spine of sphenoid bone

jugular foramen

levator palpebrae superioris m

transverse foramen, CV1

superior rectus m

occipital bone

lateral rectus m

groove for vertebral a

inferior oblique m

rectus capitis
posterior minor m

maxillary sinus

lateral pterygoid m

transverse sinus

buccal fat pad

rectus capitis
posterior major m

tendon of temporalis m

orbicularis oris m

splenius capitis m

buccinator m

submuscular recess and longus colli m

tensor veli palatini m

obliquus capitis inferior m

medial pterygoid m

joint capsules, CV2, CV3, CV4

styloglossus m

multifidus m

stylopharyngeus m

trapezius m

stylohyoid m

levator scapulae m

central tendon of digastric m

splenius cervicis m

sternocleidomastoid m

longissimus capitis m

platysma n

transverse process, TV1

rib 1

posterior tubercles, CV4, CV5

longus colli m

transverse processes, CV6, CV7

Section L3 from midline.

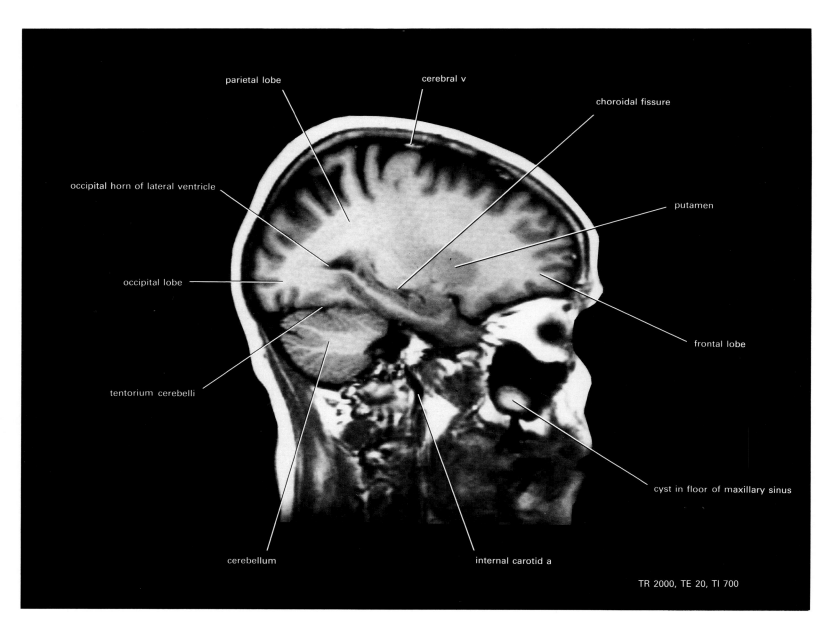

Section L3 from midline.

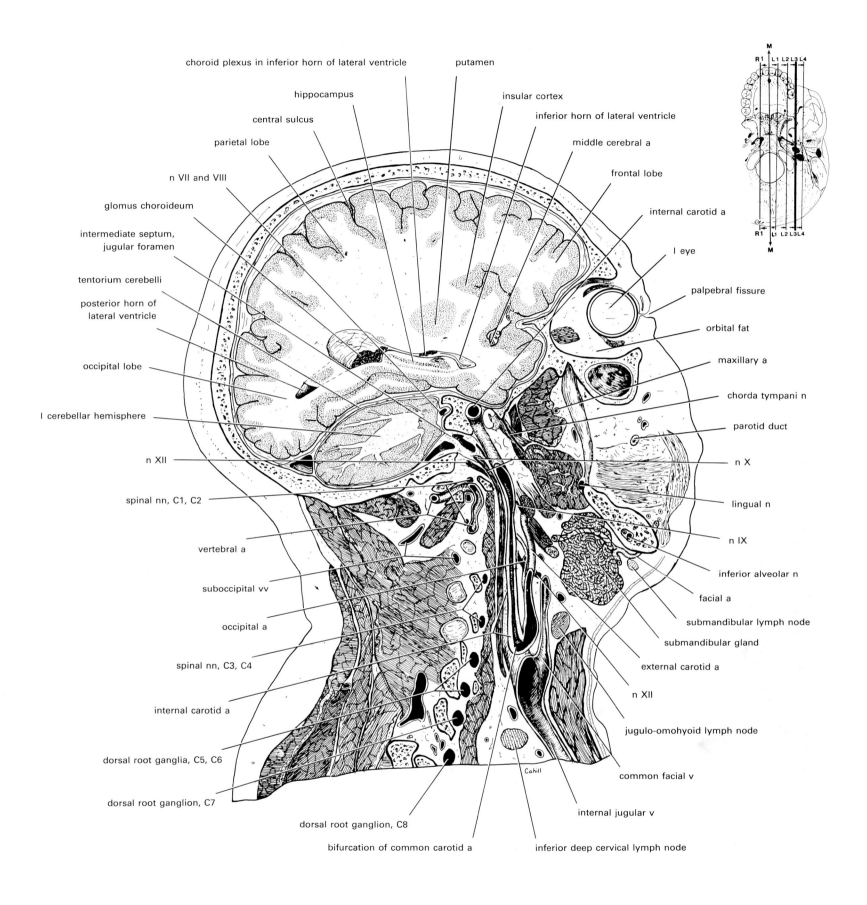

choroid plexus in inferior horn of lateral ventricle

putamen

hippocampus

insular cortex

central sulcus

inferior horn of lateral ventricle

parietal lobe

middle cerebral a

n VII and VIII

frontal lobe

glomus choroideum

internal carotid a

intermediate septum, jugular foramen

l eye

tentorium cerebelli

palpebral fissure

posterior horn of lateral ventricle

orbital fat

occipital lobe

maxillary a

chorda tympani n

l cerebellar hemisphere

parotid duct

n XII

n X

spinal nn, C1, C2

lingual n

vertebral a

n IX

suboccipital vv

inferior alveolar n

occipital a

facial a

spinal nn, C3, C4

submandibular lymph node

submandibular gland

internal carotid a

external carotid a

n XII

dorsal root ganglia, C5, C6

jugulo-omohyoid lymph node

dorsal root ganglion, C7

common facial v

dorsal root ganglion, C8

internal jugular v

bifurcation of common carotid a

inferior deep cervical lymph node

Cahill

Section L3 from midline.

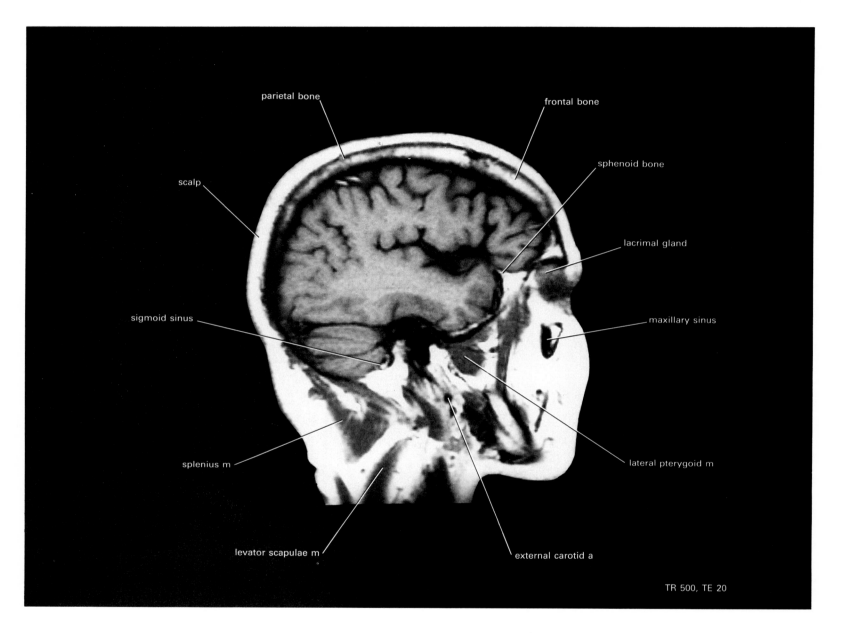

Section L4 from midline.

BONES AND MUSCLES

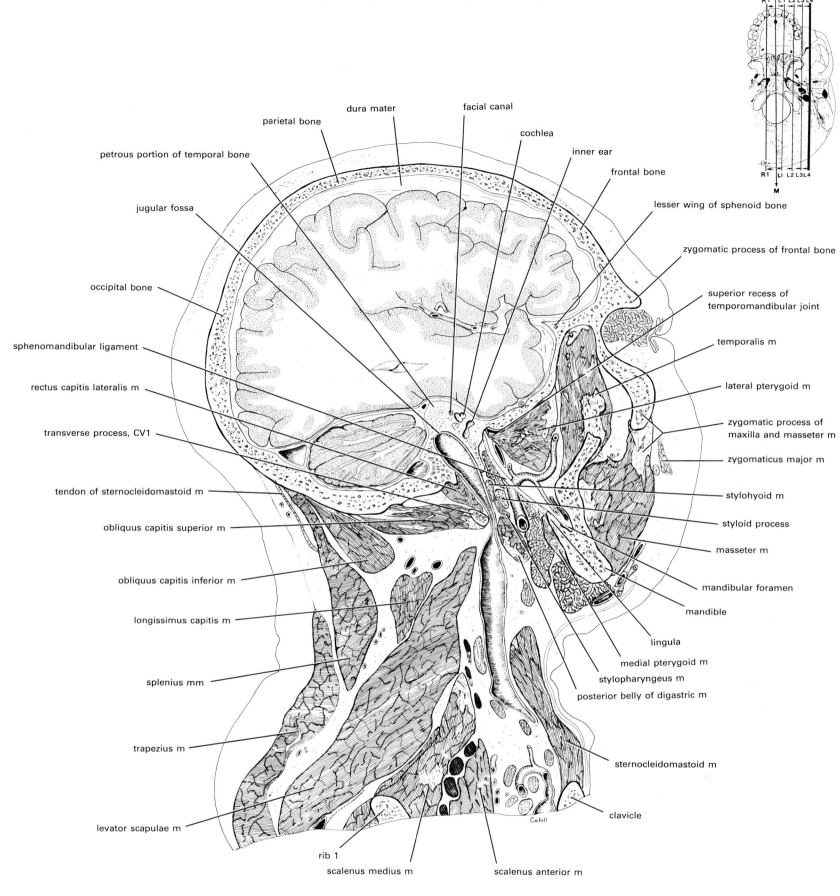

petrous portion of temporal bone

parietal bone

dura mater

facial canal

cochlea

inner ear

frontal bone

jugular fossa

lesser wing of sphenoid bone

zygomatic process of frontal bone

occipital bone

superior recess of temporomandibular joint

sphenomandibular ligament

temporalis m

rectus capitis lateralis m

lateral pterygoid m

transverse process, CV1

zygomatic process of maxilla and masseter m

zygomaticus major m

tendon of sternocleidomastoid m

stylohyoid m

obliquus capitis superior m

styloid process

masseter m

obliquus capitis inferior m

mandibular foramen

mandible

longissimus capitis m

lingula

medial pterygoid m

splenius mm

stylopharyngeus m

posterior belly of digastric m

trapezius m

sternocleidomastoid m

levator scapulae m

clavicle

rib 1

scalenus medius m

scalenus anterior m

Cahill

Section L4 from midline.

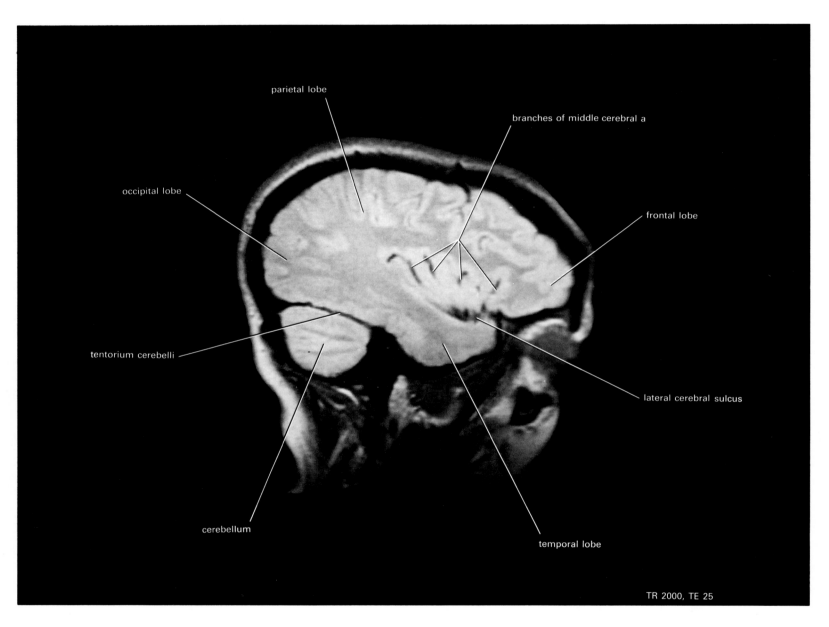

Section L4 from midline.

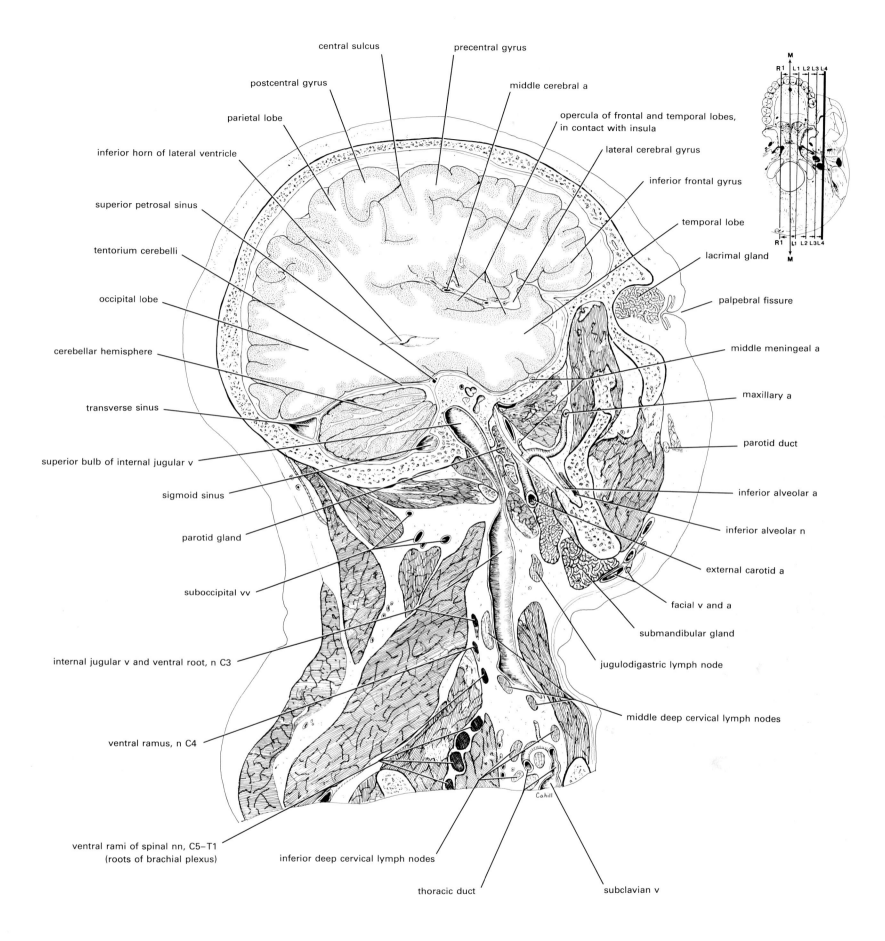

central sulcus

precentral gyrus

postcentral gyrus

middle cerebral a

parietal lobe

opercula of frontal and temporal lobes, in contact with insula

inferior horn of lateral ventricle

lateral cerebral gyrus

superior petrosal sinus

inferior frontal gyrus

tentorium cerebelli

temporal lobe

lacrimal gland

occipital lobe

palpebral fissure

cerebellar hemisphere

middle meningeal a

transverse sinus

maxillary a

superior bulb of internal jugular v

parotid duct

sigmoid sinus

inferior alveolar a

parotid gland

inferior alveolar n

suboccipital vv

external carotid a

internal jugular v and ventral root, n C3

facial v and a

ventral ramus, n C4

submandibular gland

jugulodigastric lymph node

middle deep cervical lymph nodes

ventral rami of spinal nn, C5–T1
(roots of brachial plexus)

inferior deep cervical lymph nodes

thoracic duct

subclavian v

Cahill

Section L4 from midline.

The Head and Neck in Coronal Planes

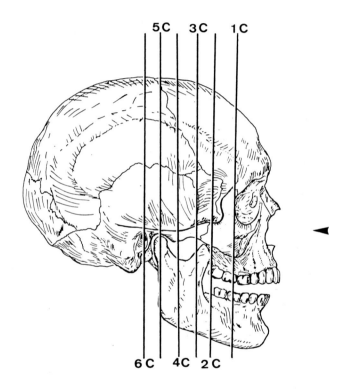

BONES

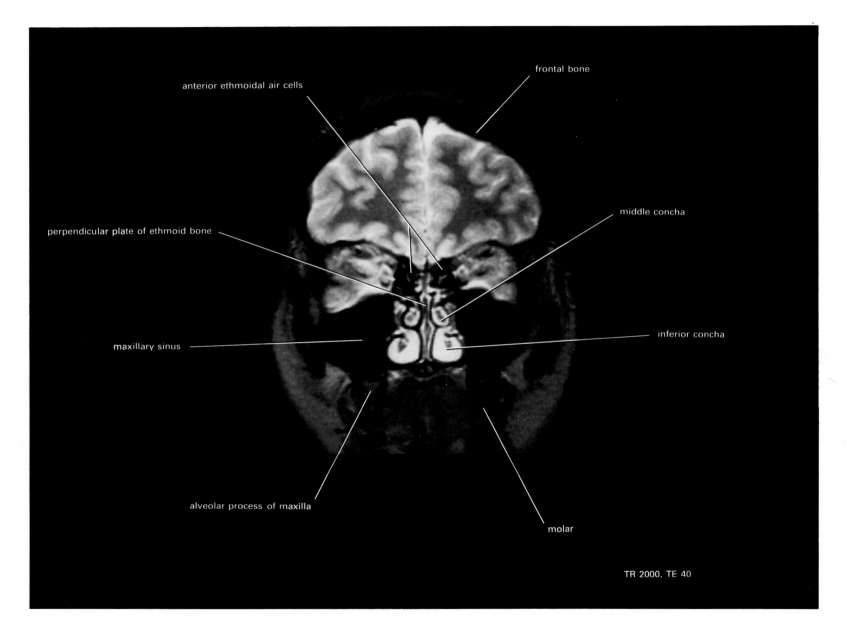

anterior ethmoidal air cells

frontal bone

perpendicular plate of ethmoid bone

middle concha

maxillary sinus

inferior concha

alveolar process of maxilla

molar

TR 2000, TE 40

Section 1C from front.

BONES

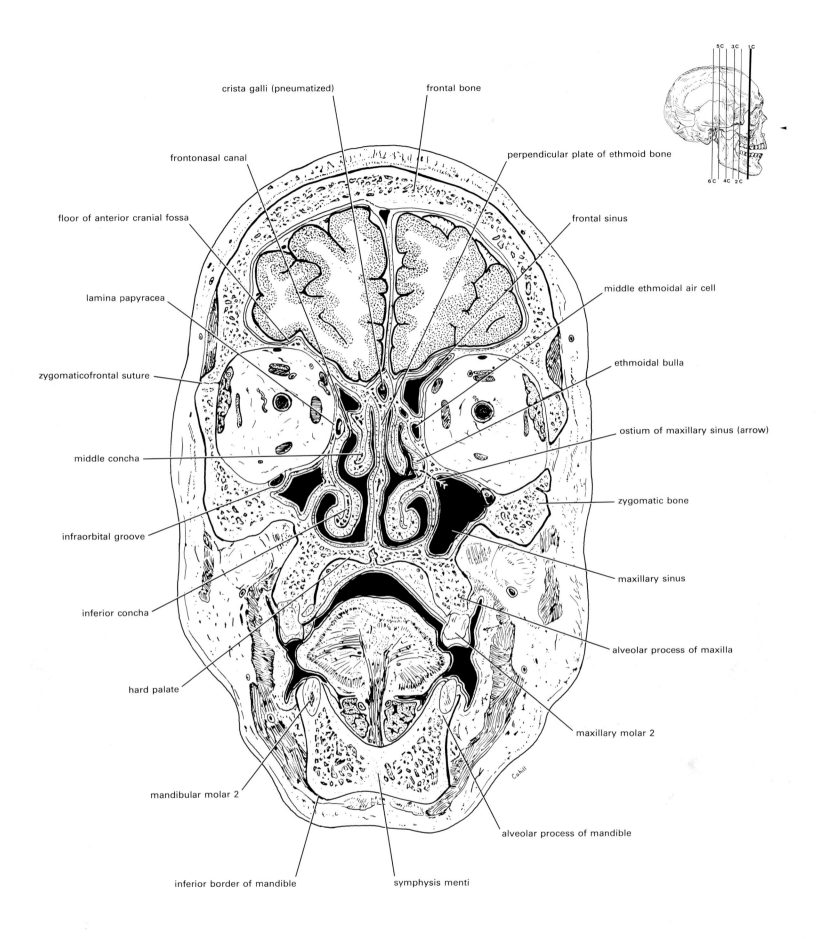

crista galli (pneumatized)

frontal bone

frontonasal canal

perpendicular plate of ethmoid bone

floor of anterior cranial fossa

frontal sinus

middle ethmoidal air cell

lamina papyracea

zygomaticofrontal suture

ethmoidal bulla

middle concha

ostium of maxillary sinus (arrow)

zygomatic bone

infraorbital groove

maxillary sinus

inferior concha

alveolar process of maxilla

hard palate

maxillary molar 2

mandibular molar 2

alveolar process of mandible

inferior border of mandible

symphysis menti

Section 1C from front.

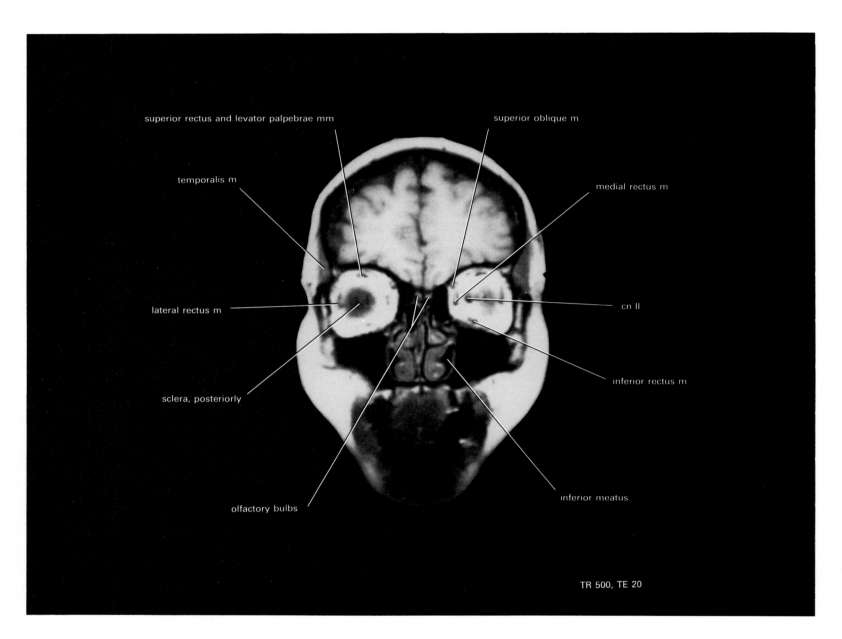

superior rectus and levator palpebrae mm

superior oblique m

temporalis m

medial rectus m

lateral rectus m

cn II

sclera, posteriorly

inferior rectus m

olfactory bulbs

inferior meatus

TR 500, TE 20

Section 1C from front.

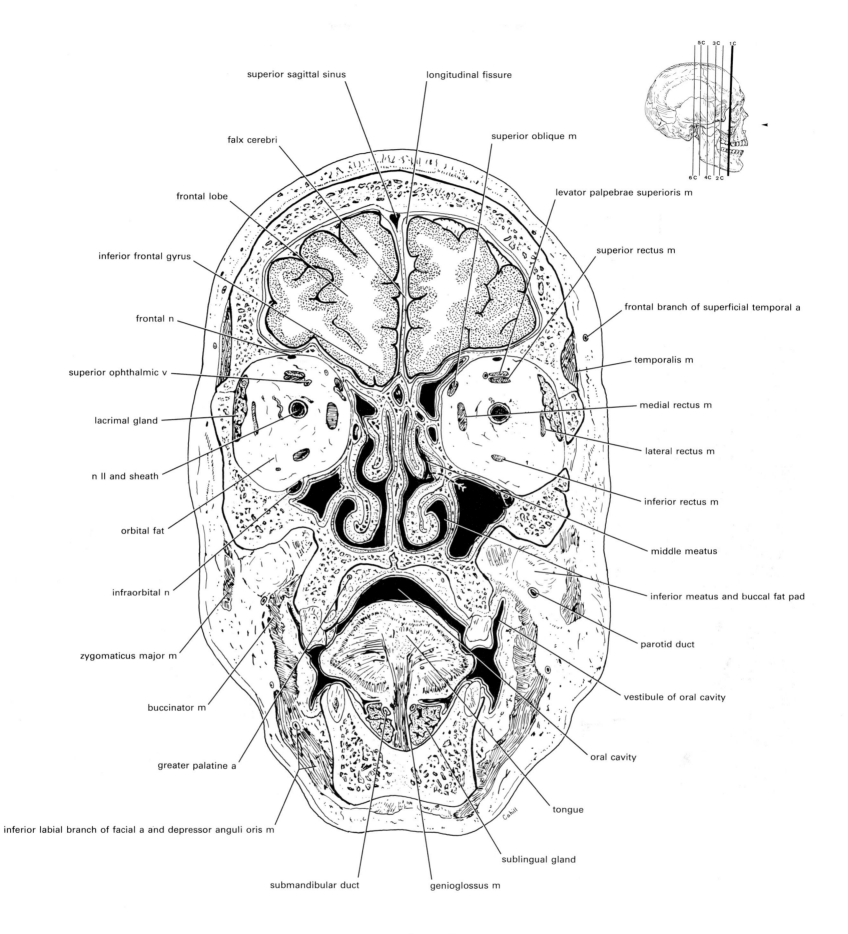

superior sagittal sinus

longitudinal fissure

falx cerebri

superior oblique m

frontal lobe

levator palpebrae superioris m

inferior frontal gyrus

superior rectus m

frontal n

frontal branch of superficial temporal a

superior ophthalmic v

temporalis m

lacrimal gland

medial rectus m

n II and sheath

lateral rectus m

orbital fat

inferior rectus m

middle meatus

infraorbital n

inferior meatus and buccal fat pad

zygomaticus major m

parotid duct

buccinator m

vestibule of oral cavity

greater palatine a

oral cavity

inferior labial branch of facial a and depressor anguli oris m

tongue

sublingual gland

submandibular duct

genioglossus m

Section 1C from front.

BONES

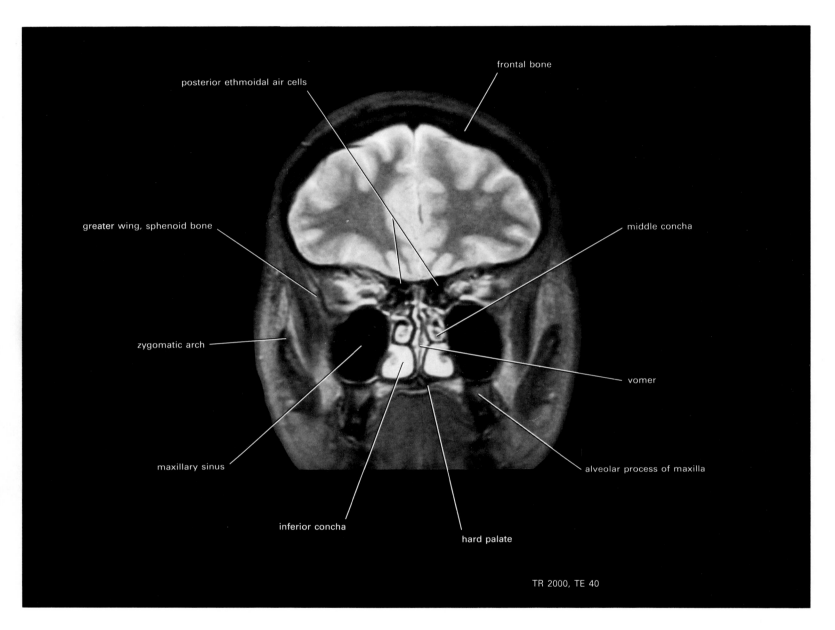

Section 2C from front.

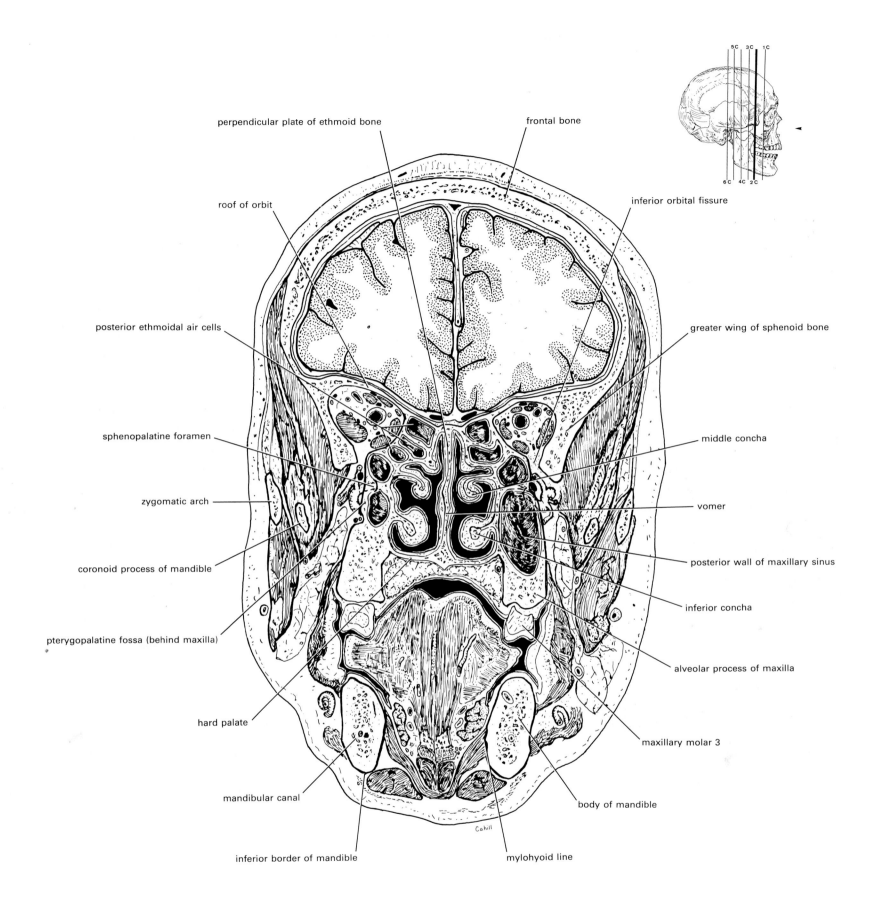

perpendicular plate of ethmoid bone

frontal bone

roof of orbit

inferior orbital fissure

posterior ethmoidal air cells

greater wing of sphenoid bone

sphenopalatine foramen

middle concha

zygomatic arch

vomer

coronoid process of mandible

posterior wall of maxillary sinus

inferior concha

pterygopalatine fossa (behind maxilla)

alveolar process of maxilla

hard palate

maxillary molar 3

mandibular canal

body of mandible

inferior border of mandible

mylohyoid line

Cahill

Section 2C from front.

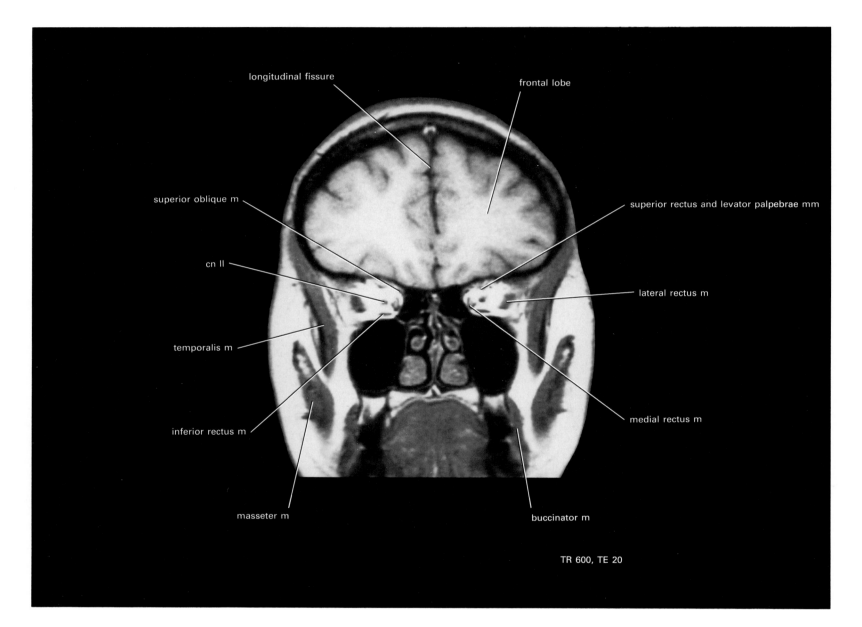

longitudinal fissure

frontal lobe

superior oblique m

superior rectus and levator palpebrae mm

cn ll

lateral rectus m

temporalis m

inferior rectus m

medial rectus m

masseter m

buccinator m

TR 600, TE 20

Section 2C from front.

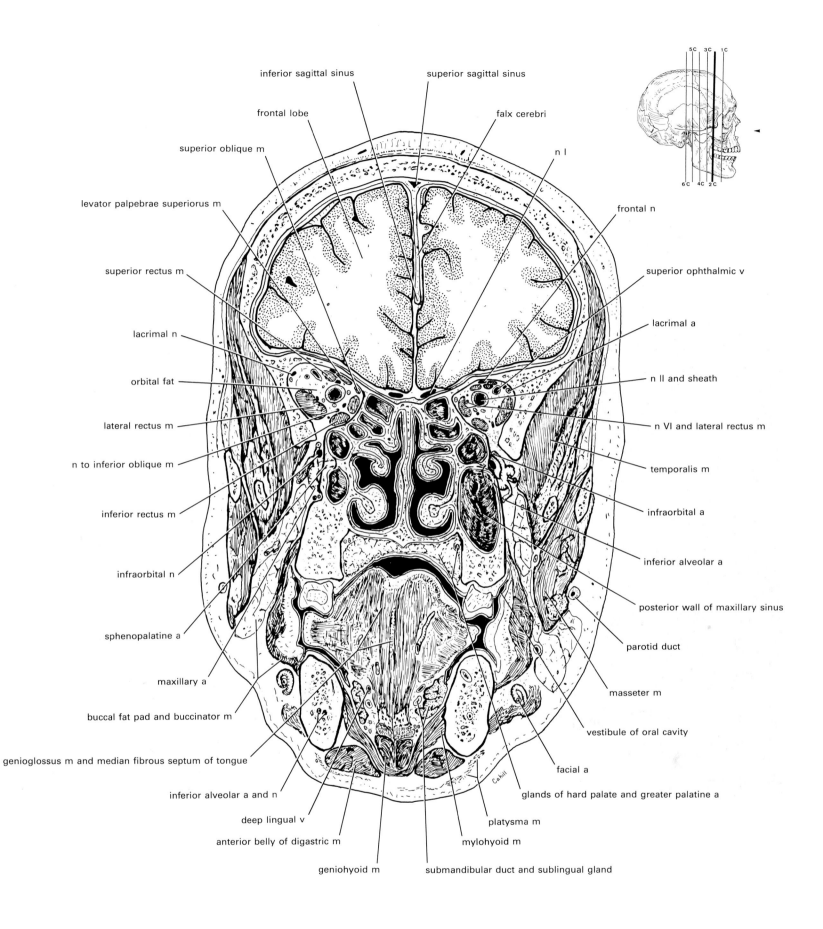

inferior sagittal sinus

superior sagittal sinus

frontal lobe

falx cerebri

superior oblique m

n I

levator palpebrae superiorus m

frontal n

superior rectus m

superior ophthalmic v

lacrimal n

lacrimal a

orbital fat

n II and sheath

lateral rectus m

n VI and lateral rectus m

n to inferior oblique m

temporalis m

inferior rectus m

infraorbital a

infraorbital n

inferior alveolar a

sphenopalatine a

posterior wall of maxillary sinus

maxillary a

parotid duct

buccal fat pad and buccinator m

masseter m

genioglossus m and median fibrous septum of tongue

vestibule of oral cavity

facial a

inferior alveolar a and n

glands of hard palate and greater palatine a

deep lingual v

platysma m

anterior belly of digastric m

mylohyoid m

geniohyoid m

submandibular duct and sublingual gland

Section 2C from front.

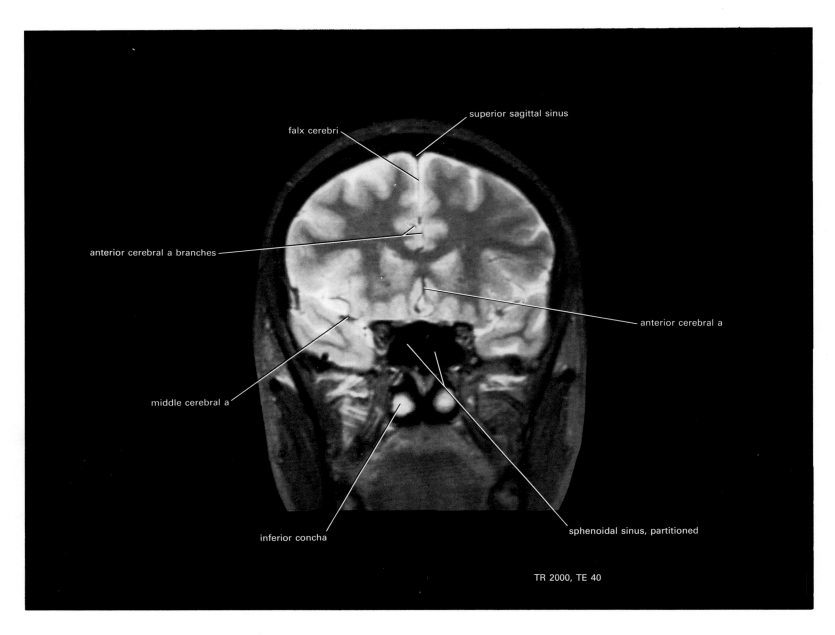

falx cerebri

superior sagittal sinus

anterior cerebral a branches

anterior cerebral a

middle cerebral a

sphenoidal sinus, partitioned

inferior concha

TR 2000, TE 40

Section 3C from front.

BONES

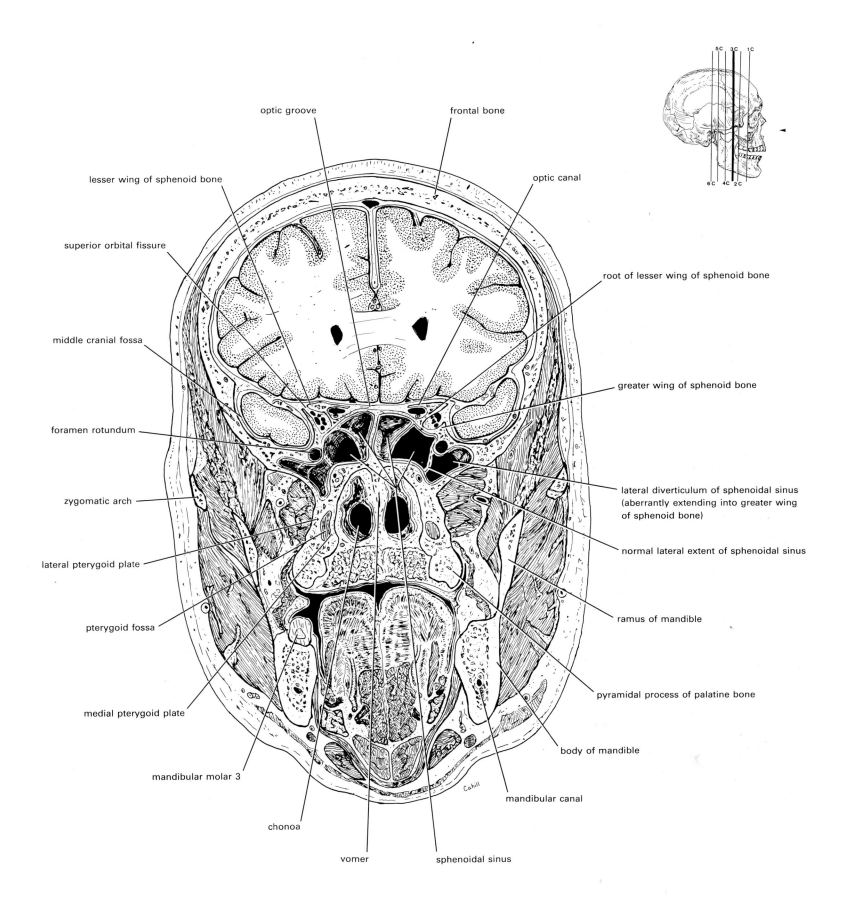

optic groove

frontal bone

lesser wing of sphenoid bone

optic canal

superior orbital fissure

root of lesser wing of sphenoid bone

middle cranial fossa

greater wing of sphenoid bone

foramen rotundum

zygomatic arch

lateral diverticulum of sphenoidal sinus (aberrantly extending into greater wing of sphenoid bone)

normal lateral extent of sphenoidal sinus

lateral pterygoid plate

ramus of mandible

pterygoid fossa

medial pterygoid plate

pyramidal process of palatine bone

body of mandible

mandibular molar 3

mandibular canal

chonoa

vomer

sphenoidal sinus

Section 3C from front.

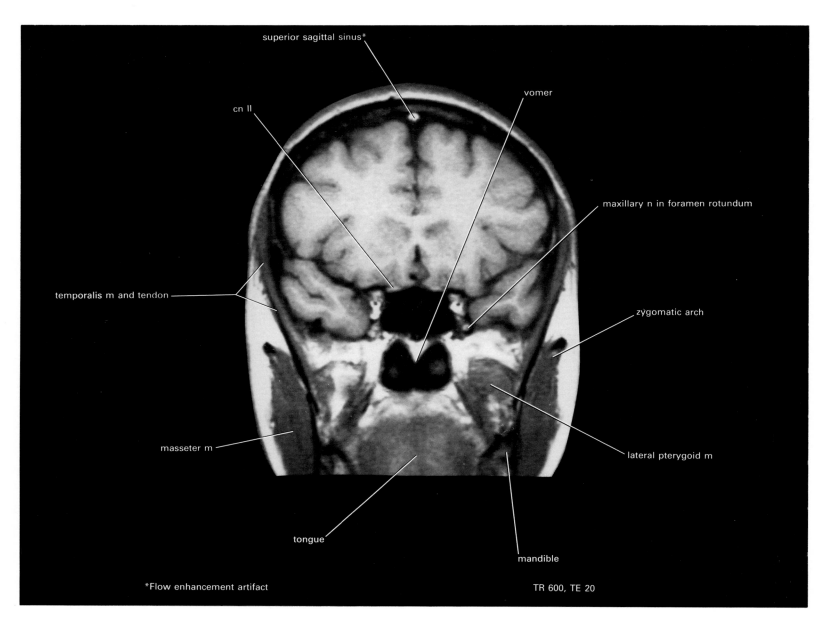

Section 3C from front.

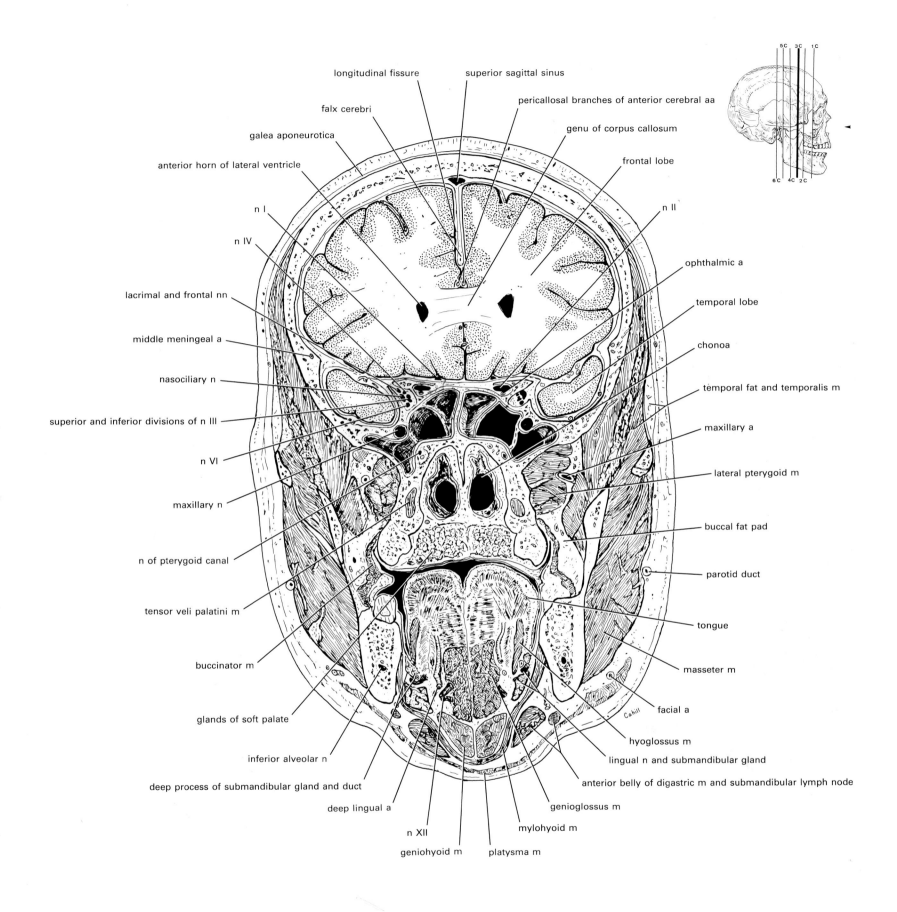

longitudinal fissure

superior sagittal sinus

falx cerebri

pericallosal branches of anterior cerebral aa

galea aponeurotica

genu of corpus callosum

anterior horn of lateral ventricle

frontal lobe

n I

n II

n IV

ophthalmic a

lacrimal and frontal nn

temporal lobe

middle meningeal a

chonoa

nasociliary n

temporal fat and temporalis m

superior and inferior divisions of n III

maxillary a

n VI

lateral pterygoid m

maxillary n

buccal fat pad

n of pterygoid canal

parotid duct

tensor veli palatini m

tongue

buccinator m

masseter m

glands of soft palate

facial a

inferior alveolar n

hyoglossus m

lingual n and submandibular gland

deep process of submandibular gland and duct

anterior belly of digastric m and submandibular lymph node

deep lingual a

genioglossus m

n XII

mylohyoid m

geniohyoid m

platysma m

5C 3C 1C

6C 4C 2C

Cahill

Section 3C from front.

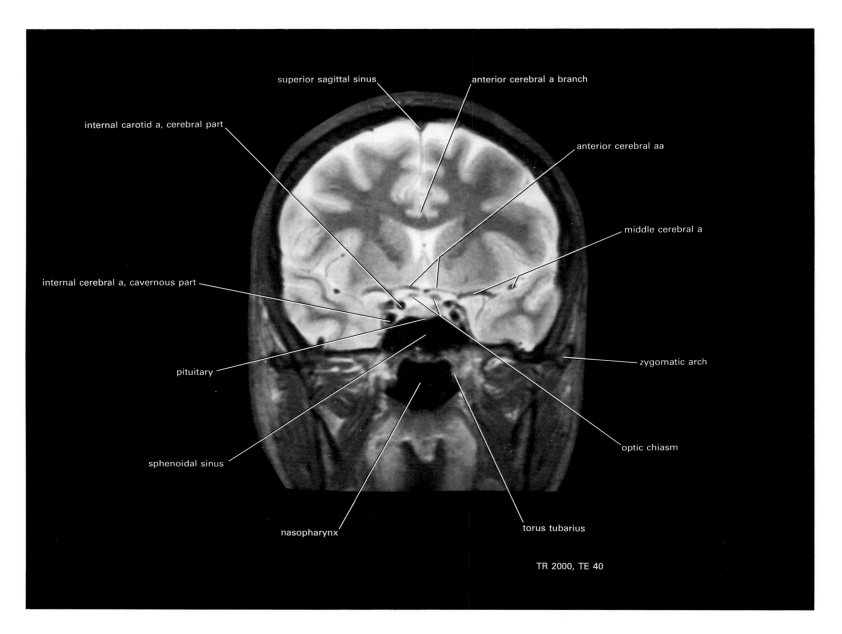

superior sagittal sinus

anterior cerebral a branch

internal carotid a, cerebral part

anterior cerebral aa

middle cerebral a

internal cerebral a, cavernous part

zygomatic arch

pituitary

optic chiasm

sphenoidal sinus

nasopharynx

torus tubarius

TR 2000, TE 40

Section 4C from front.

BONES, VESSELS, AND VISCERA

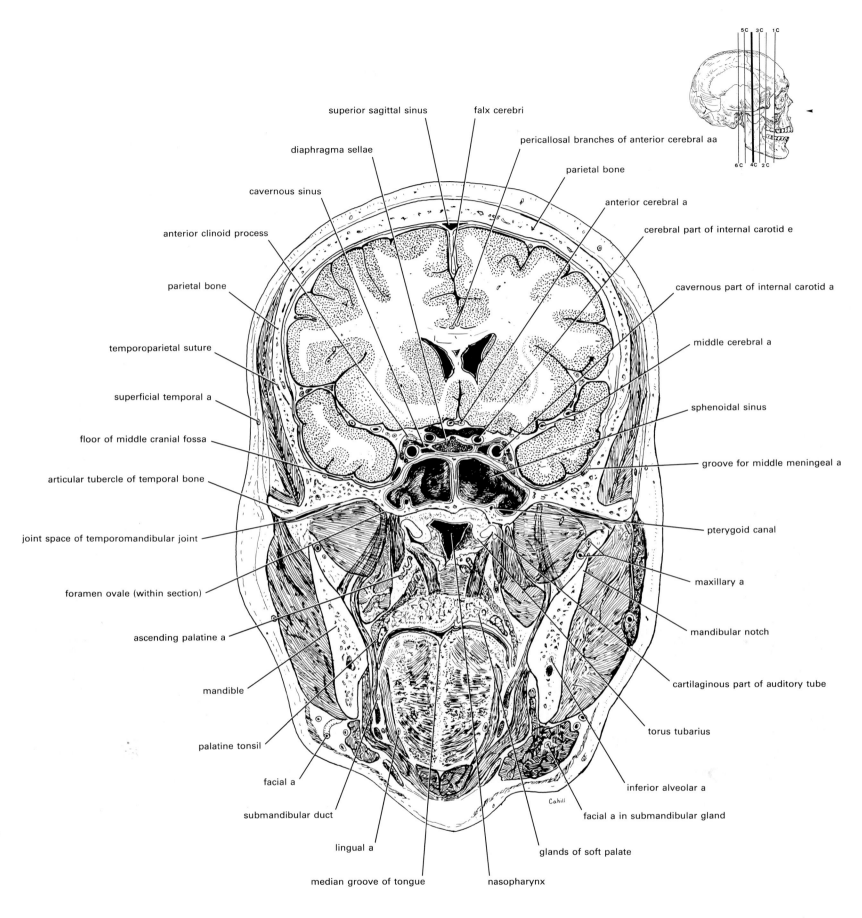

superior sagittal sinus

falx cerebri

diaphragma sellae

pericallosal branches of anterior cerebral aa

cavernous sinus

parietal bone

anterior clinoid process

anterior cerebral a

cerebral part of internal carotid e

parietal bone

cavernous part of internal carotid a

temporoparietal suture

middle cerebral a

superficial temporal a

sphenoidal sinus

floor of middle cranial fossa

groove for middle meningeal a

articular tubercle of temporal bone

joint space of temporomandibular joint

pterygoid canal

foramen ovale (within section)

maxillary a

ascending palatine a

mandibular notch

mandible

cartilaginous part of auditory tube

palatine tonsil

torus tubarius

facial a

inferior alveolar a

submandibular duct

facial a in submandibular gland

lingual a

glands of soft palate

median groove of tongue

nasopharynx

Cahill

Section 4C from front.

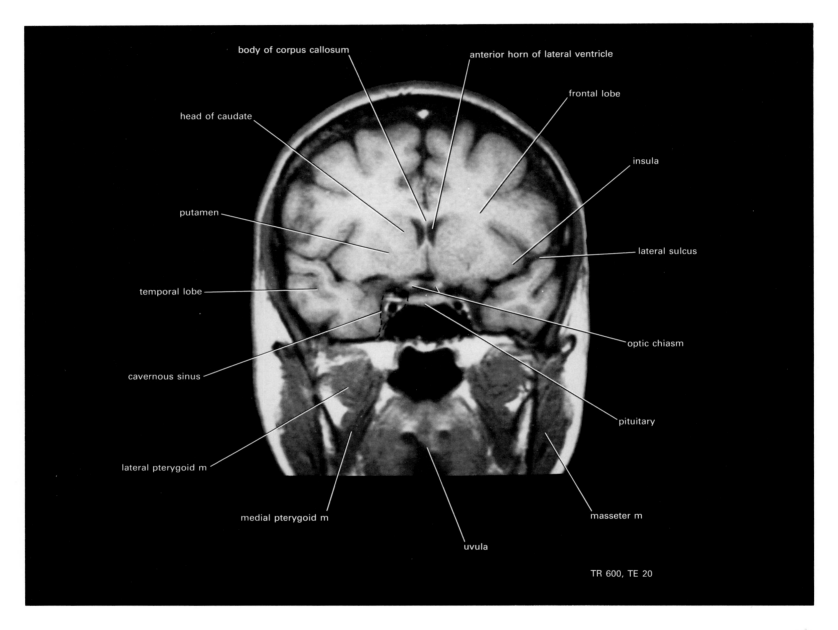

body of corpus callosum

anterior horn of lateral ventricle

frontal lobe

head of caudate

insula

putamen

lateral sulcus

temporal lobe

optic chiasm

cavernous sinus

pituitary

lateral pterygoid m

medial pterygoid m

masseter m

uvula

TR 600, TE 20

Section 4C from front.

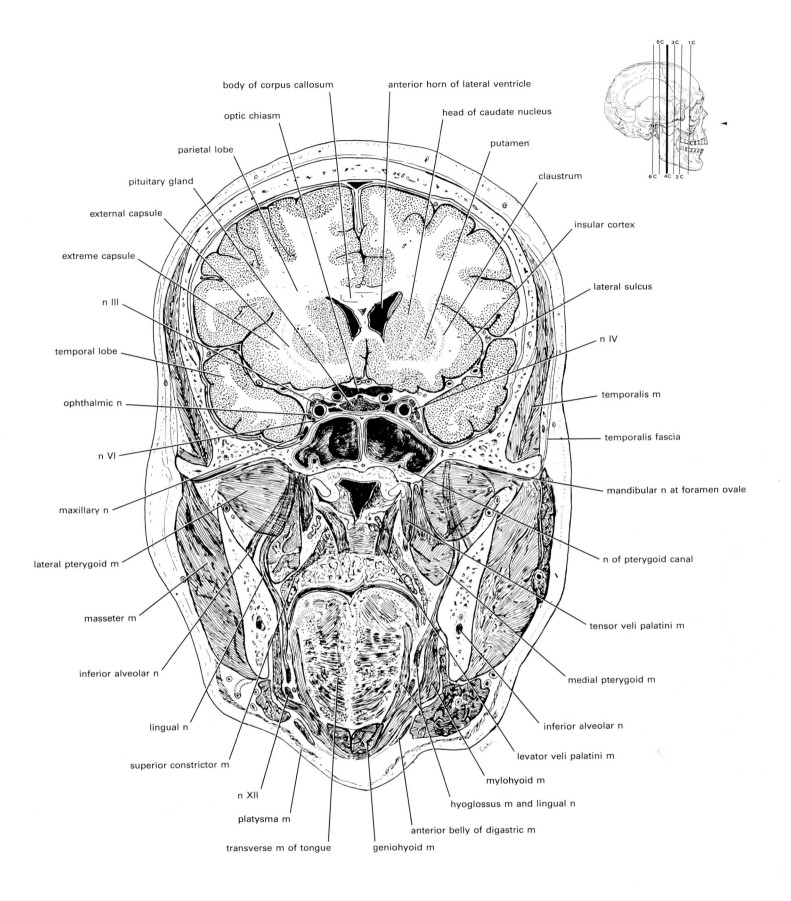

body of corpus callosum

anterior horn of lateral ventricle

optic chiasm

head of caudate nucleus

parietal lobe

putamen

pituitary gland

claustrum

external capsule

insular cortex

extreme capsule

lateral sulcus

n III

n IV

temporal lobe

temporalis m

ophthalmic n

temporalis fascia

n VI

mandibular n at foramen ovale

maxillary n

n of pterygoid canal

lateral pterygoid m

tensor veli palatini m

masseter m

medial pterygoid m

inferior alveolar n

inferior alveolar n

lingual n

levator veli palatini m

superior constrictor m

mylohyoid m

n XII

hyoglossus m and lingual n

platysma m

anterior belly of digastric m

transverse m of tongue

geniohyoid m

Section 4C from front.

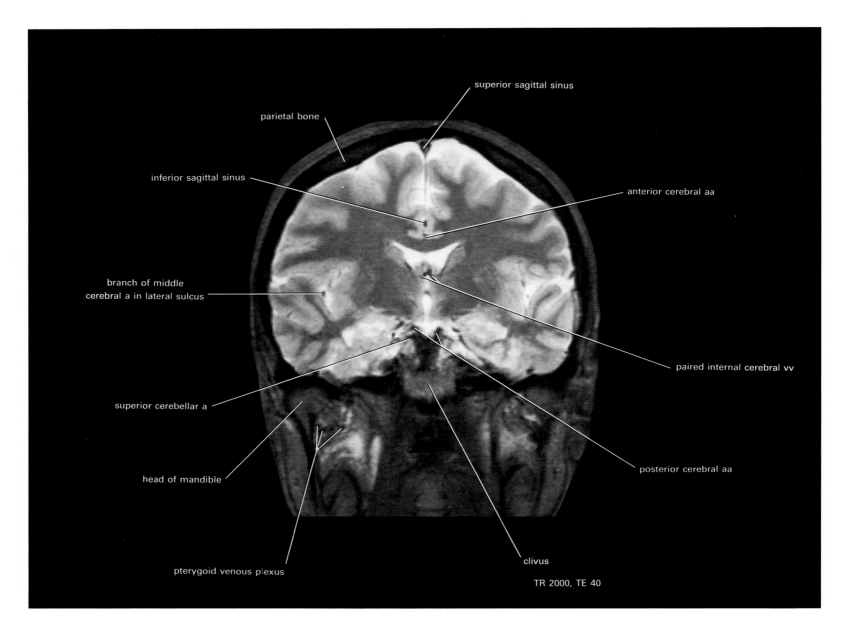

Section 5C from front.

BONES, VESSELS, AND VISCERA

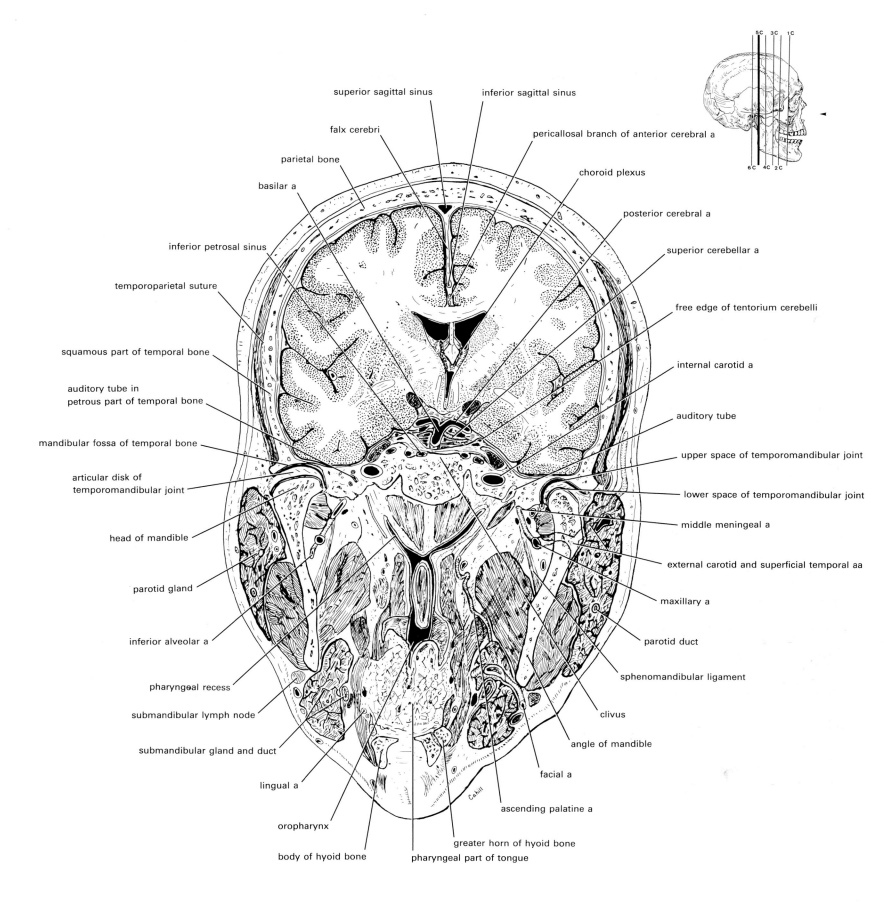

superior sagittal sinus

inferior sagittal sinus

falx cerebri

pericallosal branch of anterior cerebral a

parietal bone

choroid plexus

basilar a

posterior cerebral a

inferior petrosal sinus

superior cerebellar a

temporoparietal suture

free edge of tentorium cerebelli

squamous part of temporal bone

internal carotid a

auditory tube in petrous part of temporal bone

auditory tube

mandibular fossa of temporal bone

upper space of temporomandibular joint

articular disk of temporomandibular joint

lower space of temporomandibular joint

middle meningeal a

head of mandible

external carotid and superficial temporal aa

parotid gland

maxillary a

inferior alveolar a

parotid duct

pharyngeal recess

sphenomandibular ligament

submandibular lymph node

clivus

submandibular gland and duct

angle of mandible

lingual a

facial a

oropharynx

ascending palatine a

body of hyoid bone

greater horn of hyoid bone

pharyngeal part of tongue

Section 5C from front.

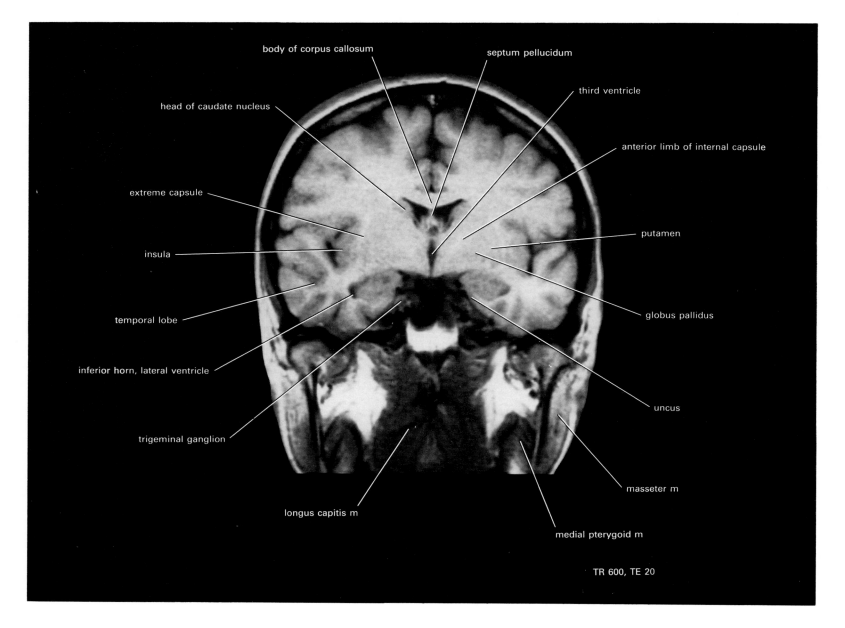

Section 5C from front.

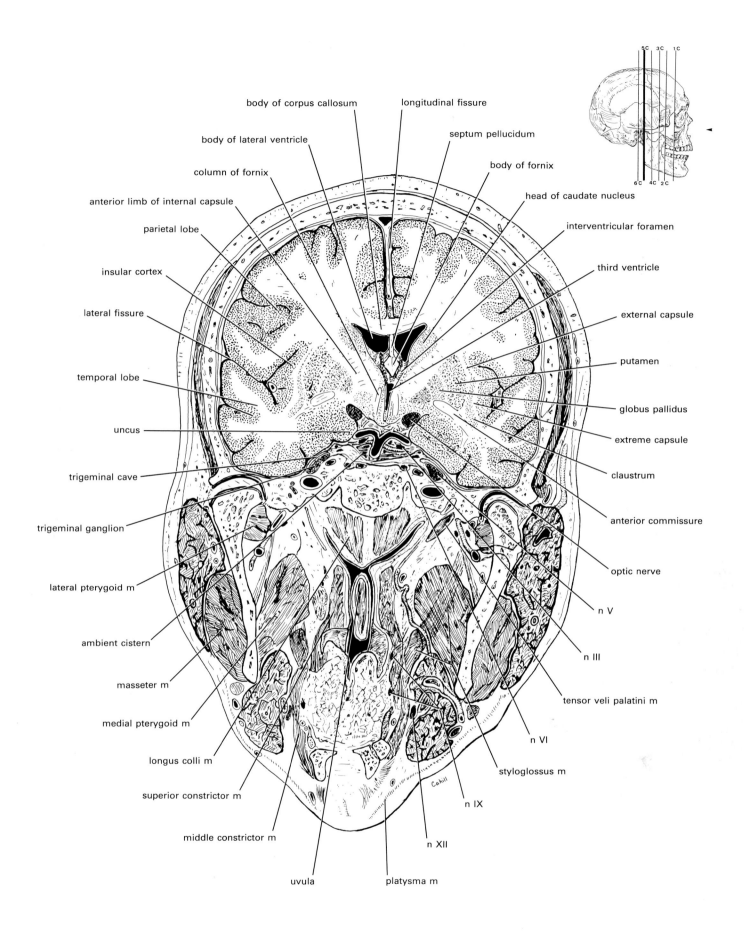

body of corpus callosum

longitudinal fissure

body of lateral ventricle

septum pellucidum

column of fornix

body of fornix

anterior limb of internal capsule

head of caudate nucleus

parietal lobe

interventricular foramen

insular cortex

third ventricle

lateral fissure

external capsule

putamen

temporal lobe

globus pallidus

uncus

extreme capsule

trigeminal cave

claustrum

trigeminal ganglion

anterior commissure

lateral pterygoid m

optic nerve

ambient cistern

n V

masseter m

n III

medial pterygoid m

tensor veli palatini m

longus colli m

n VI

superior constrictor m

styloglossus m

middle constrictor m

n IX

uvula

n XII

platysma m

Cahill

Section 5C from front.

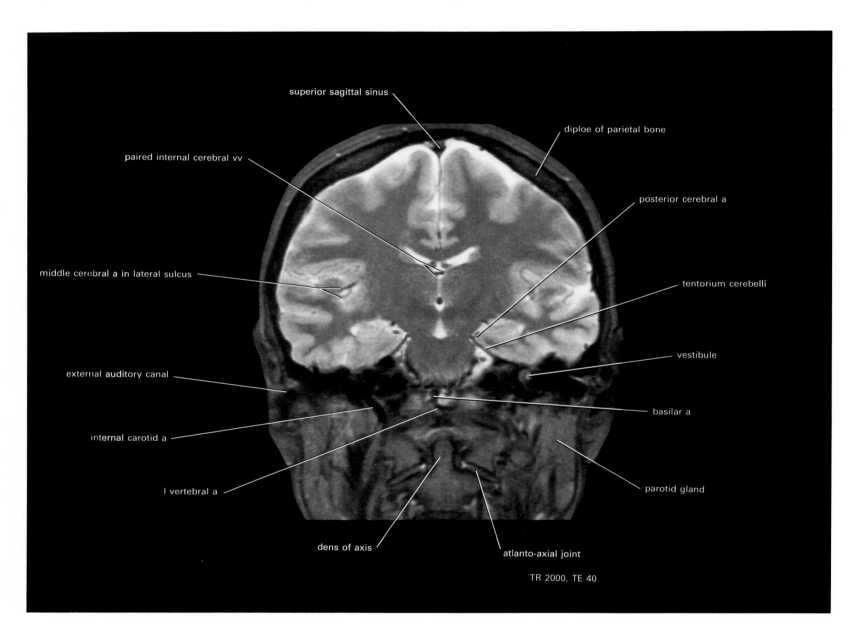

superior sagittal sinus

diploe of parietal bone

paired internal cerebral vv

posterior cerebral a

middle cerebral a in lateral sulcus

tentorium cerebelli

vestibule

external auditory canal

basilar a

internal carotid a

parotid gland

l vertebral a

dens of axis

atlanto-axial joint

TR 2000, TE 40

Section 6C from front.

BONES, VESSELS, AND VISCERA

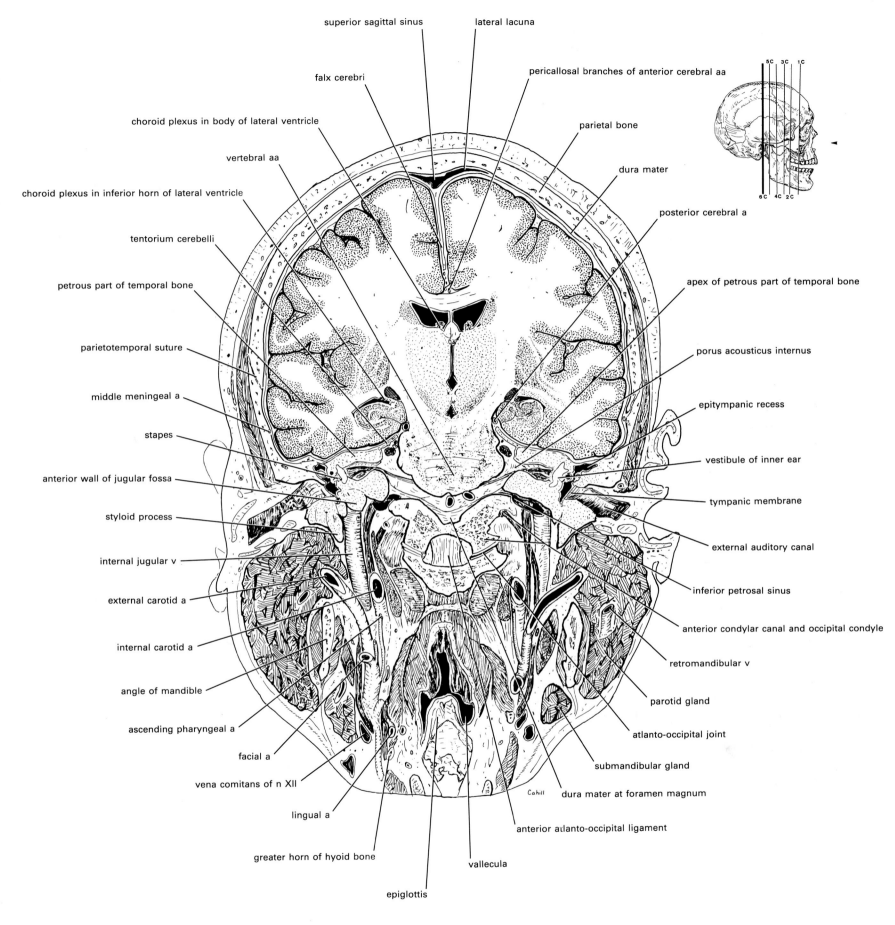

superior sagittal sinus

lateral lacuna

pericallosal branches of anterior cerebral aa

falx cerebri

parietal bone

choroid plexus in body of lateral ventricle

dura mater

vertebral aa

posterior cerebral a

choroid plexus in inferior horn of lateral ventricle

apex of petrous part of temporal bone

tentorium cerebelli

petrous part of temporal bone

porus acousticus internus

parietotemporal suture

epitympanic recess

middle meningeal a

vestibule of inner ear

stapes

tympanic membrane

anterior wall of jugular fossa

external auditory canal

styloid process

inferior petrosal sinus

internal jugular v

anterior condylar canal and occipital condyle

external carotid a

retromandibular v

internal carotid a

parotid gland

angle of mandible

atlanto-occipital joint

ascending pharyngeal a

submandibular gland

facial a

dura mater at foramen magnum

vena comitans of n XII

anterior atlanto-occipital ligament

lingual a

greater horn of hyoid bone

vallecula

epiglottis

Cahill

Section 6C from front.

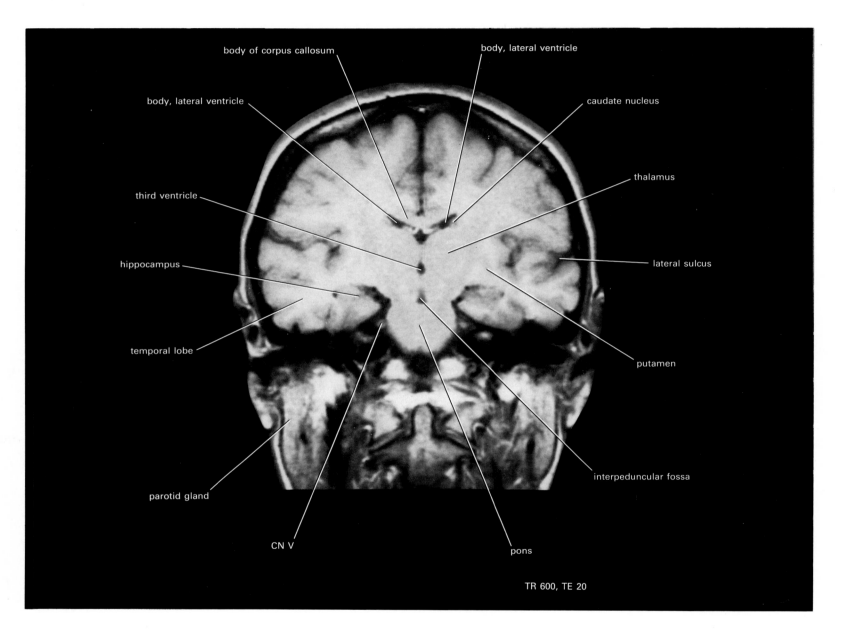

Section 6C from front.

NERVOUS SYSTEM AND MUSCLES

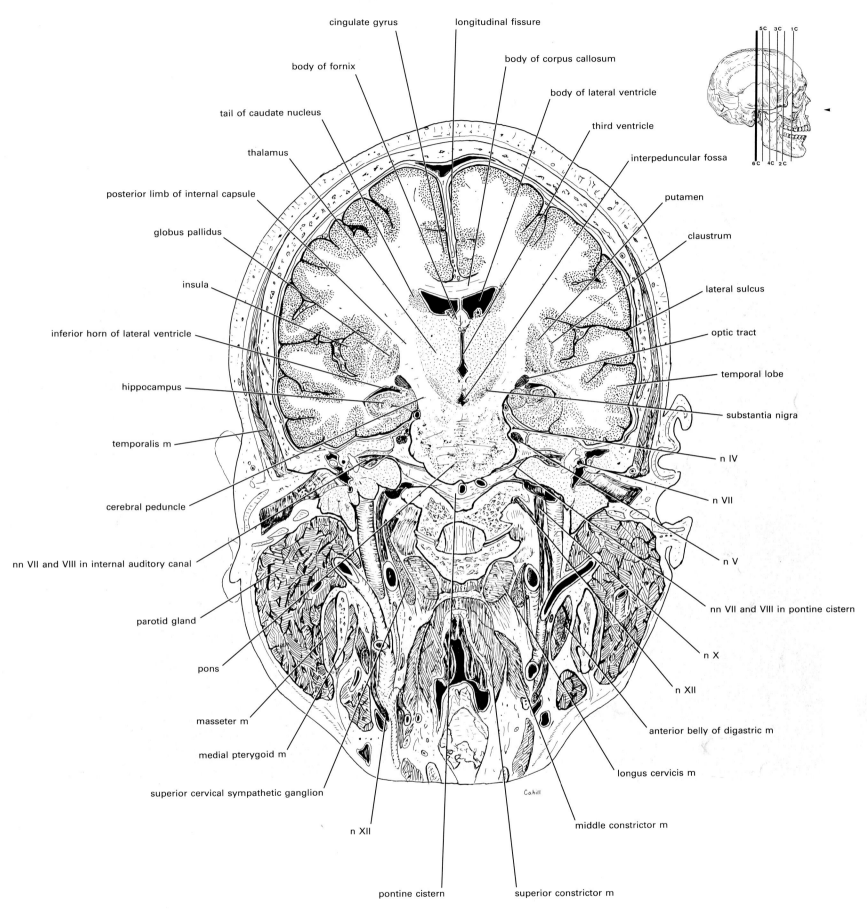

cingulate gyrus

longitudinal fissure

body of fornix

body of corpus callosum

body of lateral ventricle

tail of caudate nucleus

third ventricle

thalamus

interpeduncular fossa

posterior limb of internal capsule

putamen

globus pallidus

claustrum

insula

lateral sulcus

inferior horn of lateral ventricle

optic tract

temporal lobe

hippocampus

substantia nigra

temporalis m

n IV

cerebral peduncle

n VII

nn VII and VIII in internal auditory canal

n V

parotid gland

nn VII and VIII in pontine cistern

pons

n X

masseter m

n XII

medial pterygoid m

anterior belly of digastric m

superior cervical sympathetic ganglion

longus cervicis m

n XII

middle constrictor m

pontine cistern

superior constrictor m

Cahill

Section 6C from front.

Index